6-17-22

Cargo Containers

Their Stowage, Handling and Movement

Cargo Containers

Their Stowage, Handling and Movement

By

HERMAN D. TABAK

Cornell Maritime Press, Inc.

CAMBRIDGE   1970   MARYLAND

Standard Book Number 87033-136-1

Library of Congress Catalog Card Number: 78-100658

Printed in the United States of America

1539449

To Alice Tabak, My Dear Wife

Contents

Comments by Prominent Transportation Experts

As a member and Chairman of the Committee on Commerce of the United States Senate I have had the privilege of being a close observer and to some degree a participant in the development of our national transportation system. When that system is considered from a technical standpoint it is fascinating. However, when one views transportation as a system essential to nearly every diverse undertaking of mankind, then the importance of that system and its proper development take on characteristics of a vital national and international attribute.

Our transportation system has undergone many profound changes. Perhaps the most dramatic example is represented by the relatively few years that have transpired between the flight of the Kitty Hawk and the landing of two American spacemen on the surface of the moon. But the development of the container system is a unique change in our transportation system and one that surely warrants special attention and study. Containerization is the technical outgrowth of the understanding that all forms of transportation have the common purpose of moving cargo in the most efficient, expeditious and safe manner. The use of containers compels closer integration of our multi-mode transportation system, and a more integrated system will best serve the public and the transportation industry.

The preparation of *Cargo Containers—Their Storage, Handling and Movement* was an ambitious and challenging undertaking. The author has met that challenge successfully and the result is a comprehensive and detailed study of the many aspects of the cargo container revolution. Mr. Tabak's contribution to all who wish to pursue the many ramifications of this unique transportation development is extensive.

WARREN G. MAGNUSON, Chairman
United States Senate Committee on Commerce

The shipping of cargo in containers is one of the most significant and exciting developments experienced in the transportation industry in many years. Although the concept of containerization is not new, its great potential for broad applications is just beginning to develop in a meaningful way.

Improved efficiency in cargo handling, more economical terminal handling, reduced requirements for terminal facilities, and improved protection for cargo against weather, damage and pilferage—all of these are made possible by the development of container usage. Interchangeability between modes of transportation and improved utilization of equipment are additional benefits.

Containerization may very well be the key to a whole new era in transportation.

As is frequently the experience with innovations, containerization has problems relating to design, standardization, administration, documentation, financing, ownership and other matters for which acceptable answers must be found.

Cargo Containers—Their Stowage, Handling and Movement is a discussion of the growth of containerization, of the problems and of the efforts to solve these problems. I welcome publication of this book as a useful contribution to a fast-growing concept that has a great potential in our transportation economy.

C. E. CRIPPEN, President
Chicago, Milwaukee, St. Paul and
Pacific Railroad Company

To the uninitiated, containerization, like Topsy, seems to have "just growed." In the three years before this writing, interest in the intermodal transportation of the standardized box has flared in all parts of the world and among all types of carriers.

No single component of the transportation complex has been more affected by nor more responsive to the container concept than the steamship industry. The principal American-flag operators and many of their foreign-flag competitors have invested heavily of their corporate funds to obtain the specialized ships, the terminals, the road equipment and the containers themselves, which this far-reaching program requires if it is to bring maximum benefit to shippers and carriers alike.

Perhaps the most exciting challenge which containerization offers is to the *status quo.* Old line methods and old line thinking must give way before this new giant. New laws, new documentation, new interline agreements and new handling methods are being devised every day. The emphasis on intermodularity requires the virtual elimination of parochial thinking at every level of the transportation industry.

In this volume, H. D. Tabak has done much of the spadework for this new thinking. He gives us the why and the how of the development of containerization and a detailed account of the state of the art today. There is something here of value for the president and the shipping clerk, for the sales director and the cargo mate, for the terminal manager and the longshoreman—in short, for everyone concerned with the physical distribution of goods.

MANUEL DIAZ, President
American Export Isbrandtsen Lines, Inc.

Transportation is like a river.

This river has been flowing for as long as man has had the need and desire to move himself or any article from one place to another. Indeed, one of the means of identifying the existence of a civilization is the recognition that a group developed the ability to transport people and goods.

As a river builds by the convergence of its tributaries, and as it is used by the divergence of useful streams, the methods of transportation, parts of a total system have developed, replacing obsolete ways with more efficient systems which form the mainstream, joining together the world's economic entities.

Man's own labor was replaced by the beast. Carts or wagons were added for greater efficiency only to be superceded by the machines of the age of steam and internal combustion. The footpath has developed into the greatest highway system known to man in order that people may move freely and serve the producer and user.

The fullest swelling of this stream has, in America, created the most useful, complex and efficient system in the world contributing immensely to our industrial leadership, but more needs to be, and will be done. No system of transport can come into being or continue to exist unless it serves and grows with the economic needs of its user, the public. This is the genius of the American system—independent, competitive modes of transport each striving on its own and interdependently to continue to satisfy the ever-changing needs of the public. For this reason each mode constantly must search for the means to further improve its efficiency and competitive position.

Containers are as old as man, from the water gourd to the woven reed basket, to the oxcart, to the truck. Containers are as new as today, holding men and products as they cross a continent, circle the world, or penetrate the vast reaches of outer space. Some serve sophisticated, specific purposes; some have a multiplicity of general uses, yet each has its own unique efficiency.

The trucking industry uses a vast variety of containers in almost every conceivable way—from ordinary wheeled trailers to inflatable rubber bags, from small cardboard cartons to large steel and aluminum boxes lifted by crane or rolled on and off ocean-going ships carrying any commodity industry might need.

And so a new brook enters the mainstream—containerization in ever more specialized, yet more widely used form, furthers the desired end that the public shall receive a better, more efficient transportation service.

LEE R. SOLLENBARGER, President
Transcon Lines

Preface

I have always admired the human ingenuity that is continually devising new and more efficient ways of harnessing motion for the global transport of cargo from source to user. Until the advent of the cargo container, the ship was the master of the entire movement, the determining factor as to the packaging, the routing, the handling, the storage, and even the production timing. But a new era dawned, and entire new concepts are coming forth. The ship finds itself only one segment of the entire movement of the cargo. The "ship" is uncertain—can she still set the terms? Can she still control? Where does the railroad fit into the overall movement? Where does the truck? Who controls the cargo? Who controls the container? The impact of transport by intermodal container has spilled over into the highways, the railroads and is even starting to reach out for the airways. The impact has gone beyond the transport mode alone, for involved now is packaging, distribution, warehousing, marketing, and even plant location. The ripples go far, but the stone has been cast and this book was written in an attempt to answer the need for basic information on *containers* and their rapidly growing role in the transport of today and of tomorrow.

Obviously, in putting together a book of this type, one must stop at a given point. While there are basics in containerization whose premise will continue to remain valid, there will continue to be new developments in equipment, in documentation, in handling, etc., for this is a dynamic field and human ingenuity will continue to work for improvement. So that this book may continue to serve as a guide, therefore, it will be periodically updated.

An endeavor such as this is never the product of one man alone. The cooperation and assistance that I have received from all to whom I have turned has been outstanding, and my heartfelt thanks go to them. I would be remiss, however, if I did not offer a very special thank you to George J. Abagnolo, Manager of Hapag/Lloyd Container Services, to Anthony L. Horstman, Assistant Director of Flexi-Van International, Penn-Central Railroad, and to Mr. David Davis, Assistant Administrator to the Executive Vice-President of Wilson Freight Company. Each gave freely of his knowledge and experience in his field.

H. D. T.

1

Cargo Containers by Land and Sea

History of Containers

The term "containerization" is a static one and denotes nothing more than the use of a container into which something is put. This concept of packaging is not new. Materials have been placed in boxes, barrels and bottles for hundreds of years. Cargo, whether grouped or loose, was first handled by men who would literally hand an article from one to the other. Then material was moved on small trucks. More recently, ship-mounted winches and booms or small capacity quay cranes were used. Technically, these methods could be called containerization. However, the significance of the term "containerization" has changed greatly since the turn of this century.

In the years following World War II it was realized that improved handling of general cargo in and out and within the ship was an economic necessity. Consequently, during the 1950's, a great deal of money and effort was spent in research of the problem. Thorough detailed studies were made of existing methods of handling break-bulk cargo, palletization, forklift operation, improved cargo gear, hatch configuration, roll-on/roll-off ships, containers, and so on. This research clearly brought out the costliness and inefficiency of existing cargo-handling methods and pointed to various means by which improvements could be effected.

Modern containerization and the major breakaway from conventional cargo-handling systems dates from 1957 when Pan Atlantic Steamship Company, a forerunner of the present Sea-Land Service, Inc., installed container cells in the hold of the first of six conventional cargo ships and installed special purpose cranes aboard ships to load and unload 8' square, 35' long, 25 long ton containers. The containers were built with extra structural reinforcing to permit them to be stacked four high in the ship, and to be lifted at the top four corners by a lifting device attached to the crane. New devices were developed to connect and lash the containers on deck for the sea voyage. The gantry crane and its lifting device were built with the ability to place a container into any ship cell and to reach over the ship side to load and unload highway trailers on the quay. Each item of equipment was compatible with the other to form an integrated system. There were no industrial, national, or worldwide standards at this time. The first system was self-contained and could only expand through extension of its own facilities and not through use of others in a normal cargo interchange.

Early in 1959, Matson Navigation Company, serving Hawaii from the West Coast of the United States, inaugurated a major container operation also using equipment designed especially for handling containerized cargo.

Matson started its container operation by carrying containers as deck cargo. But, within a very short time, the cellular-hold container ship *Hawaiian Citizen* was placed in service. The Matson system uses 22½-ton capacity containers, 8′ x 8½′ x 24′. The special corner fittings, latching devices, lifting beams and cargo-lashing hardware are Matson's own design. Because five or six ships were to be used and only two ports were initially on the call list, port-based loading-unloading cranes offered the greatest economy and efficiency.

At the same time Matson moved to containerization, Grace Lines converted two ships, *Santa Eliana* and *Santa Leonore*, to cellular design to transport containers from the East Coast of the United States to South America. The system used 20-ton, 8′ x 8′ x 17′ containers with a third type of lifting and latching device which differed from the Sea-Land and Matson devices. Prompted by increasing competition from the highway trucking industry, the railroads developed their own containerization systems, a development which closely paralleled that of the shipping industry. The original method was that of the piggyback system by which the wheeled highway semi-trailers were carried on flatcars. From this, special containers separable from the highway chassis were developed. Both systems are in use by the railroads today.

STANDARDIZATION OF SIZES

Problems of standardization become obvious when Sea-Land, Grace Lines and Matson were each operating a full container system with different sized containers, different maximum weight containers, and different types of lifting and latching devices. The lack of standardization prevented interchangeability of containers between trade routes and further complicated the problems of inland shipping by truck or rail. Under sponsorship of the American Material Handling Society and the American Society of Mechanical Engineers, the American Standards Association (ASA) adopted standards for container sizes in 1961 and for container strength and fittings in 1962. In late 1965 the International Organization for Standardization (ISO) tentatively adopted the ASA standards in all aspects except that strength standards would be based on four high instead of six. The container standards adopted by the ASA in 1961-62 differed from all existing types in use in the marine industry at the time. Most of these non-standard container systems still remain in use today and will probably remain so indefinitely because of heavy financial commitments made toward their use and because of specialized uses which render them ideal for the purpose.

While the external dimensions of the containers have become standardized, the internal ones have not. Towards this end, a coordinating panel has been set up by the British Standards Institute under the chairmanship of a member of the Ministry of Transport. This panel will advise and expedite agreement on the national and international aspects of the dimensional coordination necessary for economic multipurpose transportation by freight containers so far as this concerns unit loads, packages, pallets, etc. It will eventually put forth proposals to the International Freight Containers Committee for the acceptance of internal constants for containers. Two working groups of the ISO Technical Committee for Freight Containers (T.C.104) have recommended internal dimensions for Series I General Purpose Containers in which no specific provisions have been made for thermal insulation. The recommended dimensions are: Width 2,300 mm (90½″) Height 2,195 mm (86½″), Length 185 mm (7¼″) less than the minimum external length. These dimensions are subject to the ratification of the parent committee.

The American Bureau of Shipping, in another step forward, has published a guide for the certification of dry cargo containers. This guide sets up procedure for manufacturing standards. Upon adherence to these standards, which follow as far as possible international standards of construction and testing, after due attestation by a Bureau Surveyor, American Bureau of Shipping certification will be given and an emblem affixed to the containers.

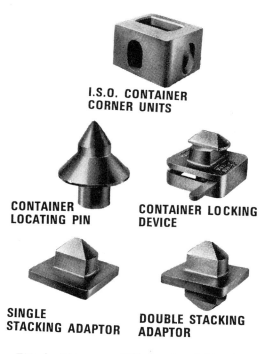

I.S.O. CONTAINER CORNER UNITS

CONTAINER LOCATING PIN

CONTAINER LOCKING DEVICE

SINGLE STACKING ADAPTOR

DOUBLE STACKING ADAPTOR

Fig. 1. ISO corner fitting and matching fittings for holding and locking container onto transport equipment, "Holtite."
(Holt Williams & Co. , Ltd.)

In addition, Lloyds Register of Shipping is issuing its emblem for freight containers certified by the Society. Lloyds Register certifies containers on a series production basis. This means that the containers are built at a factory approved by the Society in accordance with the Society's "Requirements for the Construction and Certification of Freight Containers."

The important step forward that was taken, in conjunction with the sizes, was the establishment of standards for container "corner fittings" so that interlocking, intermoving and locking would be no problem whether on chassis, rail or shipboard. These standards were approved by the Technical Committee of ISO meeting in Moscow in June 1967 and now await approval by the member countries. This corner fitting for the container and the matching fittings for holding and locking onto ancillary equipment are shown in Fig. 1.

The identification and marking subcommittee of the MH-5 sectional committee of the United States of America Standards Institute is well on its way to putting a report on standard container identification markings in final form for consideration by the USASI.

A set of proposed container markings was expected to be ready soon for circulation and consideration by subcommittee members and assuming acceptance or recommendation, for submission later to higher levels of USASI and ultimately to the International Standards Organization. A draft resolution has already been prepared and submitted to the member bodies as of December, 1966.

Standard sizes and corner fittings make the containers easier to handle here and abroad, assuming, of course, that as many carriers in as many countries as possible carry out the standardization process to the fullest. It makes for the ready interchange of containers and, hopefully, would mean that shippers and receivers would get the most favorable rates.

Identification and marking of containers fits in this general package.

Data processing is becoming increasingly important. By using a simple standard system, containers can be located faster and kept track of better, rather than having all carriers and shippers mark their own containers any way they choose.

There still is no standard domestic U.S. marking system, but this could come once an international standard is adopted and particularly when more and more basically domestic companies reach outward into the foreign trade.

Basically, what the subcommittee has done so far is to fairly well gain agreement among its members to a four-letter company identifying system, followed by six digits, which the owning company may use to number the container and for any other purposes it may require.

Thus, the first line might look like this: ABCD 000123. Below that would be the type and size of the container. The type would be in two digits, beginning with "20" (denoting a merchandise dry cargo demountable container), up to "90," standing for a demountable tank. Provision would be made for variations within each type to be assigned a number as they come into use. The size is signified by two digits, ranging from "03" for 8′ x 8′ containers, five feet in length, to "40" for containers 40 feet in length.

The type and size, then, painted just below the company's name and container identity number, might come out like this: 20 03.

The gross and tare weight, in pounds and kilograms, would be entered next below the type and size and, lastly, the country of origin would appear in an abbreviated code form.

So far it has been agreed to suggest requiring the first line, that is, the owner's name and container identification number, on the rear left door panel and the upper forward one half of the container on both sides. The type, size, weights and country of origin only are being suggested for the rear.

The standard markings are to be applied with a good quality, durable paint, in colors to contrast with the container. Current thinking is that letters may be required to be about four inches high.

The one remaining marking in this standard is the socalled "TATA plate" which was intended to include all the minimum international required information, plus the inside volume in cubic feet and meters, the manufacturer's name, location, serial number of the container, and the month and year of its manufacture. That metal plate was to be fastened to the lower forward curbside corner of the container. Size has not been tentatively agreed on but the thinking is that it would be about 8 by 12 inches.

Not only have many objections and special needs been handled, but in working out the proposed markings—very much like the experience in working out any other such system where government and industry work cooperatively—an almost endless number of special circumstances arise. One was the need to leave free the right-hand end panel for special markings required by European railroads.

Definitions of Freight Containers

A freight container is an article of transport equipment: a) of a permanent character and accordingly strong enough to be suitable for repeated use; b) especially designed to facilitate the carriage of goods, by one or more modes of transport, without intermediate reloading; c) fitted with devices permitting its ready handling, particularly its transfer from one mode of transport to another; d) so designed to be easy to fill and empty; e) having an internal volume of 1m³ (35.3 ft.³) or more.

The term freight container includes neither vehicles nor conventional packing.

Descriptions. Containers are divided by ISO into three groups. Groups I and II are demountable cargo containers that can be mounted and secured on open top carrier equipment. Group III containers cover stowable cargo containers that can be stowed efficiently inside standard Groups I and II demountable cargo containers, railroad box or gondola cars, modern volume highway vans or air cargo planes. The sizes of each series must be interchangeable in or on carrier equipment and universally interchangeable between modes of transportation.

Group I Containers:

	Size	Displacement (Max.)	Inside Capacity (Min.)	Maximum Gross Wt.	Maximum Payload
1A	40 × 8 × 8 ft.	2560 cu. ft.	2090 cu. ft.	67,200 #	—
1B	30 × 8 × 8 ft.	1916 cu. ft.	1560 cu. ft.	56,000 #	—
1C	20 × 8 × 8 ft.	1272 cu. ft.	1040 cu. ft.	44,800 #	—
1D	10 × 8 × 8 ft.	628 cu. ft.	490 cu. ft.	22,400 #	—
1E	6'8" × 8 × 8 ft.	413 cu. ft.	329 cu. ft.	15,680 #	—
1F	5' × 8 × 8 ft.	307 cu. ft.	248 cu. ft.	11,200 #	—

Group II Containers:

2A	9' 7" × 7' 7" × 6'11"	—	—	15,680 #	—
2B	7'11" × 6'11" × 6'11"	—	—	15,680 #	—
2C	4' 9" × 7' 7" × 6'11"	—	—	15,680 #	—

Some Non-Standard Containers (inside cube and weights can vary somewhat depending on when manufactured).

	Size			Displacement		Inside Capacity		Maximum Gross Wt.	Maximum Payload
	40'	× 8'	× 8'6"	2752	cu. ft.	2415	cu. ft.	68,030 #	60,000 #
	35'	× 8'	× 8'6"	2408	cu. ft.	2088	cu. ft.	—	45,000 #
	24'	× 8'	× 8'6-1/2"	1632	cu. ft.	1415	cu. ft.	50,000 #	46,200 #
	20'	× 8'	× 8'6"	1360	cu. ft.	1185	cu. ft.	49,860 #	44,800 #
	17'	× 8'	× 8'	1088	cu. ft.	914	cu. ft.	35,840 #	32,040 #
(Conex)	8'6"	× 6'3"	× 6'10-1/2"	365	cu. ft.	295	cu. ft.	10,543 #	9,000 #
(Dravo)	7'10-1/2" × 6'10-3/4"	× 6'7"	358	cu. ft.	283	cu. ft.	15,100 #	13,200 #
(Jeta)	7'9"	× 6'5"	× 6'10-1/2"	343	cu. ft.	280	cu. ft.	13,550 #	12,000 #
(Half Conex)	4'3"	× 6'3"	× 6'10"	184	cu. ft.	135	cu. ft.	10,050 #	9,000 #
(10 cbm.)		—		—		353	cu. ft.	11,101 #	8,960 #
(8 cbm.)		—		—		274	cu. ft.	11,700 #	10,000 #
(7 cbm.)		—		—		247	cu. ft.	10,587 #	8,960 #
(5 cbm.)		—		—		177	cu. ft.	10,261 #	8,960 #
	9'4.7"	× 7'10"	× 7'8"	564	cu. ft.	499.5	cu. ft.	22,950 #	20,000 #
	8'10.7"	× 7'4"	× 6'7"	429.3	cu. ft.	374	cu. ft.	22,500 #	20,000 #
	8'6.1"	× 7'4"	× 6'10-1/4"	410	cu. ft.	345	cu. ft.	14,000 #	12,000 #
	7'10"	× 7'	× 6'9"	370	cu. ft.	336	cu. ft.	11,450 #	10,000 #
	7'10"	× 7'	× 6'10"	375	cu. ft.	325	cu. ft.	11,590 #	10,000 #
	7'10"	× 7'	× 6'10"	375	cu. ft.	320	cu. ft.	22,150 #	20,000 #
	8'6"	× 7'	× 6'10"	406.5	cu. ft.	341	cu. ft.	11,700 #	10,000 #
	8'	× 7'	× 7'3"	406	cu. ft.	324	cu. ft.	9,700 #	8,000 #
	8'1"	× 7'6"	× 7'3"	444	cu. ft.	388	cu. ft.	13,450 #	12,000 #
	8'1"	× 7'6"	× 7'3"	444	cu. ft.	388	cu. ft.	19,500 #	18,000 #
	5'11-3/8" × 7'11-1/2"	× 6'9"	286	cu. ft.	242	cu. ft.	13,000 #	12,000 #
	6'1"	× 4'2"	× 6'5"	163	cu. ft.	144	cu. ft.	6,780 #	6,000 #
	7'2"	× 4'5"	× 6'10"	216	cu. ft.	180	cu. ft.	21,395 #	20,000 #

Two Sizes of Collapsible Containers Are:

Size	Displacement	Inside Capacity	Maximum Gross Wt.	Maximum Payload
4' × 3'4" × 5'4"	71 cu. ft.	63 cu. ft.	2,455 #	2,240 #
3'6" × 4'9" × 3'6"*	—	46.7 cu. ft.	1,342 #	1,118 #

*16 of these units can be carried in a standard 20-ft. container. When folded, these containers can be stacked to occupy one-fifth of their erected volume.

An additional type of pallet container is designed for carrying liquids. When open and filled, it measures 61-1/4" long × 46-1/2" wide × 41-3/4" high. The inner bag, made of Neoprene Hycaflex, Butyl Nitrile or Nitrile drinking water quality, are deflated when empty, enabling the container's hinged side panels to fold together so that the bag is protected and the container is one-third of its original height and weighs approximately 480 pounds. Its liquid capacity is 300 gallons.

Collapsible Containers:

These containers are sometimes called "pallet" containers and usually have a built-in pallet. They are usually manufactured to specific order and can be made of steel, wood, pressed board, or cardboard. They can be used for holding and shipping the cargo as is, or loaded with their cargo into the larger 20- and 40-ft. containers.

Some companies are already manufacturing collapsible containers in modules designed for use within the Group I containers.

Wire Mesh Containers:

This "cage" type container is made in varied sizes but is used mainly in domestic movements.

The U.S.A. Standards Committee MH–5, U.S.A. representative of the International Standards Organization, has not yet submitted to the U.S.A. Standards Institute for approval as a U.S.A. Standard, the ISO recommendations and draft recommendations for Freight containers. The USA Standard MH 5.1—1965 on corner fittings has been temporarily suspended for revision. However, the USA Standard Committee endorses for general use, particularly in International Distribution Systems, the following excerpts from ISO recommendations and draft recommendations:

ISO Recommendation R668 (formerly Draft Recommendation No. 804)
    Dimensions and Ratings of Freight Containers
ISO Recommendation R830 (formerly Draft Recommendation No. 1055)
    Terms and Definitions of Freight Containers
Draft ISO Recommendation 1019
    Specification of Corner Fittings for Series I Freight
    Containers (IA, IB, IC, ID)
Draft ISO Recommendation 1496
    Specification and Testing of Series I Freight Containers

CONTAINER CONSTRUCTION

Container construction is as varied as the sizes and devices used in their handling. The container is a relatively thin-skinned box built around a framework consisting of four corner posts connected by a header and sill at each end and tied together longitudinally by top rails and lower rails. The corner posts are designed as columns to withstand stacking under dynamic conditions experienced aboard ship. The horizontal sill which has a skin is most rigid, while the end containing the doors must depend on the end frame for its resistance to transverse racking unless specially designed doors are installed. Without special doors, the design of the door end frame and the connections between the horizontal members and the corner posts must provide the strength to resist the racking effect caused by ship motion or when stacked eccentrically. This is particularly important for stowage above deck where lashings are used instead of cell guides. Corner fittings are located at the top and bottom of each corner post for lifting and securing. Top rails and lower rails tie the end frames together to form the box.

The floor of the container consists of cross-members on about 12″ centers which transmit the load carried in the container to the lower rails. The cross-members are sheathed with load-bearing material. The side walls tied to the lower rails and top rails give the container its longitudinal strength as a girder. The roof, of light construction, consists of transverse stiffeners or bows on which is laid a light skin strong enough to support two men. The bows maintain the top rail straight and prevent its deflection under load.

The high load-bearing corner-post construction is the principal feature of the container which makes it different from normal over-the-road vans.

One important source of cargo damage claims is torn bags and boxes. Tears and rips are often caused by snagging on damaged container body liners. The liner, initially smooth, is under periodic attack by lift trucks and sharp-cornered cargo crates. Sheet metal and wood cannot withstand the abuse of accidental, but frequent, impact.

Metal dents, then tears; plywood splinters, delaminates. The resulting knife-sharp points can ruin a load of cargo. The need for periodic inspection and maintenance can ruin an operating budget.

A promising solution is a means of covering the surface either before or after the damage has occurred. Covering damage-prone surfaces with more of the same material can only be temporary. A tougher surface is needed; it must be smooth, impact-resistant and easy to install.

The proven toughness of resin-bonded glass fiber sheet can solve this performance problem, as fiberglas materials have solved similar performance problems as components for truck bodies, transit car bodies, automobiles, appliances, and aircraft.

In the container damage problem, the simple requirements of smooth flat surface can be met by compression molding fiberglas mat with polyester resin. The tough glass fibers produce a surface that brushes off light impact and absorbs heavy impact with a resilient give-and-bounce-back action. Every material has a maximum strength, but fiberglas laminates are outstanding even beyond their rupture limit. The unique laminate structure confines actual breakage to a small area which then can be repaired simply with the same laminate material. The repaired section can be as tough as the original laminate.

A fiberglas laminate is easily drilled for mechanical fasteners. A simple adhesive system may be preferred, eliminating any projection of fastener heads. The fiberglas/polyester laminate is compatible with most adhesive systems.

Seams can be a nuisance in installation and a source of contamination when foodstuffs are carried. Development of a proprietary continuous molding process resulted in the introduction of continuous liner. Produced in any length, it is normally shipped in lengths up to 40'. The resilient nature of the laminate allows it to be coiled for shipping, then uncoiled for flat installation. Thermal expansion and contraction can make forty-foot lengths of dissimilar materials tear apart. A liner of polyethylene installed full length when the temperature was 80° would contract about 4" when the temperature reached −20°. The difference between this movement and the minor contraction of a liner made of separate steel panels would break fasteners or surface.

The fiberglas-polyester laminate has a low expansion coefficient (similar to aluminum, closer to steel than other plastics). The relative contraction over 100° temperature range is insignificant in most applications.

Despite the many different types of containers that are gradually coming into use, the majority of containers will be strictly dry-freight boxes suitable for the carriage of general cargoes in both directions. Containers are manufactured from aluminum, steel, plywood, fiberglas and combinations of these.

Each type of container has its own particular place in transportation. All have advantages and disadvantages depending upon the kind of operation which requires them. Accordingly each type has its champions and its critics.

The design of a serviceable yet economical container is a complex problem. For maximum results, the designer must have a thorough knowledge of materials, fabricating methods and service requirements—all seasoned

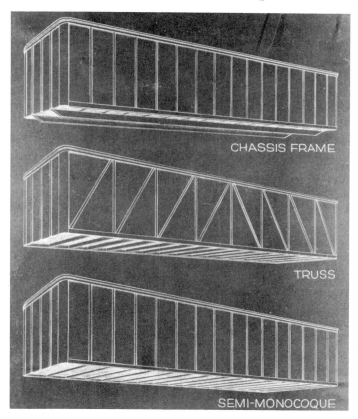

Fig. 2. Different types of container construction. (Aluminum Corp. of America)

with experience. Reputable container manufacturers have the required knowledge and experience and, for best results, container design is usually placed in their hands. From a design standpoint, containers are highway van trailers without wheels. Container designers, therefore, draw heavily upon experience gained during the past 30 years in which highway van trailers have been a vital factor in the American transportation picture. There are three basic methods of construction which could be considered for container design and all have been explored by container manufacturers. The first and oldest design utilizes a chassis frame consisting of longitudinal load-bearing members to which a boxlike body is attached. The second type of construction consists of building the body sides as trusses, covered by light gauge panels—the panels are not intended to carry stress. The third method is the monocoque or skin-stressed type. The philosophy behind this development is that as long as the body must be provided to protect the cargo, it might also be utilized in supporting the load; for it is obvious that a properly built body with sides approximately 8 ft. high would be much stronger than a shallow chassis. Of the three types, the monocoque, or more accurately semimonocoque as applied to container construction, has proved to be most efficient (lightest) and most economical to fabricate.

Once side-panel design has been established, other container components can be engineered by applying common structural design formulas.

Aluminum Containers

Cross-members. Conventional container flooring systems consist of cross-members which act as beams in supporting the lading and transferring this load to the side walls. The cross-members are usually covered with flooring that is laid longitudinally.

It has been found satisfactory to design cross-members on the basis of simple beam action, with maximum deflection limited to about 1 in. In most instances the critical load to be carried will not be that of the lading but, instead, that of the lift truck used in the loading operation. Unless specifications indicate otherwise, it is wise to anticipate lift truck front-axle loads of 12,000 lbs.

Flooring. In a 40-ft. container, a standard 1¼-in. aluminum floor can save more than 500 lbs. when compared to wood flooring and also offer additional advantages. Aluminum floors are stronger and provide a splinter-free surface of exceptional durability. For refrigerated units, extruded aluminum floors are most useful because they can be designed to provide space for air circulation under the load and for water runoff, and because they are impervious to moisture.

Side Sills and Posts. The unique attributes of the extrusion process make aluminum one of the logical choices for side sills and side posts. With the extrusion process, structural sections of virtually any cross section can be produced. Metal can be placed where it is required for strength or desired for convenience. Aluminum is the only material generally acceptable for container applications that can be easily extruded. Also, aluminum's high impact resistance is particularly desirable because side sills and side posts are vulnerable to damage.

Corner-post Design. At the front corners, where depth of post in the transverse direction is not objectionable, aluminum posts can be used to effect an appreciable weight saving. At the rear, the transverse depth of posts is important because it is usually this depth which governs maximum clear-door width for noninsulated containers. If a very shallow post is required to provide a maximum rear-door width, aluminum offers little weight advantage. On the other hand, if maximum door width is not essential, a deeper aluminum post can be used and weight can be saved.

The benefits of aluminum corner-post construction that may be considered are as follows:

1. Complicated corner-post cross sections can be extruded in one piece, minimizing fabricating costs.

2. Other aluminum components can be welded to aluminum corner posts; aluminum corner fittings save weight.

3. Corner posts are exposed to scraping and it is virtually impossible to keep paint on posts made of steel. The rusting that occurs is both unsightly and detrimental to the strength of the members.

Side Panel and Roof Sheet. All aluminum alloys commonly considered for container side sheet and roof sheet have a high degree of resistance to marine and industrial atmospheric corrosion. When necessary, corrosion resistance may be improved by using alloy core having surface layers of high-purity aluminum or of an alloy that is resistant to corrosion and anodic to the core alloy. If a pit forms, it penetrates only as far as the core alloy, where its further progress is stopped by the electrolytic action between the cladding alloy and the base metal. This prevents significant loss of sheet properties due to corrosion and virtually eliminates the possibility of leakage through the side or roof sheet due to corrosion. A second advantage of aluminum sheet is that it is available in extremely wide widths, making possible such design features as one-piece roofs, sides and ends. The one-piece roof, secured only around the periphery, eliminates a multitude of potential leak points down the length of the roof and can actually reduce total cost by eliminating the labor involved in making a series of riveted or lock-seamed roof joints.

One-piece side sheets are used for special bulk containers to obtain smooth interiors free from unnecessary sheet seams which could entrap cargo contaminants. One-piece construction can also be used at the front end of units to minimize the potential leak hazard in this area which is subjected to high wind loads in over-the-highway operation.

Insulated Containers. In general, insulated containers are heavier than noninsulated units and are subject to more severe corrosion hazards. Inherent light weight and resistance to corrosion make aluminum especially well suited for this type of service.

Pickup Hardware—Corner Fittings. The general advantages of aluminum construction—lighter weight and lower maintenance costs—are applicable for corner fittings, too. Aluminum corner castings also provide for attachment by welding to an otherwise all-aluminum unit. When needed, properly designed inserts may be incorporated in aluminum fittings (usually castings) to provide increased abrasion resistance. Aluminum castings are standard items with companies producing container corner fittings.

It is also possible to rivet or bolt steel castings to aluminum members, but precautions must be taken to eliminate dissimilar metal contact. Such precautions are not necessary for inserts that are cast in place in aluminum because the bond line is impervious to moisture.

Straddle Pickup. The extrusion process makes it a simple matter to design side sills that provide a ledge to accommodate the lifting bars of a straddle lift. The aluminum sill does not need a protective wear surface in this instance because the area of contact is comparatively large and there is very little relative motion between the pickup device and the containers.

Forklift pockets. Forklift pockets can be designed in aluminum for additional weight saving.

There are at least four possible joining methods that can be considered for aluminum container construction: riveting, welding, adhesive bonding and mechanical bonding.

Experience indicates that riveted construction is the most suitable and least expensive for general-purpose dry-cargo and refrigerated units.

The lower sheet gauge limit for economical production welding with consumable-electrode equipment is about .080 in. Welded construction is desirable for bulk containers that require heavy side sheet to withstand bulging loads and smooth interiors to prevent cargo contamination.

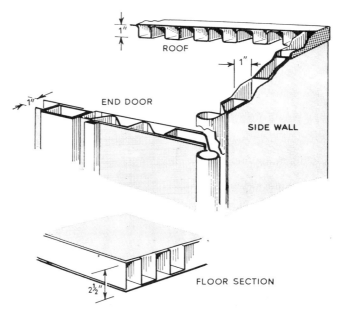

Fig. 3. York "corrugated principle" steel container construction. (Containerisation International, Oct. , 1967)

Adhesive bonding offers great promise for the future but must still be considered experimental and used accordingly.

Mechanical-bonding techniques, such as lock-seams, snap-lock or roll-lock extruded sections, are all limited by their inability to carry shear loads. As with adhesive bonding, such approaches must be considered strictly experimental at this point.

Steel Containers

The traditional objections to the use of steel for container construction have been weight and corrosion and these are rapidly being overcome. The weight factor, which naturally tends to be reflected in a reduced payload, has been overcome in the adoption of different techniques of construction.

One company has devised a lightweight container based on the use of a continuous steel double-wall which, they claim, provides strength and stiffness at low weight and needs no framing. The York Trailer Company of England has managed to reduce the unloaded weight of a 20-ft. ISO type container to 37 cwt, claiming that this is only 7 cwt heavier than containers made out of other types of material. York's all-steel container is constructed on the corrugated cardboard carton principle which gives strength while reducing the amount of steel. Framing is eliminated as well as the necessity for cross-members under the flooring.

Rust, abrasion and corrosion are overcome with the modern antirust and corrosion primers now on the market. Galvanized steels have all the qualities needed for long, trouble-free service. Zinc is an excellent, economical protective coating; an effective, low-cost agent for the protection of the base metal against corrosion. Prepainted sheets offer a custom, precoated steel ready for fabrication. These sheets have a finish that is uniform in color, gloss, texture, and thickness, with excellent paint adhesion and durability. Prepainted steels utilize a two-coat system. This allows more precise color matches as well as increasing corrosion resistance, improving appearance control, assuring uniform film coverage, improving paint adherence, minimizing pinholes or paint voids in the finish and permitting two types of coatings to be used. Prepainted steels are fabricated by the same processes that are used to fabricate rare metals. They may be stamped, pierced, drawn, formed, blanked, or fastened by all conventional fabricating methods except welding. For handling containers that demand absolute freedom from contamination—namely, tank-type containers—stainless steels offer excellent possibilities. They cost more initially but, in exchange, offer less maintenance, uninterrupted performance and long service life. Stainless tank containers can offer three times the corrosion resistance of other steels and other metals, can give higher strength with the result of a lighter weight and competitively priced unit, and resistance to high and low temperatures without losing significant strength or ductility.

From the standpoint of damage, especially in the interim period where conventional ships and handling gear will continue to be used, the advantages of the strength of steel are not to be overlooked. Collisions with "foreign bodies" such as the forks of a lift truck, or the ship's or other vehicle's superstructure, will be inevitable, often causing severe puncturing of aluminum body work. Containers made of corrugated steel tend to have "foreign bodies" glance off the corrugations. Should a steel container wall be pierced or dented, repairs and maintenance present little difficulty. Steel plating is easy to come by and even in out-of-the-way places, garages have welding equipment that can be utilized in case of an emergency.

If the container concept is taken to its logical conclusion, a container may be in Alaska one month and pass through the steaming heat of the Red Sea the next. These terrific extremes of temperature can create havoc with a structure of dissimilar metals, due to the different coefficient of expansion. These extremes can also encourage the loosening of riveted joints. A distinct advantage claimed for steel, provided that the floor is made of corrugated steel sheeting and depending on the sealing of the doors, is the watertight sealing as a result of the method of construction possible with steel, where no rivets are needed. It is through loose rivet holes that dampness and dust can seep, causing contamination and possibly the complete loss of a load.

Plywood Containers, Fiberglas Containers and Combinations

By definition, softwood plywood is a flat panel built up generally of an odd number of thin sheets or veneers of wood in which the grain direction of each ply or layer is at right angles to the one adjacent to it. These veneers are united under pressure by a bonding agent to create a laminated panel with an adhesive bond as strong as, or stronger than, the wood.

Inherent properties in such an engineered panel, brought about through the contiguous layering of veneers in odd numbers, are equalization of strains, minimization of shrinkage and warping, and the prevention of splitting. Further, plywood is available, easily worked, is designable for engineered applications, and is stronger in bending load-carrying capability and bending stiffness than the same weight of steel, aluminum or magnesium.

Plywood can be readily combined with other materials to produce a product with superior characteristics to either of the individual components. For example, such products as the fiberglas reinforced plastic/plywood (FRP) combination in cargo containers and the ply-metal configurations used for truck bodies and van doors. Fiberglas is a basic material with many forms for many uses. The material form examined here for cargo container uses is fiberglas reinforced plastic—FRP. The combination of fiberglas yarns and strands with various chemical resins produces a structural material with a happy combination of properties for containers.

In Group II demountable cargo containers, plywood units with steel frames have been used in quantity for many years. In these oldtimers, the plywood is, of course, directly exposed to the elements and subjected to the usual abuses of handling. Even in this case, however, these mostly nonstandard, smaller units have performed admirably, showing a significant maintenance advantage over similar sized containers of other materials. Although new for containers, this combination material is time proven as a durable, truly seagoing material, since it has been widely used in commercial, private and military boat hulls for more than 20 years. FRP/plywood containers are also easy to repair if damage occurs. Unskilled labor in any port of the world or aboard ship can replace punctured or chipped areas (by far, the most common damage occurrences) with a minimum of tools in a matter of minutes; no unloading, welding, riveting or emptying of contents is required.

FRP/plywood containers are not subject to electrolysis or corrosion; plywood and fiberglas-reinforced plastics are immune to damaging electrolysis and do not act as a medium for galvanic action or other deterioration caused by salt air, salt water, or caustic industrial atmosphere. FRP/plywood containers have smooth, uncluttered walls; there are no joints, ribs, bracing, hat sections, channels, or other framing configurations to take up space, trap dirt, damage goods or support contamination.

In addition, FRP/plywood containers have greater cube efficiency than most other containers. Due to the same thin, uncluttered, structural walls, there is more usable space inside the container. This often runs from 7–10% more than other containers. Bigger pay load in the same displacement means more money to the carrier.

Considerations of both cost and design have so far prevented an all-fiberglas product from actively joining the materials race in containers for overland and sea use. One of the intriguing characteristics of this basic material, however, is its ability to combine with other materials to do a job that neither could do alone.

In the area of Group III stowable cargo containers, more familiar to us as pallet bins, plywood is playing an ever-increasing role. In agriculture alone, nearly two million reusable plywood pallet bins are in service nation-wide, more than any other type of bin used for agricultural commodities. The reasons, briefly: low initial cost, high durability (some of the containers have been in use now for 10 years), high strength and rigidity; low weight/capacity ratio, ease of fabrication and maintenance, and standard quality and availability. Many more of these Group III containers are, of course, used in industry and vary considerably as to their makeup and intended use. The same advantages, however, apply.

The Veenema & Wiegers (V & W) Company manufactures a combination aluminum/fiberglas/plywood container, utilizing the best properties of each.

TYPES OF CONTAINERS

Containers mold the cargo to be carried into preset dimensions and shapes for efficient handling aboard the ship, on the railcar or on the highway. There are numerous types of cargoes and a number of various containers have been developed to especially suit them.

The commonest container is the general dry-freight type which can be loaded with a wide variety of packaged, bagged, or cartoned goods. Other types available are open-top, insulated, insulated and heated, insulated and refrigerated, ventilated, liquid, shallow liquid, bulk, watertight bulk, top-loading and end-loading bulk, open-bulk trays, automobile carriers, open pallets, livestock pens and side-opening types. (*See* Figs. 4 & 4A.)

Other Types

While the purpose of this book is to discuss intermodal containers, it is interesting to note, in passing, that wherever there is a need, special-purpose equipment will always be devised. This is true of containers as much as anything else. Collapsible rubber containers turn open or closed trucks, trailers and vans into dual-purpose tank trucks.

Uniroyal rubber containers are currently saving money for food processors and many other companies that ship liquid cargoes like petroleum products, paper chemicals, textile chemicals, urethane chemicals, liquid sugars, alcohols and liquid fertilizers because these containers allow them to haul liquids in one direction and general cargo in the other.

Fig 4. (upper left) Insulated Container; (upper center, top) Tank Container; (upper center, lower) Dry Cargo 20-foot Container; (upper right) Livestock Container; (center left) Dry Cargo 40-foot Container; (center right, top) Platform, Tray or Pallet Type Container; (center right, lower) Car-haul Container; (lower left) Tank Containers; (lower right) Open-top Container. (Freuhof Division, Freuhof Corp.)

Fig. 4A. Side opening container. (Duramin Engr. Co., Ltd.)

Fig. 5. Collapsible rubber containers. Can be used in open or closed trailers.
(U. S. Rubber - "Uniroyal")

Despite the size of these rubber containers (they range from 1,500 to 4,570 gals. capacity), no special preparation in setting them up is necessary. In 15 minutes, two men can turn an ordinary flatbed truck, van, reefer, or even a ship or barge, into a liquid carrier, and at the end of the run, turn it back again. Their most obvious advantages over rigid drums are their much lighter weight and the fact that when empty they collapse to only 15% of their filled size. Other benefits include important cost savings and greater speed in handling. In realizing the importance of "two-way movements" to container economics, this may well be one way of increasing the versatility of the dry-cargo container.

Still in the experimental stage are insulated marine containers whose cargo temperatures have been pulled down by normal refrigeration methods, or by the introduction of low-temperature nitrogen within the container.

The Canadian Pacific Railroad, in designing an insulated container, decided on the use of plastics. A frameless, one-piece molded all-plastic insulated container was chosen as the basic unit to design around in the development of an integrated intermodal temperature-controlled container. The container was built of 2½" rigid foam core (polyurethane) sandwiched between an inner and an outer skin of polyester resin fiberglas laminate bonded together by a series of honeycomb-type stiffenings 12 inches apart to the inner and outer polyester skin. This plastic sandwich provides the unusually low overall heat transmission coefficient for the entire container of less than 50 b.t.u.'s per hour per degree Fahrenheit temperature difference. The advantages of great rigidity, high impact resistance, immunity to corrosion and contamination and unusual thermal qualities were the features desired. It was still necessary to provide a small enough refrigeration power-pack and components when installed on the 18-foot-long container that would be within the international standard 20-foot length. A compact refrigeration power-pack was designed, including evaporator, condenser, compressor, aircooled diesel-electric generator, sea-water heat exchanger, fuel tank and all automatic controls to fit into an 18-inch-long compartment across the front of the container. The power-pack operates on the aircooled diesel-electric generator while on road, rail or on the deck of a ship, and is operated off the ship's electric circuit when stowed below deck, then the condenser can be cooled by means of a sea-water condenser. The Lentz system of reverse airflow envelope-cooling was installed to add greater thermal efficiency to the already excellent insulating qualities of the plastic container. This principle of envelope-cooling has proved to be eminently successful.

Refrigeration methods employed in most vehicles fall roughly into two main classes:

Mechanical. (a) Evaporator coil and circulation fan; (b) eutectic plates charged mechanically.

Dispensable Chemicals. (a) Dry ice (solid CO_2); (b) liquid CO_2; (c) liquid nitrogen (N_2); (d) liquid air.

Currently one of the most popular methods is the first method of mechanical refrigeration which involves the use of the Thermo King unit. Low maintenance costs and a choice of no less than five power sources—gasoline, propane, diesel, hydraulic and electrical—are the reasons for this popularity.

In addition, running costs are generally much lower than any other method.

Short Sea Crossing. As far as the U.K. is concerned, international road haulage has developed with the growth of roll-on/roll-off ferries, but only a handful of the many services offering are, or will be, operating their own containers. There is, of course, the extremely successful Danish Bacon container service between Esbjerg and Grimsby operated by the United Steamship Company for the Danish Bacon Factories' Export Association. Initially 400 refrigerated semi-trailer vans were in use on a shuttle service, making the sea crossing in a specially designed 2,100-ton double-deck refrigerated vessel (*Somerset*), hauled at either end for a fleet of 70 tractive units. In this case the containers are plugged into the ship's electrical system to maintain a constant temperature of 35°F. using the Prestcold system. During the relatively brief road journeys in the U.K. and Denmark, the specially insulated containers keep the temperature inside between 35°F. and 40°F.

Another short-sea container operator whose vessels make regular crossings of the North Sea—this time to Gothenburg in Sweden—is the England Sweden Line, which offers shippers 350 steel container vans and 900 flats, plus normal below-deck accommodation for freight vehicles and other unit loads. ESL recently introduced a batch of refrigerated containers which are proving popular among shippers. Designed to maintain temperatures of between 55°F. to minus 10°F., these 20-ft. ISO type, Duramin-manufactured, frameless glass-fiber-coated containers have an internal capacity of 830 cu. ft. Refrigeration is provided by a Thermo King unit carried at the after end of the container, the principal source of power being electricity with the alternative of gasoline. An abundant supply of plug-in points on board the vessels enables the temperature to be maintained by the ship's electrical system, and on the parking areas ashore, some of the concrete plinths (on which the containers are parked before transportation to or from the ships) are fitted with plugs. Once the containers transfer to road vehicles, low temperatures are achieved and maintained by the Thermo King gasoline-powered unit. If refrigerated traffic is not offering, ESL uses the containers for the conveyance of ordinary general cargo.

In the long-voyage operation emanating from England, the thinking follows the course of involving two systems, the ship's specially designed refrigeration plant for the long sea journey plus a detachable mechanical unit for the land journeys. One of the problems to be solved in a U.K./Australia refrigerated service will be the different requirements of the cargo. The refrigerated cargoes themselves will alternate between the carriage of meat, butter and fruit from Australia, while in the other direction the need for refrigeration will be at a minimum initially. One of the problems is that meat and butter need to be kept at a constant cold temperature, but fruit requires a different treatment, having to be maintained within a fairly fine range of temperatures. Research on this is going on from the operational as well as economic standpoint in many areas of the world simultaneously.

In the United States a liquid nitrogen system has proved successful in the movement of perishables from the interior of the U.S. to the interior of the European continent. Behind this is Oxytrol, a low-oxygen, controlled atmosphere system marketed in the U.S. by Occidental Petroleum Company and abroad by its international affiliate, International Ore & Fertilizer Corporation. The Oxytrol system consists of an oxygen sensor which monitors the level of oxygen inside the shipping compartment, an electronic analyzer to control the release of nitrogen gas into the system, super-insulated liquid nitrogen storage containers and other cryogenic hardware necessary to perform system functions. The Oxytrol process is an adjunct to existing methods of refrigerating highway trailers and ocean-going container vans. Developed several years ago, the technique was only recently made available to international shipping. The system holds promise of revolutionizing international movements of perishables, creating new markets and expanding old ones.

Already containers and trailer vans loaded with beef from Texas, peaches from South Carolina, and celery, broccoli and strawberries from California have been placed aboard ocean-going ships at the Port of New York for delivery to markets in Europe and Great Britain. The crisp freshness with which these commodities have reached their overseas destinations was not possible via surface transport until a short time ago.

The Oxytrol process is based on well-known scientific facts; oxygen and certain other gases in the air promote spoilage and foods maintained in an oxygen-free atmosphere may be preserved. What Oxytrol does, therefore, is to substitute nitrogen for air in the controlled atmosphere within highway trailers or containers to enable produce to remain in these refrigerated units for longer periods of time and still retain their freshness.

Oxytrol installations consist of individual units attached to trailers or recessed in containers. Each van contains extremely sensitive oxygen sensors which control release of nitrogen from a 100-gallon tank. To maintain a low oxygen level (1–2%), the atmosphere within each van is changed six or seven times a day. As air diffuses into the container, causing the oxygen level to rise, nitrogen is sprayed into the van, reducing the oxygen level and flushing out carbon dioxide. A transatlantic voyage requires about $20 worth of liquid nitrogen per van at current retail prices.

Initial international shipments in Oxytrol-equipped refrigerator vans have been remarkably successful. Celery, broccoli and strawberries consigned to London reached market in perfect condition. As regards celery; this shipment had been loaded in a trailer in California, shipped overland to New York and transported by sea to Britain—16 days en route. Produce men usually consider four days the maximum shipping time for strawberries; yet, when the strawberries in this trailer were unloaded in London, they were field-fresh.

For the first time, commercially marketable fresh beef from Texas is reaching German markets. Formerly, the only practical method of transporting beef such distances was in a frozen state. Now, however, the Estes Packing Company, Fort Worth, and the Texas Farm Bureau are delivering many tons of refrigerated beef weekly for German tables. The meat is loaded into containers in Texas, transported via Sea-Land's coastal service to the Port of New York and then transshipped onto another container ship for the transatlantic journey.

The same marketing advantages reaching Europe are being felt in North America. Via the Port of New York—San Juan route, fresher and sweeter pineapples from Puerto Rico are arriving for the New York market.

It is perhaps a happy coincidence that international container shipping is undergoing a revolutionary expansion at the time Oxytrol is becoming commercially available. Oxytrol is closely tied to container or trailer shipping, for it is easier to control the atmosphere in a closed van than in the hold of a ship. Furthermore, it is a marked advantage to maintain an uninterrupted controlled atmosphere for the produce or meat from origin to destination, a feat almost impossible if the commodities are carried in shipboard refrigerated areas.

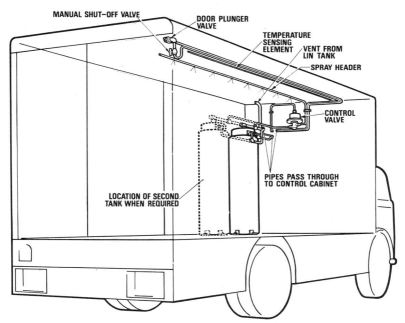

Fig. 6. Typical installation of liquid nitrogen refrigeration, "Cryo-Guard." (Air Products, Ltd.)

In addition to the obvious marketing advantages offered in extending shipping times, Oxytrol, through certain biological quirks, creates interesting and advantageous side effects to the shipped commodities. One is the process referred to as "setting." Once the processes of deterioration are retarded by low-oxygen/nitrogen atmosphere, they are slow to resume and never continue at the same rate. This means longer "shelf life" at retail outlets.

Other economic advantages accrue to shippers using Oxytrol-equipped trailers and containers. For instance, lettuce, tomatoes, melons, grapes and other fresh produce can be packed into cartons in the field and then loaded directly into controlled-atmosphere vans. Once packed, the van receives a burst of injected nitrogen which reduces the oxygen level immediately and begins to slow down respiration, a spoilage process in plants. At the same time, refrigeration equipment begins to decrease the atmospheric temperature within the van.

Reduction of the respiration rate of fresh fruits and vegetables reduces demands on a van's refrigeration system. In a standard van about 80% of the heat dispersed is generated by the cargo itself through respiration. By cutting internal heat production, both the weight and cost of refrigeration units may be halved and perishables may be more closely packed in the van.

Having demonstrated the feasibility of Oxytrol in domestic shipments, companies recently began to pioneer international applications. However, as the Oxytrol system proves itself, carriers are arranging to make installations of the units on their own equipment. An Oxytrol installation leases for about $4 a day.

In the United Kingdom, a similar nitrogen gas system for trailers and containers, called "Cryo-Guard," is manufactured by Air Products in New Malden, Surrey. This system consists of a vacuum-insulated, liquid nitrogen storage tank with a capacity of 375 lbs. of liquid nitrogen usually installed inside of the container (in ISO containers) but can be mounted outside, piping to distribute the liquid nitrogen to the cargo and control equipment to maintain the desired temperature. Installation does not require any special skills and can be completed by a good pipefitter in less than a day. There is no special setting-up or test procedure, and as soon as installation is complete the liquid nitrogen tank can be filled and the container is in service. There is no need to provide air-circulation space around the cargo, since nitrogen gas penetrates every crevice of the container, giving a uniform temperature regardless of how tightly the cargo has been packed. Mechanical systems of refrigeration are generally accepted to require a minimum circulation space of 2″ under the load, 18″ above it, ¾″ at the sides and 2″ at the ends, as well as requiring some form of delivery duct from the cooler. The liquid nitrogen in the tank constantly emits a small amount of gas, sufficient to maintain a pressure adequate to force the liquid up into the spray header. The gas also provides pressure to actuate the control valve so that the system is entirely self-reliant. There are no pumps or other moving parts and the system needs no connection to the vehicle, or to the container's electrical system or to any auxiliary power source.

Controlled-atmosphere shipping, one of the newest innovations in international commerce, augurs well to be the forerunner of a great new wave of perishable exports and imports.

Bulk containers are especially designed to carry such commodities as grains, powders and granular commodities in bulk. When granular chemicals are carried, thorough cleaning is necessary to prevent contamination of the next cargo carried. These containers are emptied at their destination by hydraulic equipment which tilts the container endwise, permitting discharge through the rear doors. Heavy bulk materials such as ore concentrates are carried in shallow trays which limit the volume per tray and thus the weight. Open pallets conforming to the base dimensions of standard containers are used for carrying smaller palletized loads or they may be used as a "master pallet." Their height may be half a normal container or full height, or they may be fitted with removable corner posts that may be of lengths selected to suit the cargo height.

Many other forms are possible and may include any type seen on the highways or on the rails.

A feature which has been in use is a coupling system in which two 20-ft. containers, or one 30-ft. and one 10-ft. container, can be joined together to form a 40-ft. container. The space for the coupling requires three inches. ASA standard containers of the 10-, 20-, and 30-ft. lengths are less than their nominal lengths partly for this reason. The coupling is inserted in the end

Fig. 7. Twin 20-foot containers coupled as single traveling unit. (Strick Corp.)

circular openings of the bottom corner fittings of abutting containers and twist-locked in place. This coupling remains in tension. A top-fitting serves as a compression member. Specially designed chassis are used to transport the coupled containers over the road.

INTERMODALITY AND THE TRANSMODALIST

With the advent of the Group I standard containers, containerization entered its latest phase, that of being a mode of transport in itself. This quality of intermodality enables this container to become a truck by putting it on a chassis or bogie and hooking up a tractor. Lift it off, put it on a rail-car, add a locomotive, and it becomes a train. Lift it off and put it into the hatch or cell of a ship and it becomes one of the holds of the ship; it becomes the ship.

This new flexibility, more than any other single factor, served to expand the "international trade" consciousness of manufacturers and distributors all over the world, but in particular, in the United States. While the advantages were originally exploited by the steamships on a port-to-port basis, container transport companies came into the picture offering the service of moving cargo in containers. Frequently these companies were truckers or freight forwarders, even railroads. They had access to containers, their own, or leased, or belonging to steamship companies, arranging for the different segments of the needed transfer, yet not owning the ships nor necessarily controlling all the segments.

Individual freight forwarders, naturally concerned about their participation in the container market, have banded together in groups, operating container depots and offering their customers consolidation services and break-bulk services. In the U.S. these groups have formed separate operating companies as non-vessel operating common carriers which would enable them to comply with the Federal Maritime Commission regulations whereby a freight forwarder cannot act as an NVO for his own cargoes. However, some freight forwarders do not operate under a separate tariff but only charge for consolidation and delivery to the pier where the regular steamship bill of lading is issued and the steamship rates apply. Others go even farther and route the cargoes directly to the piers or to the steamship consolidation facility. Some freight forwarder groups formed for the purpose of handling cargo in containers are: Conship, Hamburg/Bremen; Confracht, Vienna; Soteco, Paris; Amcon (Cooperative Vereniging Container Consolidatie Groep U.A., Amsterdam; Contrex (Container Combinatie Van Expediteurs), Rotterdam; ACF (Antwerp Container Forwarding Cooperation), Antwerp; MGL (Mersey Groupage Ltd.), Liverpool.

Other arrangements in the handling of cargoes for containerized movements are made by "consolidators," who receive LCL cargo in their warehouses, load containers and deliver full containers to the piers. In reverse, they take over the containers from the piers (which consist of LCL cargoes loaded by another consolidator overseas), bring them to their warehouse where they break bulk and then arrange distribution. Container Express, Ltd., of London, and World Warehouse, in New York, are examples of this type of operation. Many times, freight forwarders, not members of the larger groups and not having large enough volume, will utilize the services of a consolidator. As another alternative, groups of shippers will unite and form a group, as 148 Connecticut manufacturers have done under the name of Charter Oaks Cooperative Shippers Association. Lastly, the NVOCC, exists. The NVO performs all of the above, functioning as a transport company. Some NVO's are: United Cargo Corp., Gilbert International and REA Express. In the U.S. the entire system of handling consolidated containers off waterfront facilities and piers without longshore labor is creating many problems for consolidators.

Companies, operating to and from the United States, are licensed by the Federal Maritime Commission as Non-Vessel Operating Common Carriers by Water in International Trade (NVOCC's). They file independent tariffs and they started the trend away from the seaport being the beginning and the end of the international transport of the cargo and toward the concept that the cargo starts at the manufacturer's plant and ends at the customer's warehouse. The carrier who moves cargo on this door-to-door basis, with intermodal equipment assuming the responsibility for all the segments, issuing one bill of lading for the entire movement (although, at the present time, if he is not a trucker with ICC rights in the particular inland area, he will have to let the ICC carrier issue his own lading, referring to the fact, as well as the terms of movement in his own lading), is now called a transmodalist.

The transmodalist fills a unique area in the transport of cargo. He moves it, and though oft-times owning the containers, yet he does not own the means of their movement. He arranges "through" rates, yet they are not always the sum total of all the pieces. That he fills a need is recognized by the Federal Maritime Commission in officially regulating the existence of the NVOCC. While the original intent of the NVOCC regulation did not necessarily envision this carrier as one that would serve for other than essentially port-to-port water carriage, the concept of NVOCC as a building block for door-to-door transport through the interaction of land and sea modes has taken hold and grown and has been one of the leading means of expanding the role of the intermodal container.

A popular misconception has been that the total of rates comprise only the U.S. inland cost, the ocean cost, and the overseas inland cost. Ascertaining the true costs involves a thorough analysis and usually brings to light factors in which, quite often, the ocean rate plays a minor rather than a major role.

Packaging is the first cost to consider. As mentioned elsewhere, there are many considerations involved in establishing the proper packaging, but in the use of wooden casing for packaging, the costs of the wood, the storage of the wood, the labor in construction, the cost of transport of the wood throughout the inland, sea and overseas movement, the duty on the wood (since in most instances it is considered as part of the "first cost" of the merchandise), and then finally the cost of disposal require careful consideration. Selling terms may be such that the shipper relieves himself of responsibility at the port and passes it on to the consignee, yet the *cargo* bears the cost and, in the final analysis, *total* costs determine the selling price and whether it is competitive. The same can be said for drummed cargo, since it is possible, when shipping by container, to utilize a lower gauge metal, since drums of varying gauges are readily available. Though export packaging may not be eliminated, it can often be materially reduced. Other costs to be taken into consideration are the "pier delivery" charges, unloading charges, possible storage charges, possible detention charges. In addition, on U.S. inland movements costs vary with quantity. It is very rare when manufacturers have their production geared to "container loads." There is usually an "overflow." The cost for moving this LCL will vary.

Stabilization of the rate picture through the establishment of a single rate factor, reduction (if not elimination) of pilferage and damage through reduced handlings and exposures, the single supervision and responsibility factor through worldwide organization and through bills of lading, the ability to consolidate and break cargo, handling LCL's with the same ease as container loads, have all pointed up the fact that the NVOCC has filled a very real need and has earned a place in containerized transport.

OWNERSHIP, LEASING, AND METHODS OF USAGE

Today's largest users of containers are the steamship lines. Some lines continue to use non-standard container sizes because of captive service operations, but a majority of the industry is purchasing new standardized equipment and/or converting present equipment to standard maritime freight service. The trucking industry has been relatively slow in accepting Group I containers. One reason is that a large-scale shift to containerization requires large investments in new equipment and loading facilities that few motor carriers can afford. Railroad use of containers is steadily increasing and a large percentage of United States railroads is presently using Group I containers. Very few shippers own containers, notable exceptions being the U.S. Government, leasing companies and transmodalists. Most carrier-used containers are owned or are on long-term lease.

While most Group I containers still remain in the hands of steamship companies, leasing companies are playing an increasingly larger role in the available supply. In Europe, Contrans (a German company formed in 1952 by the Port Authorities of Bremen and Hamburg, German Railways, the Federal Association of Forwarders, Hamburg American Line and North German Lloyd), serves as one of the large container leasing and rental organizations in Europe. Presently owned by the German Railway, Hapag-Lloyd and the OETKER organization, Contrans operates 3,000 containers of 5–10 cubic-meter volume and about 200 ISO units.

Sea Containers, Ltd. (a London-based company, Mercantile Engineering Ltd., in Grays, Essex, England) operates a container leasing service. Users can lease containers for two minimum periods of 3 and 5 years which, under certain circumstances, can be extended. For the three-year period, lessees pay 36 monthly payments; each five-year leasing period participants pay 60 monthly installments, each month in advance, at the rate of 2.202% of the delivery cost. If it is desired to extend the minimum leasing periods, on the 37th and 51st months, respectively, 0.05% on the original delivery price will be charged. In the United States, Interpool, an REA Express subsidiary, leases containers as part of a cargo rail system in the U.S. This system is described in more detail in Chapter 5. Other American leasing companies are Rentco (a division of Freuhof), in New Jersey; XTRA, in Boston; North American International, in Indiana; Container Transport International and Integrated Container Service, in New York, among others. The Integrated Container Service plan differs from that of some of its competitors in that each member is responsible for the equipment it has and must hold the container until it can pass it on to another member. This is sometimes described as the "hot-potato" system. The function of the pool is to supply the equipment initially and to make the hand-over function as smooth as possible by providing up-to-date information on the whereabouts of the boxes through a locator system. The Container Transport International

Plan permits the user to turn in the container without resigning from the pool, offers a bonus for avoiding interchange and provides for a penalty payment for turn-ins. North American International, a leading supplier of small non-ISO containers, has lease rates as low as 12½¢ per day.

There are containers available on lease for large shippers whose volume of transfer keeps the containers in constant use. While the trend has been toward the use of smaller, non-standard sizes for this purpose, the standard 20-footer is also being added.

Steamship companies will make their containers available to shippers for their use, with a varying range of charges depending on the area of the world, the steamship conference involved, and the size of the shipper. Steamship companies have not yet developed a consistent policy on allowances for use of nonsteamship-owned containers, but are feeling their way along in this area. Some offer a flat amount; others offer a percentage; still others, a per-diem allowance.

LIMITATIONS: GEOGRAPHIC, TYPE OF CARGO, LACK OF TWO-WAY MOVEMENTS

While this new transport mode has created a revolution in shipping, there are certain conditions, which, though they may change in the future, presently put a limiting factor on the container's uses. Certain areas of the world are not ready to move into containerization. Many of the underdeveloped countries of Africa and Asia fall into this category. But many countries not considered underdeveloped, such as South and Central American countries, are not yet ready to handle regular container traffic. For the start of such traffic, steamship service is essential; however, service with ships that can carry but a few containers on deck or the small non-standard containers is also meaningless. In addition to service with partial or full-container ships, port facilities are necessary with the required lifting equipment, storage areas, rail sidings, chassis and tractors. Inland railroads and highways must be developed. Above all, the cargo must be available and this requires a sophisticated manufacturing economy—it requires participation by the "have" nations that have the monies and know-how in international trade.

For the immediate future, most transport companies are selective about the type of cargoes they transport in containers. Some commodities, especially those that move in bulk, have a low rating and therefore are not considered desirable. Some commodities have noxious qualities, such as the odor of hides, which could contaminate other cargoes in the container. "Label" cargoes are not carried.

The two-way movement of cargo on a regular basis is a desirable, if not essential, part of the economics of container movement. The costs of moving empty containers one way can make the entire operation prohibitive and some steamship companies will accept low-paying freights if only to cover the costs of moving containers into position for a desirable one-way movement. This is done, however, when the potential for development of two-way traffic exists.

The limitations are few and will further diminish when availability of equipment and ships increase.

2

Documentation

The Bill of Lading in international commerce simultaneously acts as a contract of carriage, as a receipt and as a document of title. The Ocean Bill of Lading took on particular importance in dealing with precontainerized movements as the terms of sale usually involved the transfer of title at the domestic or foreign port. An additional function was therefore added to this ocean bill of lading as it was used when documentary credits were involved, for presentation to a bank as evidence of shipment. The bank then accepted these, together with other supporting documents (usually insurance documents and commercial invoices), and made payment against the credit authorized for the merchandise in question.

In the movement of cargo by steamship container (i.e., where cargo is loaded and discharged on the steamship piers) we are still dealing essentially with an ocean bill of lading. The only area that poses a problem is the method used for establishing liability limits covering damage or loss. At the present time there is a $500.00 limitation for any package or "unit" of cargo, established in the Bills of Lading Convention of 1924, and known as *The Hague Rules*.

With a view toward developing uniformity of the rules governing the obligations and liabilities of shipowners under a bill of lading, a code of rules was drawn up by the Maritime Law Committee of the International Law Association, in 1921. These rules, *The Hague Rules*, technically established the liabilities and rights of shippers and ocean carriers. They were reissued with some amendments in 1924, and were then placed before the International Conference on Maritime Law, at Brussels, in 1924. The delegates at the Conference unanimously recommended the adoption of a draft convention embodying *The Hague Rules* as the basis of a convention between the participating countries. In accordance with these recommendations, the rules have been given statutory force with certain amendments in various countries, and form the basis of the *Carriage of Goods by Sea Acts* of Great Britain and the United States.

A problem arises in the interpretation basing the $500.00 limitation on a "package" of cargo regardless of its size or contents. This interpretation considers a container a single package even if that container be a 40-foot-long container. Action is being considered to correct this in the near future.

Just as the conception of the cargo movements has expanded beyond that of only the port-to-port portion, so has the need for a covering bill of lading changed. No longer is the ocean bill of lading enough. A door-to-door bill of lading covering the complete movement is necessary. Although freight forwarders, manufacturers and others have been issuing their own bills of lading, these are but memos and have no legal significance, either for purposes of title or for purposes of compliance with Federal Maritime Commission regulations. As mentioned earlier, the "document" covering the movement of the cargo, the Bill of Lading, must fulfill the following requirements:

1. It must be a certificate of title and account for the cargo in detail;

2. It must be transferable and thus be able to serve as a financing instrument on the basis of which the financiers will, if necessary, be able to exercise their proprietary or security rights;

3. It must give the holder the right to claim compensation in case of loss, damage or wrong delivery from the company that issued the document.

Indeed, as we well know, the banks can be told what documents to accept in a Letter of Credit, but anything short of a recognized Bill of Lading is degraded to the status of a mere receipt and lacks the requirements for financing. Only recognized common carriers can issue Bills of Lading and the Non-Vessel Operating Common Carrier by Water in Foreign Trade can issue a Bill of Lading to an inland foreign destination. Title is vested in the owner of such Bills of Lading without interruption or fragmentation to final destination. The NVO assumes complete responsibility as the single carrier. However, a necessary change must take place in the execution of a Bill of Lading when the loading is carried out on the shipper's premises. In such instances, the shipper can no longer hold the carrier responsible for issuing a receipt for weight or count. The Bill of Lading will merely be noted "subject to shipper's load and count." The shipper will then have to answer directly to his consignee as in the case of questions of non-delivery of part of the shipment when no signs of entry are evident on the exterior of the container.

The U.S. *Bill of Lading Act of 1916*, known as the *Pomerene Act*, states in Section 21 (in part):

That when package freight or bulk freight is loaded by a shipper and the goods are described in a bill of lading merely by a statement of marks or labels upon them or upon packages containing them, or by a statement that the goods are said to be goods of a certain kind or quantity, or in a certain condition, or it is stated in the bill of lading that packages are said to contain goods of a certain kind or quantity or in a certain condition, or that the contents or condition of the contents of packages are unknown, or words of like purport are contained in the bill of lading, such statements, if true, shall not make liable the carrier issuing the bill of lading, although the goods are not of the kind or quantity or in the condition which the marks or labels upon them indicate, or of the kind or quantity or in the condition they were said to be by the consignor. The carrier may also by inserting in the bill of lading the words "Shipper's weight, load, and count," or other words of like purport, indicate that the goods were loaded by the shipper and the description of them made by him; and if such statement be true, the carrier shall not be liable for damages caused by the improper loading or by the non-receipt or by the misdescription of the goods described in the bill of lading.

The rate structure in present FMC-filed NVO tariffs must show port-to-port rates. Any other rates in the tariff, including overseas inland rates, may be shown in addition to these. However, to show the inland rates in the U.S., the Interstate Commerce Commission becomes involved, as well. Thus, at the present time, in order for a through, or door-to-door Bill of Lading to be issued, so as to be able to quote a single door-to-door rate, a tariff must be filed with the FMC and ICC.

Separate permission, however, must be obtained from both bodies in order to do so. Things may be changed when the ICC, either in conjunction with the FMC or independently, will recognize intermodal carriers and authorize them to quote through rates even though they are not truckers, after attesting to the fact that such rates applicable to the U.S. inland portion are not lower than ICC-filed rates. To date, a certain degree of progress has been achieved in this area. Various provisions of the *Interstate Commerce Act* already permit the establishment of joint coordinated intermodal service between carriers of the several modes subject to the commission's jurisdiction. This, at present, does not provide for the transmodalist carrier nor for the interrelation of the regulatory functions between the ICC and the FMC. The future, however, will eventually see pressures of the realities of the situation force a combination of all the transportation regulatory bodies into a single entity. The *Trade Simplification Act of 1968* attempts to solve this. This bill would authorize and foster joint rates for international transportation of property and would facilitate the transportation of such property. The 1968 Congress did not consider the bill and attempts are being made to bring it to the floor in 1969.

This bill is designed basically for the American shipper; it would satisfy a number of basic needs. It would remove the obstacles to publication of through, single-factor tariffs. It would put to rest any reservations the regulatory agencies may have about their power to accept intermodal tariffs for filing. This is considered to be of primary importance, since an exact knowledge by the shipper—in advance—of what his transportation costs will be is one of the keys to a flourishing international trade.

Related to the need for simple and concise tariff quotations is the need to eliminate the present chaos in commodity descriptions. This bill attempts to promote uniform commodity descriptions in tariffs.

In addition, the bill would make an initial assault on the paperwork problem by permitting all the carriers participating in a through movement to issue a single through bill of lading for the entire trip, door-to-door. Such documents, voluntarily and cooperatively developed by carriers, would satisfy the desire for paper whose negotiability is recognized by the banking community here and abroad, and would make the benefits of financing with such paper available from the beginning of a shipment's inland movement, prior to the customary and traditional issuance of the ocean carrier's bill of lading.

This legislation also would deal indirectly with the serious impediment which stems from the absence of a simple and uniform system of carrier liability. Variations in the substantive rights provided by numerous domestic and foreign laws and international conventions, are so wide that a quick, comprehensive legislative overhaul would not be practical at present.

Therefore, this bill tries to create a mechanism that will allow competitive market forces to work toward uniform liability. It will provide the impetus for carriers to assume full responsibility for loss or damage, and to reflect their costs in the joint rate. This is already being done by some Part IV freight forwarders.

Finally, in view of the high cost of containers and related equipment, maximum utilization of that equipment is a prerequisite to an efficient intermodal transportation system. For that reason, the bill would permit carriers to interchange equipment.

The fundamental approach of the bill is voluntary and permissive. Carriers and shippers would be permitted to establish and use joint rates and through routes. They would not be *required* to do so, nor would they be *required* to devise simplified, through bills of lading. In the same vein, no new type of carrier would be created nor certificated by the bill, and all existing types of carriers would be free to enter into joint rates with all other types of carriers.

It is not the intention of the bill to extend or alter the present rights of Part IV freight forwarders to make special contracts with their underlying carriers. A Part IV freight forwarder would merely be authorized to participate as a party to a joint through rate arrangement in providing one leg of a through movement.

None of the regulatory agencies would have jurisdiction over the whole intermodal rate or intercarrier agreement, but the proposed arrangements would have to receive the approval of each relevant regulatory agency under the laws, regulations and interpretations which the agency now applies to rates or practices involving single-mode shipments.

Transporters for hire—that is, carriers operating between points outside the United States—are authorized by the bill to participate in joint rates.

The *Trade Simplification Act* would work as follows:

After receiving the approval of their respective regulatory agencies, a group of carriers who desired to establish joint rates would meet together and come up with a proposal including the joint rate; the division of revenues among the various carriers; the apportionment of liability; procedures for the interchange of equipment, and so on. Each carrier would then file the joint rate tariff with its respective regulatory agency.

Let us assume that a trucking company, an ocean carrier and a foreign surface carrier have agreed upon a joint rate. Suppose further that the door-to-door joint rate will be $100 per ton and that the trucking company would receive $30 and the steamship company, $50. The trucker would file the joint rate with the ICC. The commission might also require the trucking company to identify the $30 division it will receive. The commission could request that this division be filed for informational purposes or actually shown in the tariff.

The same requirements would apply to the steamship company before the Federal Maritime Commission. Similarly, if the joint rate involved an air carrier, the relationship between that carrier and the CAB would be the same. A joint rate would depend upon the approval of all the regulatory agencies involved, but no agency would be given jurisdiction over the entire rate.

It is not the intention of this bill to require transporters for hire to file divisions of rates with the regulatory agencies either for informational purposes or in tariffs. Each regulatory agency would retain the authority it now has over carriers which operate under its jurisdiction.

Each agency would be authorized to treat a joint-rate agreement just as it would an agreement entered into by competing carriers, all of whom are under its jurisdiction.

Antitrust immunity would be conferred on joint-rate arrangements when each regulatory agency involved approved such arrangements under the legal standards each agency administers.

Each agency would have authority to approve or disapprove participation in such discussions by carriers subject to its jurisdiction. Agreements would be permitted between a single carrier or group of carriers in one mode and a single carrier or group of carriers in another mode. The agencies would cooperate among themselves in issuing any necessary uniform regulations to be observed by carriers entering into joint-rate agreements.

In addition, the U.S. Government has formed a transportation facilitation committee consisting of a group of government and transportation industry officials who will direct wide ranging efforts to eliminate unnecessary documentation, processing procedures, inspection and clearance.

The Task Force on Intermodal Transport

A series of task forces are assigned specific areas of responsibility. The Task Force on Intermodal Transport has the following work groups.

Surface Intermodal Container Systems, whose objectives are:

To promote and support efforts to achieve national and international acceptance of those features of intermodal transport which will ease flow of containers moving via surface means through interchange terminals and transshipping points, between modes and across national frontiers. This is to include those elements of size, structural strength, lifting and securing devices, and similar considerations, that will aid and abet compatible interchange between the various means of transport and by the trading countries will yet not impede technological development nor the initiative of manufacturers and owners.

To determine appropriate container safety provisions required to assure safety of transport equipment and crews and to study feasibility of action with principal U.S. manufacturers of containers, underwriters and transport operators to formulate detailed proposals to achieve simplified type-approval plans for meeting domestic or foreign safety requirements and international custom regulations.

To delineate and work for the adoption of the commonly agreed means of marking, certifying and, if necessary, registering, intermodal containers so that pre-arrival notice can expedite clearance through interchange points and across international borders.

To arrange for and conduct necessary experimental or pilot operation of intercontinental container movement via surface means to examine the effectiveness of procedures, physical handling and equipment requirements, as well as to isolate detrimental aspects and areas warranting attention.

Intermodal Statutory and Tariff Provisions, whose objectives are:

To develop, propose and coordinate legislative and/or administrative changes which will permit, for international movement, carriers to jointly publish through (single-factor) freight rates, specify joint routes and interchange agreements, and to jointly assume responsibility from origin to destination under through bills of lading.

By consultation with regulatory agencies, commercial and government shipping activities, carriers and rate publishing associations, to define and promote the acceptance by all modes of: (1) a common basis of expressing rates and charges which will give due recognition to weight and cube relationships, and (2) maximum uniformity or at least comparability of the rules portion of joint through-rate tariffs so as to facilitate comparison and use.

To consider methods of international rate-making employed by various modes of transportation and rate control by government agencies, so as to prepare recommendations in respect to correlation, coordination, exemption or other facets promoting intermodal operations. This should include study of current methods available for handling shippers' requests for joint through (intermodal) rates, and carrier or rate conference consideration of shipper request, for the purpose of improving, simplifying and unifying procedures and expediting publication. To promote simplified tariff provisions for rates and charges of intercontinental containerized cargo with ease of dissemination, so that shippers and others involved in foreign trade can readily ascertain complete charges for total door-to-door delivery cost.

To develop data on agreements or provisions specified by carrier conferences, government commissions or international treaties regarding carrier liability for loss or damage provided in different degrees and on different bases by the respective means of transport employed in joint through (intercontinental) transportation.

To determine results of pending court cases concerning limitations on carrier liability for container and actions related to review by Comite Maritime International of the Draft Convention (for combined international transport) drafted by the International Institute of Unifications of Private Law, in Rome; as related to the carrier liability provisions of the International Goods Railway Convention (C.I.M.) and the International Carriage of Goods by Road (C.M.R.).

Through information developed as outlined above, and the necessary consultation and coordination with interested agencies, to define and clarify carrier liability to be expected; and when deemed feasible, to promote maximum uniformity or compatibility among the respective modal carriers participating in joint through movement under single contractual arrangement (through bill of lading) between shipper and carrier.

To arrange for economic study of open cargo policy and other types of insurance coverage and average costs available for carriers and shippers, as well as for passengers moving internationally to include liability indemnity.

To develop information on minimum insurance requirements governing international transportation and appropriate recommendations to protect public interest.

To obtain data on claims contributing to rate level of premium, and to promulgate corrective measures which would tend to reduce the claim incidence.

Intermodal Inspection Procedures, whose objectives are:

To identify and propose programs for resolution of impediments in the transportation processes due to existing legal, regulatory or operating practices governing inspection and declaration procedures.

To evaluate government and industrial interests affected by inspection processes, and develop the most efficient and effective solutions to problem areas under consideration in international trades.

To examine specialized inspection procedures that will reciprocally obviate the necessity to open sealed containers at frontier points for routine sampling and inspection, and permit the temporary admission of containers and chassis.

Intermodal Statistics and Coding, whose objectives are:

To develop a system for statistical compilation of data related to intercontinental traffic moving in containerized or unitized loads to provide information in respect to the direction and magnitude of such traffic based, to the extent feasible, on the consensus of using agency needs. To devise a system encompassing and/or amplifying data collection now performed by modal administrations, regulatory agencies and international organizations with the purpose of minimizing any duplication.

To identify problems and coordinate technological solution to intermodal application of commodity descriptions and codes; to promote international use of standard commodity descriptions and codes.

To determine most effective means for common commodity identification for carriers, shippers, and other interests and promote adoption of techniques to achieve compatible or common coding. Make an effort to evaluate the various needs of government and industry, including the different transport modes and users of statistical data, in order to develop an approach which will result in a coding system to universally or compatibly code commodities for domestic and international movements.

A voluntary, nonprofit group, called the National Committee on International Trade Documentation (NCITD), was formed in Washington, D.C., whose purpose is to simplify and improve international trade documentation and documentation procedures.

In Europe, the Institut International pour L'Unification du Droit Privé, known as *Unidroit,* has submitted to the Bureau International des Containers a draft convention providing for the introduction of a new shipping document, called *Titre de Transport Combiné* (abbreviated as *TTC*), which will cover the whole journey in the case of combined transport; e.g., overland and by sea. The TTC will have to be issued by what the convention calls the "transporteur principal," with whom the shipper makes his contractual arrangements and who assumes liability for the goods throughout the entire transport operation, i.e., also for the carriers whose services the "transporteur principal" employs for parts of the journey. Who the "transporteur principal" will be in actual practice is not defined in the convention.

According to the convention, the TTC will have the same features that are at present embodied in the bill of lading, i.e., it can be made out to order, be endorsable, and embody the holder's title to the goods. The documents for the various stages comprised in the journey as a whole, such as road and rail consignment notes, and also the marine bill of lading, will in that case merely retain importance as haulage and shipping documents. The TTC alone gives title to the goods and will—in the case of combined transport, at any rate—supersede the bill of lading as the "document of title."

There are still difficulties to be overcome before the general acceptance of the through bill of lading by banking and steamship people. Education and experience in the new usage will gradually enable the necessary changes to evolve.

1539449

OTHER EXPORT AND IMPORT DOCUMENTS

Exports. In the preparation of cargo for export, it is necessary to comply with U.S. government regulations. The general export clearance requirements follow.

1. No exporter or his agent, including any carrier, shall place or permit placing on a pier or dock or other place of loading for the purpose of exporting by water or air, load or carry or permit loading or carrying onto an exporting carrier, or present to the Customs Office for inspection and clearance for export, any commodity or technical data until:

(i) For shipments requiring a validated export license: A validated license therefor has been presented to the Customs Office, and a related duly executed Shipper's Export Declaration in the requisite number of copies covering such commodity or technical data has been presented to, and authenticated by, the Customs Office, and a copy returned to the person presenting it. (Recent liberalization of the Export Control Regulations permit certain categories of exporters to receive a single "distribution" license covering a full year's shipments.)

(ii) For shipments under a general license: A duly executed declaration, in the requisite number of copies, consistent with the provisions of an applicable general license, has been presented to, and authenticated by, the Customs Office, and a copy returned to the person presenting it. Where the filing of a declaration is not required, an oral declaration describing the commodity or technical data about to be exported and identifying the applicable general license shall be made to the Customs Office at the port of exit.

2. No carrier shall load or carry any commodity or technical data onto an exporting carrier or permit any commodity or technical data to be loaded or carried onto an exporting carrier for export by water or air, until such carrier has received its copy of the authenticated Shipper's Export Declaration.

The license provisions remain the same, but the procedures on the authentication take on a different complexion in the movement of the cargo from inland locations by container. Part of the authentication requirements now read:

"That the shipment is or will be available for inspection and has not been loaded on an exporting carrier." However, the U.S. Bureau of Customs has made special rulings for the handling of cargo that moves by container. They state:

"Merchandise for exportation laden into containers must be listed on U.S. Department of Commerce, shipper's export declaration Form 7525-V, in the same manner that would prevail if the cargo were being carried in other than a container. If the port at which the loading commences is a port authorized to authenticate SED's, the lading of the export cargo can be verified, and the container sealed for movement to the port of exportation. Generally, no further examination of the contents will be required, unless the container or seals indicate that they have been the object of tampering. If the container has entered the United States as an instrument of international trade and if in its incidental movement to an exterior port is desired to include merchandise from an intermediate port, this can be accomplished without jeopardy. In this instance the Customs officer would affix and record such other customs seals that may be required, if it is intended to avoid any delay for inspection at the exterior port.

"Assuming that merchandise from the originating port and intermediate port is sufficient to fill the container and further considering that said cargo is destined to the same country, the container will not necessarily require opening at the port of exit from the United States, if the lading of the container has been under the supervision of customs. In the event the cargo laden at the interior port is not sufficient to fill the container, or if all cargo laden is not intended to be transported to the same destination, the container would be placed in a container station at the port of exit for the purpose of further utilizing the instrument. This movement is entirely on the initiative of the carrier and is not a requirement of U.S. Customs. It is necessary that merchandise laden in the containers be detailed on vessel manifests. For example, the marks and numbers, together with a description of packages in the container according to usual name of the package, such as carton, case, bag, etc., and the actual description of the merchandise should be shown."

In addition, the Bureau of Customs, upon application, will arrange for permission to load export shipments in containers at inland ports, with authentication of the related shipper's export declaration at such inland ports. It thus becomes possible for the inland city to become an "inland port of export." The shipper's cargo is loaded, for example, in Dayton, Ohio, the export declaration is authenticated in Dayton, the foreign freight forwarder in Dayton completes the necessary documentation (the consular invoices, when required, still have to be sent on to the cities where consular offices exist), the Bill of Lading is issued in Dayton and the papers are presented for collection to a bank in Dayton.

Import. All merchandise imported into the United States from abroad is subject to custom's entry, examination and appraisement. Here, too, the Bureau of Customs states:

"Merchandise being imported in containers must be identified on the manifest of the importing carrier to indicate the description of the packages in addition to the contents of each package. The cargo may be cleared at the first port of arrival or may be transported under customs bond to an inland port of entry for customs clearance. In either event, actual examination and release may be accomplished at the pier, terminal, container stations, importer's premises, or other location, depending on the circumstances. The choice of clearance at the first port of arrival or at an inland location is that of the importer while the exact location within the port is usually a matter of agreement between customs and the importer.

"Merchandise in transit from the port of arrival to an inland port of entry is transported in a container secured with a customs seal, or, if there is not a sufficient quantity of merchandise to utilize the container fully, additional cargo may be laden therein and the sealing equipment waived. Such movement is transacted under a carrier's bond which insures that the merchandise will be delivered to the customs office at the port of destination, or the office nearest such terminus point."

It, therefore, is possible to bring cargoes all the way through to the inland container station for clearance where the importer himself can be present while the cargo is classified and duty assessed.

BANKING

International trade flows as freely as the exchange of the goods for payment permits it to flow. From the viewpoint of the merchants involved, therefore, international trade can present many problems. While the documentary credit is banking's major solution to the particular one of making and receiving payment against delivery of goods, it is important to remember that, in documentary credit, all parties concerned deal in documents, not goods. Basically it is a conditional promise of payment, assuring the seller of his money by substituting the credit-worthiness of a bank for that of the buyer, and interposing that bank's expertise on behalf of the buyer to ensure constructive delivery in the form of simplicity. In practice, however, it can display a weakness. Practice can only be as good as the practitioner, and traders are merchants, not financial experts; banks are not all international in outlook and understanding; and some countries lack banking experience and have still to develop a banking tradition. Consequently, misunderstandings may arise regarding the meaning of terms used, and disputes can develop over the obligations imposed, a state of affairs which at worst may lead to litigation, and at best may justify the voicing of doubts as to the effectiveness of documentary credits.

A leading organization in the field of international commerce is the International Chamber of Commerce. Founded in 1919, the ICC brings together producers and consumers, manufacturers and traders, bankers and insurers, carriers and transport users, and legal and economic experts from more than 70 countries. Extending across political frontiers, the ICC enables them to meet and pool their experience and forge a common policy adapted both to national and to international requirements.

"More than thirty-five active years," it has been written, "have brought the ICC world-wide recognition as a fully representative body, with realistic and objective methods of work. Governments and inter-governmental agencies increasingly turn to the ICC for advice and guidance about business requirements. The ICC's recommendations have many times been reflected in official decisions closely affecting international trade and ensuring its expansion."

Some fifty international technical commissions and committees composed of businessmen and experts appointed by member countries, with whose activities more than fifty specialized international organizations are associated, help to further the ICC's program of work. This program is divided among four main groups: 1) Economic and financial policy; 2) Production, distribution, advertising; 3) Transport and communications; and 4) Law and commercial practice.

The Uniform Customs and Practice for Documentary Credits (1962 Revision), as drawn up by the International Chamber of Commerce, has formulated a code of uniform practice which has been adopted by banks and banking associations in 173 countries and territories. This code in Section C—Documents, states:

Article 13. "All instructions to issue, confirm or advise a credit must state precisely the documents against which payment, acceptance or negotiation is to be made.

"Terms such as "first class," "well known," "qualified" and the like shall not be used to describe the issuers of any documents called for under credits and if they are incorporated in the credit terms, banks will accept documents as presented without further responsibility on their part."

Article 14. "Except as stated in Article 18, the date of the Bill of Lading, or date indicated in the reception stamp or by notation on any other document evidencing shipment or despatch, will be taken in each case to be the date of shipment or despatch of the goods."

Article 17. "Unless specifically authorized in the credit, Bills of Lading of the following nature will be rejected:

a) Bills of Lading issued by forwarding agents.

b) Bills of Lading which are issued under and are subject to the conditions of a Charter Party.

c) Bills of Lading covering shipment by sailing vessels.

However, unless otherwise specified in the credit, Bills of Lading of the following nature will be accepted:

a) "Port" or "Custody" Bills of Lading for shipments of cotton from the United States of America.

b) "Through" Bills of Lading issued by steamship companies or their agents even though they cover several modes of transport."

Article 18. "Unless otherwise specified in the credit, Bills of Lading must show that the goods are loaded on board.

Loading on board may be evidenced by an on-board Bill of Lading or by means of a notation to the effect dated and signed or initialed by the carrier or his agent, and the date of this notation shall be regarded as the date of loading on board and shipment."

Article 20. "Banks will refuse a Bill of Lading showing the stowage of goods on deck, unless specifically authorized in the credit."

The key to the entire usage of banks for making payments against documentation lies in Article 13. While the code attempts to set up guidelines, the seller states exactly what it is against which the bank is to make payment. Nothing is sacred. Everything is possible. If, therefore any change is to take place, it will be the seller who will help set the new guidelines. In Article 17 (b) usage will demand change so that any recognized transport company issuing through bills of lading covering several modes of transport will have its Bills of Lading honored. This can be done even now by the maker of the credit requesting his bank to do this.

Article 18 has been interpreted, over the years, to mean on board a vessel. But when a container is loaded at an inland point, sealed there for export by U.S. Customs, and a Bill of Lading issued at that point, the cargo is "on board" the container. The intent of on board has been fulfilled, that of having the cargo irretrievably on its way. This part of the code of practice will gradually change to conform with the evolution in usage.

In addition, the terms "on deck" and "under deck," too, are undergoing subtle change. The original intent was to offer lower insurance premiums to cargo that was loaded in an area where it might be less susceptible to damage. But, with the new container ships and even some of the converted older ships, where does the deck begin and end? There is much work yet to be done in devising the best method of protecting containers while at sea. There is much divergence in thought as to which would be the best approach in vessel construction, in container construction—and, after all is said and done, there will still be the vessel operator who will stow some containers on deck to get the additional carrying capacity. The "on-deck" or "under-deck" clauses, moreover, while applicable to the ocean bill of lading, take on a different importance when it is the through bill of lading that is involved. There are many possible hazards along the way and the reduction of any of these in a measurable, constant way should be of interest to the underwriters. It is necessary to note one steamship consortium has already taken steps to face this situation. In their bill of lading they state that they shall be entitled to carry goods on deck in containers, but that they, the steamship company, will accept liability for such loadings.

We have primarily discussed the letter of credit, but the adjustment to the usage of the intermodal through bill of lading is equally applicable to sight draft, FCIA credits, and back-to-back letters of credit.

Another steamship line, U.S. Lines, has inserted an "option" clause in its bill of lading, reading:

"Goods in containers, vans or trailers, whether packed by shipper or carrier, may be carried on deck at carrier's option and, if carried on deck, the carrier shall not be required to specially note, mark or stamp any statement of on-deck carriage on this bill of lading, any custom to the contrary notwithstanding.

"With respect to on-deck carriage all risk of loss or damage by perils inherent in or incidental to such carriage shall be borne by the shipper or consignee and the carrier shall have the benefit of all the provisions of this bill of lading except those inconsistent with the provisions of this clause, and the carrier shall have as provisions of this clause the benefit of all and the same rights, immunities, exemptions and limitations as provided for in Sec. 4 of the *Carriage of Goods by Sea Act of the United States*, except subdivisions (1), (2)(j), (2)(q), (3) and (4) thereof as if the same were set forth herein in full. In no event shall the carrier be liable for any loss or damage to on-deck carriage arising or resulting from any cause whatsoever, including unseaworthiness, unless affirmatively proved by the shipper or consignee to be due to lack of due diligence or to the fault or neglect of the carrier or those for whom it may otherwise be responsible, but the carrier shall not in any event be liable for any act, neglect or default in the navigation or the management of the ship."

There are five basic ways in which an "exporter" can receive payment for the goods he ships abroad. These methods of financing exports, in order of increasing risk to the manufacturers are:

1. *Cash Deposit (in Whole or Part) in Advance of Shipment.* In times of economic or political stress in the buyer's country, or when dollar exchange is in short supply, the seller may require such form of payment. This method is of advantage to the seller since no risk is involved, but the volume of business handled on this basis may be negligible. Only in the case of goods being made to the special requirements of the buyer can exporters justify cash in advance from foreign firms of good reputation under normal conditions.

2. *Export Letter of Credit Opened by Foreign Buyer in Favor of U.S. Seller.* A letter of credit is an instrument issued by a bank to a seller, in which the bank undertakes to pay the seller for purchases made by a foreign buyer. It describes the terms and conditions under which payment will be made.

With the exception of cash in advance, this method of financing affords the exporter the greatest protection, particularly if the letter of credit is issued in irrevocable form. It assures that he will be paid in dollars by the issuing or confirming bank upon surrender of the necessary documents and compliance with conditions specified in the letter. Export letters of credit fall into four general classes:

a. The American bank may itself issue an irrevocable letter of credit and take full responsibility of honoring the seller's drafts for account of its foreign correspondent.

b. An irrevocable letter of credit issued by a foreign bank may be confirmed by the American bank.

c. An irrevocable letter of credit may be issued by a foreign bank but without responsibility of engagement of the American bank. This form may provide for drafts to be drawn on the buyer, on the foreign bank, or on the American bank.

d. The fourth class, or the revocable letter of credit, is less frequently used and serves only as a means of arranging payment. The advising bank, on its own volition or acting on instructions from the issuing bank, may revoke or amend it at any time without notice to or consent of the beneficiary.

3. *Draft Drawn by U.S. Seller on Foreign Buyer (Documentary Drafts)*. A substantial portion of exports is financed on the basis of U.S. dollar sight or time draft (30, 60, or 90 days after sight or date) drawn by American exporters on foreign buyers. Such drafts are generally forwarded by the seller to his bank for collection, and in some cases, arrangements may be made by the seller with the collecting bank to borrow against the drafts.

Drafts are usually accompanied by a full set of shipping documents. The latter are surrendered to the buyer upon payment or acceptance, as the case may be, in order that he can obtain the shipment from the ocean carrier and clear it through customs.

When shipping against dollar drafts, however, it is important for the exporter to realize that he assumes both credit and exchange risks until the return of dollars from abroad; and the order can be cancelled prior to the shipment of the merchandise.

4. *Open Account Terms*. The open account method in export financing differs little from its practice in the domestic field. It is used in the foreign field, however, only on sales to foreign buyers of established reputation. While this is a simple method, it provides little evidence of obligation and may complicate eventual collection in the event the foreign buyer defaults.

5. *Sales on Consignment Basis*. Sales on consignment terms are generally limited to shipments to overseas branch offices or subsidiaries. If these terms are extended to agents or importers, the credit risks involved should be thoroughly understood.

As in the case of open account terms, there is little evidence of obligation, which may make collection difficult in the case of default. If exchange restrictions exist, it may not be easy to convert foreign currency into dollars for return to the United States.

In the banking aspect of containerization, the evolution also spreads geographically to the extent that the trend of the bank handling the documents also moves away from the port areas and into the inland areas of the U.S. where the cargo is placed in the containers. Inland banks are no longer provincial but are maintaining their own direct relations with overseas correspondent banks. Thus, another benefit of the containerization of cargo at, or near the point of, origin is the ability and incentive to use a local bank for collections and issuances of credits. The bank that serves the domestic needs serves the international needs as well, and the convenient relationships that exist by using one's own bank, without having to have that bank work through a port city bank, will tend to make international trade easier and make it possible for companies to consider expanding their markets.

At the present time it is too soon for any definite patterns of approach to have been evolved. All that can be said is that because of these new modes of cargo carriage the old approaches, the old terminologies, the old conceptions, can no longer be acceptable.

LIABILITY AND INSURANCE

With the rapid introduction of combined transports on a large scale by means of containers and trailers, the situation has changed radically. The legal systems covering international trade and international transports are not designed for this new type of through-transports. The reason is that the checkpoints for the goods in the conventional transport has disappeared. In commercial law the ships rail and in transport law the ship's side can no longer play their decisive role as the place where the risk passes from the seller to the buyer or the risk passes from a road or rail carrier to the sea carrier. The moment when a container ship is loading or discharging is not an occasion when one has a possibility to count the number of packages or survey the condition of the goods. The only place where this can be done is where the goods are loaded in and discharged from the container. This may happen respectively at the seller's or the buyer's warehouse or at a depot for consolidating containers in the port or in the interior of the country.

In commercial relations, a consequence of this seems to be that the traditional trade terms f.o.b. and c.i.f. will not easily fit into the pattern of container shipments. It is not only the passing of the risk at the ship's rail but also the division of the costs that presents a problem. In the relation between the seller and the buyer it will be necessary to apply a clause that places the borderline between the obligations of the parties either at the place where the goods are loaded in a container or at the place where they are discharged. Of course, the clauses "ex works" and "free delivered" can always be used, but many exporters and importers are reluctant to use them. It may be added that a splitting up of the costs between the seller and buyer at another point than the one indicated under the f.o.b. and c.i.f. clauses will affect also the customs valuation of the goods which is generally based on f.o.b. or c.i.f. values.

During the sea carriage there is a general presumption that the carrier is liable for loss or damage to the goods that has occurred during the period when the goods have been in his custody. There are, however, some very important exemptions to his responsibility. According to Article 4 in *The Hague Rules,* carrier is not liable for loss or damage to the goods which result from neglect or default in the navigation or the management of the ship or resulting from fire or from perils of the sea, etc. This means that the sea carrier is generally not liable for loss or damage to goods due to grounding or total loss of the ship. When he is liable, the compensation is limited to £200 per package or unit with the Gold Clause Agreement. (Section 4 of *The American Carriage of Goods by Sea Act* of $500 per package limitation)

The railway that may have carried the container to the port follows a different legal system, defined in the CIM convention. The liability of the railway is strict, i.e., that the railway is liable for all loss and damage to the goods during the transport. Exemption from liability is allowed only under some specified circumstances where the damage has the character of *force majeure* or the damage is due to an action of the shipper or to insufficient packing or inherent vice of the goods. The compensation is limited to £12 per kg.

If, on the other hand, the container has been carried by a truck on road, we meet a third variation of liability. According to the CMR convention the road carrier, like the railways, is liable for loss or damage to the goods and he may be free from liability under the same circumstances as the railway, but also if the damage is due to "circumstances which the carrier could not avoid and consequences of which he was unable to prevent." The compensation shall not exceed £3 per kg. of gross weight short.

If these different legal systems should work properly in a transit of goods by means of a number of successive carriers, one must have a checkpoint between each carrier where the condition of the goods can be examined, packages counted, etc., and the carrier in fault be made liable. In a container transit one will see the goods only when one opens the container at the end of the transport. With the exception of cases where the container itself is lost or damaged, it will be impossible to refer to one particular carrier damages discovered at the end of the transport. It is also evident that the shipper in a container transport will not accept—he may not even be in a position to conclude—a separate agreement with each carrier. He must be able to conclude the contract on the whole transit with one carrier only, or one operator.

There are different ways to proceed in order to solve the problems. One way would be, of course, a harmonization of the various legal systems governing the different means of transports. Maybe this will be accomplished one day, but it will take a very long time. In the meantime, it should be observed that the international conventions creating unanimous rules for each particular means of transport are of great value for the international trade and will remain so for a considerable time.

Another solution would be to create a special legal system for combined transports. This is under discussion, but also such a solution will take some time before it is generally accepted. What is needed is something that could be applied immediately.

According to the Bill of Lading of the Atlantic Container Line, the sea carrier is responsible for the goods from the time when the goods are received by him either at the sea terminal at the port of dispatch or, which is new, at any inland point in Europe or the U.S.A. The responsibility for the goods follows the legal system applicable to each particular leg of the transit. Thus the sea carriage is subject to *The Hague Rules*, a road carriage between inland points in Europe follows the CMR rules and a carriage by railway follows the CIM. Between the inland points in the U.S.A. the responsibility follows the special rules and conditions adopted by the particular carrier who performs this part of the transit. In order to decide on damages which could not be referred to any of the carriers involved in the transit, there is a special clause in the Bill of Lading (Clause 3—IV) stating the following: "When the goods have been damaged or lost during the through-transportation and it cannot be established in whose custody the goods were when the damage or loss occurred, the damage or loss shall be deemed to have occurred during the sea voyage and *The Hague Rules* shall apply."

Sea-Land, on its Puerto Rican service, issues a "Through Bill of Lading," which states: "Received, subject to the terms and conditions of the applicable bills of lading (which are referred to on the reverse side), which are hereby incorporated and made part hereof, subject to the applicable tariffs, from the shipper specified below, the goods described below, in apparent good order, except as noted hereon (contents and condition of contents of package unknown), marked, consigned and destined as indicated, which the carrier agrees to transport to destination, if on its own line, otherwise to deliver to another carrier on the route to destination (the word carrier being understood throughout this contract as meaning any person or corporation in possession of the goods, including agents and independent contractors). It is mutually agreed as to each carrier of all or any of said goods, that every service to be performed hereunder shall be subject to all the conditions, whether printed or whether contained or incorporated herein, including the terms and conditions of applicable bills of lading referred to on the reverse side hereof, which are hereby agreed to by the shipper and each party interested in the goods."

In addition, in the contract terms and conditions on the Short Form, Clause 1, states: "During the period of receipt of the goods at an inland point until the time the goods are delivered into the custody of Sea-Land Service, Inc., at the port of loading, this carriage shall be governed by and subject to the terms and conditions of the initial carrier's bill of lading and/or the I.C.C. uniform bill of lading contract, whichever is applicable, including all of the contract terms and conditions generally in use.

The responsibility pattern of the NVO is generally the same as the ACL and Sea-Land approach with the overall responsibility being assumed by the issuer of the through bill of lading. The NVO protects himself with the clausing of his bill of lading reading, as does that of United Cargo Corp., for example: "The liability of U.C.C. herein shall in no event be greater than that of the underlying carrier under its bill of lading, receipt, freight note, contract, or other shipping document, and U.C.C. shall be entitled to all of the exemptions from liability therein contained."

Thus the NVO assumes the primary responsibility for the movement giving the shipper one contract to make for any claim, while the NVO, in turn, is covered by his individual contracts with the carriers he used in each mode.

In general we have been dealing with liability and insurance from two standpoints: From the standpoint of the carrier and from the standpoint of the shipper.

To list the types involving the carrier, we have:

1. An open policy for cargo damages, available for shippers not having their own.

2. A bill of lading liability type, protecting the carrier for cargo damage up to the limitation, while the cargo is in the carrier's custody.

3. A policy covering delays and accidents to cargo while in carrier's terminals.

4. A policy for damage that the container may do to equipment that is part of a ship (while loading or discharging), a train, etc.

To list the types involving the shipper, we have had until recently only the extension of the marine insurance he now carries to an all-risk policy covering the movement door-to-door. A new alternative to those offered above is now available to the shipper. An insurance association has been formed in London, called "Through Transit Marine Mutual Assurance Association Ltd.," that is set up to insure the various risks incurred by any carrier who uses containers for the carriage of cargo by sea, on the one hand, and any through transit operator who enters into contract for the carriage of cargo by methods involving the use of more than one means of transport by sea, land, or air. Cover includes two major risks—loss or damage to cargo, and loss or damage to containers—as well as many less important ones, such as expenses and liabilities arising from the use of a container which is infected, infested or contaminated, and customs and other penalties resulting from inaccurate documentation. Third-party liabilities can be covered as well.

Insurance rates, particularly for the shipper, have been the subject of much discussion. While the eventual trend will be toward a downward revision, the improved safety to the cargo which is offered by containerization still needs to have some loose ends straightened out. The pilferage aspect is quite evident to all who have immediately realized that reducing the amount of handlings, the reduction of exposure to piers, the fact that the sealed container is a physical barrier which does not advertise the contents, all point to improvement. The ease with which entire containers can be stolen, however, calls for the tightening of "pick up" procedures and security procedures. The damage to the cargo aspect is not so evident, since there should be a marked improvement, but several factors still need tightening.

Inexperienced stowage of goods in containers is, of course, one of the greatest causes of damage. Many containers are loaded as if they were only destined to be carried over land with no allowance made for the movement of a ship or the additional risks that come when a heavy box is lifted on or off the vessel and placed on a truck or railway flatcar.

The problem here is one of education and training of personnel loading the containers, particularly when containers are loaded away from the container terminals. The packaging of the commodities constitute a big factor in their protection. Heavy wood casing does not necessarily improve the chances of protection. The important factors are proper protection and bracing of the cargo within the carton or crate. Handling equipment on most ships and in many ports puts the containers under many unusual stresses. In stowing a container the packing should be able to resist the pressure exerted by its own contents as well as cargoes nested alongside and strengthened to withstand stacking loads. There can be a considerable saving in costs through reduction, if not entire elimination of export packing, but consideration must be given as to whether the container is going on to final destination. If it does not go on, but just goes to a port or a break-bulk station, the cargo may have to be sent onward to final destination without the container. Packaging should take this into consideration. The damage aspect is further complicated by present inability to control humidity in the container. The risk of damage from condensation is most severe when travel will take the container through different temperature zones.

In order to familiarize ourselves with some of the attendant problems and conditions to which containerized cargoes may be subjected, we shall follow a piece of merchandise from the point of origin, i.e., the manufacturer, to its final discharge at destination. We shall assume the passage to be "normal"

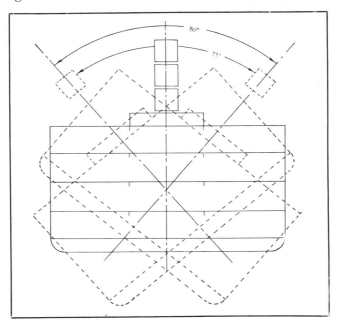

Fig. 8. Effect of vertical position on container motion. The top container moves 73 feet from side to side while a container stowed at the lower 'tween-deck centerline would only roll about its own axis.
(Containerisation International, Apr. , 1968)

or average and the article to be transported may be almost anything from a delicate watch or instrument to a bag of beans or carton of canned goods. The definition of "normal" in this context includes the possibility that the article will be rolled over, tossed, slammed, banged, dropped, rotated, slung, squeezed and lifted, all in varying degrees of severity.

All these things are going to happen as it is stowed in a container, trucked over the road at 96-112 kilometers per hour, over hill and mountain, through rain or snow, and then parked gently or otherwise in a farm at the terminal where it may either be left on wheels or handled by a straddle carrier, lift truck, side-loader or similar device, moved alongside the ship, lifted, swung and finally stowed on board. This process will be repeated between the port of discharge and final destination.

When the ship sails, the cargo is subjected to a new and continuing set of motions. (There are six ship motions: rolling, pitching, yawing, heaving, swaying, and surging. The ship may move in any one of these actions or in any combination of them at one time.) The vessel will pitch, slam and roll without ceasing and, if the weather is moderate, the cargo will be moving forward at the speed of the ship, up and down with the pitching, side to side with the rolling. Of these motions, more damage can be attributed to the rolling than to any other. It is interesting to note that acceleration forces bear a direct relationship to the vertical position of the cargo in the vessel relative to the rolling center. It can be readily seen from the diagram that the content of the top "on-deck" container is subject to far more severe motion than that in containers stowed near the rolling center, i.e., the lower 'tweendeck. (See Fig. 8.)

If it is on deck, it is going to be subjected to sea, wind and weather. It may start in 90° F. temperature, possibly experience sub-zero conditions and then be discharged in hot weather again.

The problems are not insurmountable and, once the underwriters gain more favorable experience and ships themselves gain more experience (the navigation of container ships in heavy weather is still unresolved), insurance rates will come down and stay down.

Customs

Cargo. As in the case of the cargo itself, the customs procedures, especially in the United States, have always been taken care of at ports of exit. In this respect, Europe has been more advanced than the United States in permitting goods to travel across national borders for interior clearance through the medium of TIR, an international convention for the movement of goods. While the participation in TIR will enable many exporters and importers to move containerized cargo with greater ease and thus increase the flow of commerce, NVO's have been using TIR in Europe for the inland movement of their containers. With the cooperation of U.S. Customs, the American flag NVO's have the same ease of movement in the interior of the U.S. The participation of the U.S. in TIR will expand the possibilities and at the same time, perhaps, act as a helping push toward uniting the ICC and FMC.

The main purpose of the TIR Convention is to facilitate the crossing of national frontiers without requiring the discharge of cargo from road vehicles (or containers on road vehicles) for customs inspection, payment of duties at intermediate points, or the posting of bond for transit under seal. This convention entails the establishment in each participating country of one or more guaranteeing associations who undertake to assure that all legitimate customs duties will be paid for the freight that moves under a carnet that such organization has issued.

At the present time, twenty-four countries (Austria, Belgium, Bulgaria, Czechoslovakia, Denmark, Fed. Rep. of Germany, Finland, France, Greece, Hungary, Ireland, Italy, Luxembourg, Netherlands, Norway, Poland, Portugal, Romania, Spain, Sweden, Switzerland, Turkey, United Kingdom and

Yugoslavia) participate in the TIR Convention. United States' accession to the convention has been contemplated for some time. In May 1966, the State Department proposed this action to the White House, and the President recommended favorable consideration to the U.S. Senate in June 1966. On March 1, 1967, the Senate granted its "Advice and Consent," but the instrument of accession was not deposited with the ECE until minor changes in the tariff law had been adopted. A bill (H.R. 18373) providing for U.S. accession to the convention, was reported out by the House Ways and Means Committee on July 24, 1968.

The TIR Convention, in its present form, is not applicable to most transatlantic (intercontinental) transport by containers inasmuch as the containers are generally removed from the chassis and as such are not road vehicles, nor on one. Therefore, the ECE Working Party for Customs Questions Affecting Transport, at U.S. instigation, developed a "special provision" that would make this Convention applicable, even though the containers are carried for part of the journey without being loaded on road vehicles.

European railroads have expressed their readiness to accept TIR carnets issued in the United States, and assume full responsibility whether the containers are forwarded in Europe by rail alone or have a terminal delivery by road.

While the trucking organizations have been primarily concerned for Inter-Europe transport, the ocean haul becomes involved as this service expands to intercontinental transport and EIA water carrier members may also be desirous of obtaining transit privileges available under TIR carnet procedure.

To qualify for transport under this system, trucks and containers must move with their loads under customs seal and with special documentation in the form of a TIR carnet. The carnet is a formal guarantee, usually underwritten by a responsible trucking organization, of payment of any duties, fines or penalties that may be imposed by the customs authorities for contravention of the privileges granted by the Convention. The carnet is a prescribed standard form with multiple copies that provide complete (manifest) information on the shipment.

Two forms of carnets are available. A six-voucher carnet is used for movements over a single frontier, involving only two customs administrations. A fourteen-voucher carnet is used for movements crossing more than one frontier. It is necessary for the carrier, before starting the journey, to determine the number of pages needed (one for the customshouse of departure, one for the customshouse of destination, and one for each customshouse of exit and of entry) and have the carnet filled in accordingly. These forms are obtained from the International Road Transport Union (IRU), in Geneva, Switzerland.

In practical terms the TIR carnet proves of most value in two specific situations. One is where goods shipped by truck must pass a frontier customs post en route to a customshouse at an inland destination in the same country. In this case customs inspection takes place at the inland point rather than at the frontier if the goods are covered by a valid TIR carnet.

The other situation, which perhaps is more important, is where goods shipped by truck must pass in transit through several countries en route to their destination.

Perhaps the most significant feature of the TIR carnet is the insurance protection it carries. Each carnet affords to the guaranteeing association, that purchases it through IRU, liability insurance in the amount of $46,000.00. As a result, the liability of the guaranteeing association for payment of duties, fines or penalties in case of violation of customs laws or regulations in connection with a given carnet is assumed by the insurance company. The premium for this insurance coverage is included in the price paid by the Association to IRU for the carnet form. This insurance is written by a pool of six European insurance companies which have entered into arrangements with IRU for this purpose.

The carnet system presupposes a close working relationship between the customs authorities of each contracting party to the convention and the association that issues and guarantees the TIR carnet in that country. This relationship is outlined in general terms in the convention and its annexes.

The focal point of the association procedure is the International Road Transport Union, an organization of national road transport associations which has its headquarters in Geneva. The IRU has the leading role in administering the TIR carnet system, by virtue of rather indirect language in the TIR Convention which presupposes an affiliation between each association issuing and guaranteeing carnets and an international organization. The IRU actually produces the TIR carnets, distributes them to member associations for formal issuance, and exercises general supervision over their activities.

The relationship between the IRU and each issuing and guaranteeing association affiliated with it is contractual in nature. It is established by a "deed of engagement," which is usually a unilateral undertaking by the association of specific obligations to IRU with respect to the former's participation in the TIR system. Full or associate membership in IRU is not necessary in order for an organization to act as issuing and guaranteeing association for IRU-sponsored TIR carnets.

The IRU has issued general regulations to govern the operation of the TIR carnet system by its affiliated association. In turn, each national association may issue its own regulations to govern the terms on which it will issue carnets to individual shippers, carriers, etc. The IRU does not normally review these internal regulations.

The national association issues carnets it has purchased from IRU. Its main responsibilities in this respect are to screen the applicants and to fix the period of validity of each carnet. Since effective functioning of the TIR system requires that IRU serve as a kind of clearing house for relevant information; this entails systematic reporting of normal operations, in addition to emergency reporting of other matters.

The particulars of the IRU-national association relationship are specified in a written contract. This is the special "deed of engagement" which sets forth the association's obligations to IRU in connection with the carnets it issues and guarantees.

It is important to note that the IRU is willing to convert the deed of engagement into a bilateral form of contract if an association so desires. It is also willing to include additional provisions covering any special aspects of the relationship between IRU and that particular association that may be desirable. It appears that most existing provisions of this kind are added to comply with national regulations imposed on international road transport movements by individual governments. IRU's only requirement in this respect is that the instrument agreed upon incorporate all the provisions of the standard deed of engagement.

Another area of interest concerns the relationship between the national association and the purchaser of the carnets.

The basic rule is that the national association issues carnets only to its own members and to nonmembers it has approved. Both must execute a "declaration of engagement," which is in the nature of a contract between the purchaser and the national association stipulating the former's obligations toward the latter. The Association may write into the declaration of engagement such provisions as it deems desirable. It must include, however, certain clauses required by IRU.

The applicant for carnets must furnish such specific data about his enterprise as the association may find necessary in order to properly evaluate the applicant's qualifications.

The applicant's basic commitment in the contract of declaration of engagement is to strictly observe the regulations governing the use of TIR carnets. Under the contract the purchaser also agrees to carry out certain duties in connection with customs clearance. The IRU also requires that the contract or declaration of engagement between national associations and purchasers be renewed every two years. This is a precaution against possible changes in the financial condition or other changes in his status that might make him a poor risk.

The association may grant carnets for a maximum period of two months. It is free, however, to limit the period of validity to any lesser period if it sees fit. IRU reserves the right to intervene and reduce the period of validity of a carnet issued by a national association. IRU is understood to exercise this right infrequently and only when it fears a serious breach of regulations that might jeopardize its arrangements with its insurers.

Although technically carnets are "issued" by a national road transport association to individual shippers or carriers, they are actually prepared by IRU. IRU has them made up in Switzerland under its own imprint and sells them to its members or affiliated national associations. The latter in turn sells them to individual shippers or carriers.

Guaranteeing associations affiliated with IRU in turn charge individual shippers and carriers for each carnet they issue to them to cover specific shipments of goods. The guaranteeing association is free to fix the amount it charges, within reasonable limits, the individual shipper or carrier to whom a carnet is issued.

Administrative duties of national associations involve maintaining records, keeping forms, and furnishing IRU with certain required reports. A basic requirement is the monthly report that each national association must send to IRU. The association must also report various matters to IRU in addition to the regular monthly report.

Thus a shipper may move his goods through foreign customs points en route to destination without the expense and bother of duty payments or deposits on the contents, or customs examination. When the U.S. becomes a signatory, the TIR system will enable an importer to bring his cargo from Basle, Switzerland, for example, to Indianapolis, Indiana, with the same carnet being used for each national border, even the U.S.A. It will still necessitate final customs clearance at the interior U.S. point nearest destination or at destination when arranged for. This will parallel the "in transit" entry presently made to accomplish this.

A typical example of a customs operation for cargo export from Italy operates as follows:

Documentation is prepared and submitted by broker to the customs office where inspectors are subsequently assigned to the loading terminal or plant, if the cargo is not being brought to the customs station. Inspectors come in the afternoon and there is a team of three, regardless of the amount of cargo: a Chief, a Clerk, and a Finance Officer. After officials arrive, cargo is inspected, loaded, and shipment sealed. The papers are then returned subsequently to the broker together with a bill for the charges. There is, however, a "bolette" which goes with the cargo papers to the border where another customs operation takes place. If the cargo is exported through an Italian port, then this second operation takes place at the port. The operation involves a check of documents and stamping of the bolette which is returned to the broker who uses this bolette as a basis for applying for an "export rebate." Such monies are a rebate given by the government in varying percentages on most commodities that are exported and represent refund of duties on any merchandise not originating in Italy, or refund of domestic taxes or other export incentives.

It is not necessary to have this customs operation done at the point of origin of this cargo. For example, tile can be loaded at Teramo, shoes in Firenze, bicycle saddles in Padova and the customs operation can be done in Padova, or even at the border. It only remains for the last point of loading to make up the carnet TIR.

To summarize, customs procedures do exist, and are constantly being made easier so that the movement of containers from door to door can flow smoothly.

Containers. The container itself becomes a customs consideration in its function as a housing for the cargo. All governments recognize, however, the impracticality of considering the container dutiable and they have set up registration or "matriculation" procedures so that the containers can move freely between borders.

For example, the regulations in Japan state that unless containers have previously been landed in Japan and previously registered and cleared with customs, they have to be registered and temporary landing permits obtained before loading can take place. All non-Japanese containers are so affected. In order to register, a list of containers must be submitted with history of the container together with dollar value in yen at time of purchase.

In Latin America there is not yet any clear awareness of a need to adapt the current laws and regulations to the massive utilization of containers in the International Carriage of Merchandise. Some of the regulations are:

Current provisions in the LAFTA (Latin American Free Trade Association) countries on the customs treatment of containers and of merchandise shipped in them.

Such provisions are known to exist in Argentina, Brazil, Chile, Colombia and Peru. The other two countries in the Plate basin, Paraguay and Uraguay, are not known to have enacted any specific rules on containerization. In Argentina, Law 17.347 authorizes the executive branch to regulate the use of containers in the domestic and international transport of imported and exported merchandise by sea, air and/or land. This law defines a container as a transport conveyance possessing basically the same characteristics described in the same law. Decree 1.071 enacted with the foregoing law on July 18, 1967, directed that a commission presided by the Secretary of State for Transport propose to the Executive Branch, within 30 days of its establishment, the implementing regulations for that law. On April 20, 1967, the General Customs Administration had adopted Resolution 2992 to lay down provisional rules for the handling of merchandise entering or leaving the country in containers pending the enactment of the aforementioned law and implementing regulations.

Basically, Resolution 2992 established that the agents for ships which brought containerized merchandise for importation into the country had to apply for the temporary admission thereof to the customs authority, which was to grant it on condition that the applicant assume responsibility for the return of those containers abroad. The same resolution stated the required technical characteristics of containers which were later incorporated almost without change in Law 17.347.

For purposes of importing containerized merchandise, Resolution 2992 directed that the importer was to apply to customs in writing for the transfer of the goods to his warehouse and had to show proof of ownership before they were transferred. The resolution allowed the examination and clearance of containerized goods to be effected in the warehouse of the interested firm if it was recognizd as such by the customs, and the goods could not be disposed of before they had been cleared. If examination showed the goods to have been falsely declared, they were interdicted and immediately removed to the customs warehouse at the expense of the interested firm.

In an export operation, the interested party had to apply for a temporary exit permit for the containers if they were of domestic origin and the permit was granted under the sole responsibility of the applicant. If a container was in the country under temporary admission, this status terminated at the moment of its shipment abroad. The merchandise to be exported in a container could be checked and accommodated in it on the premises of the exporting firm under a loading permit.

Brazil has had a law on containerization slightly longer that Argentina. Law 4.907 of December 30, 1965 regulates the use of containers in Brazilian foreign trade on the principle that these receptacles qualify for temporary admission status and hence are not dutiable. It also establishes the basic principle of cargo unitization and combined transport by allowing the container through passage to the customs of destination in the interior of the country without being opened or controlled by the customs of entry at the port where it is put ashore. On this point the law states: "The container in international traffic shall be exempt from all import and consumption taxes and all other Federal imposts, including those for renovation of the merchant marine and for the improvement of ports, and enjoys temporary exemption from customs duties," and "a container carrying foreign merchandise may be cleared from the customshouse of original entry in transit to any customs station in the interior of the country where the merchandise is to be finally cleared for consumption." Also, "a container carrying merchandise destined for export may be cleared for exit from any place in the country where there is a functioning customs station."

This law, though more permissive than that of Argentina, does not allow the container to be delivered uncleared to the point or place where it is to be unloaded. The regulations for this law are contained in *Decree No.* P.59.316 of Sept. 28, 1966. Also, a resolution of the Director of Customs of Brazil provides the instructions to be followed by those interested in using containers, especially in view of the rules and other legal regulations approved by Decree-law No. 37 of Nov. 18, 1966. The regulations contained in the resolution of the Director of Customs of Brazil merely develop the basic principles underlying Law 4.907 and state the rules which apply to exports and imports and the requirements to be satisfied with regard to each. In the case of imports, customs may clear containers that are either carrying foreign merchandise or empty, for temporary admission and authorize their free transit within the country. As a fiscal safeguard the owner or user of a container must post a bond with the customs, unless the carrier assumes responsibility for its security, in which case the bond need not be posted. The customs may consent to inspection of the merchandise in the warehouse of the importer only if prior authorization has been obtained, which is subject to compliance with certain conditions. In this case, fiscal agents must perform the inspection within two working days after the import taxes have been paid or after the date on which exemption from those taxes is granted. Special rules are provided to apply in controversies arising in respect of merchandise transported in containers and to allow the forwarding of a closed container carrying uncleared merchandise for inspection and release at the destination in the interior of the country provided that there is at that place a customs station authorized to carry out those formalities. Such customs station must notify the customshouse of original entry of the results of its action. The forwarding of closed containers carrying foreign merchandise to the interior of the country for inspection and release at the destination, even if no customs station is functioning there, is possible but only with express authorization from the Department of the Customs. For purposes of authorizing the through passage of closed containers carrying uncleared merchandise for clearing at destination, the customshouse of original entry must officially, i.e., automatically, seal such containers at the moment when they are first unloaded in the country. If, on inspection of the merchandise, the seals are found to have been broken, the officiating customshouse must institute investigation proceedings and call in the appropriate police authorities.

In specification of the situation described in the foregoing paragraph, the domestic carrier of such containers is answerable for the unimpaired condition of the seals and security devices en route from the customs of original entry to their destination, and this answerability may be transferred to the consignee. In the case of exports, the customs authority may, with proper safeguards, allow the containers to be loaded on the premises of the producer or exporter. Containers may be forwarded directly to the customshouse of exit from any place where there is an authorized customs station. The customshouse of exit confines its control to checking the documents and security devices, including the seals affixed by the customs at the place where the containers were loaded. In cases in which seals and security devices are found to have been tampered, the rules in force for similar cases in imports shall apply.

The above rules are provisional and will be complemented with others later on in order to cover all practical situations in the carriage of containerized goods.

In Colombia, a General Customs Regulation has been issued which, in Chapter LXII on the transportation of uncleared goods covered by the Customs Code, authorized the administrators of the customs at Santa Marta and Buenaventura to permit the direct transfer to cars of the National Railways of merchandise arriving at those ports in closed containers, and they are also empowered to authorize the forwarding of such merchandise to its destination for clearing at that place as specified in advance on the import license. For purposes of the foregoing, such containers may not be passed through to the interior of the country until the legal and regulation documents required for the unloading and carriage of *goods in transit* have been tendered. Once a container has been unloaded, the customs of original entry must take all appropriate measures to guarantee the security of the merchandise during its transit to the customs of final entry, by sealing the containers both on the flatcars and in the freight cars in which they are carried and providing, if necessary, special surveillance en route by customs personnel.

It is established in these regulations that merchandise transported in containers may only be cleared for entry at customs in cities with direct rail connections to the ports of Buenaventura and Santa Marta. Containers must be unloaded from the freight cars directly into the special warehouses placed for the purpose at the disposal of the customs by the National Railways, which are to be used for this purpose alone; moreover, these warehouses are guarded and controlled directly by the customs and are hence to be equipped to serve as customs areas. Merchandise transported in containers is appraised in the manner required by law in these warehouses, from which it may not be withdrawn without presentation of proof that the duties have been paid and the legal clearing formalities complied with. Containerized merchandise may only be transported by the National Railways of Colombia, to which end this enterprise must post a special bond with the National Customs Administration. Also, the customshouse of original entry may, upon receipt of the containers by the National Railways, state the time limit within which the latter must have transported and delivered them to the customshouse of final entry. The National Customs Administration may restrict the transit of containers in accordance with the preceding rules and may even authorize their carriage by means of transportation other than the National Railways of Colombia.

To assure compliance with the requirement that containerized merchandise be carried only by the National Railways of Colombia, there is a contract between this enterprise and the Flota Mercante Grancolombiana S.A. specifying the rights and obligations of both parties. The system set up by the Colombian customs legislation is more restrictive than those of Argentina and Brazil not only as to the ports through which containers laden with foreign merchandise may enter the country, but also as to the means of transport and the place where the customs clearing formalities are to be carried out.

Article 145 (o) of the Customs Ordinance of Chile establishes that metal "Dravo" containers and other such containers which steamship companies use to facilitate transport and protect merchandise from the port of shipment to the domicile or warehouse of the owner or consignee in the country, may enter the country on a temporary basis. As the Chilean customs legislation contains no explicit provisions on the customs treatment of containerized merchandise, it may be inferred that such merchandise is subject to the general rules in force. Despite the lack of more detailed rules on the use of containers and the like, it is possible to deduce that the Chilean customs legislation allows the laden container to proceed without breaking bulk from the port of shipment to the domicile or warehouse of the owner or consignee in the country, which is obviously an appreciable advantage. No reference is made, however, to the type of document by which temporary entry of the container is to be requested nor the formalities to be concluded for forwarding the container from the customshouse of original entry to the location of the warehouse where it is to be unloaded.

In Peru, Supreme Decree No. 82-H of May 22, 1964, lays down the general rules for the use of containers and lift-vans in imports and exports, and for the transfer of merchandise to private warehouses after inspection in customs. The General Superintendency of Customs has been given the responsibility of establishing in each case such quantitative and qualitative controls as may best serve the fiscal interest. In its discharge of that responsibility, the Superintendency issued the appropriate regulations in its resolution of December 11, 1964. Actually, this resolution refers only to inspection of merchandise imported in containers or lift-vans in the customshouses of Callao and the Lima-Callao International Airport by customs inspectors and checkers, who are even required to inspect the entirely empty receptacle, which presupposes the unloading of the merchandise on the customs premises. When the inspection has been completed, the merchandise is returned to the container or lift-van, and a customs tag signed and stamped by the inspecting officials is sealed to the security device. Guards at the exits from these two customshouses must make sure that the signed and stamped control ticket is attached to the lift-van or container and that the seal has not been broken. Merchandise slated for export is also inspected in the customs precincts; the container or lift-van is emptied and refilled and a customs ticket signed and stamped by the inspecting officials is sealed to the security device. The Peruvian system is actually quite restrictive and affords, in practice, no special facility for the transport of containerized goods.

The United States, at present, states in its regulations:

A. "There are five methods whereby containers may enter the United States."

1. *Vessel Equipment.* Containers which are considered to be equipment of a vessel may be unloaded from the vessel, emptied, loaded, and laden aboard the vessel again. No special formalities are required in this instance.

2. *Instrument of International Traffic.* Under this provision the container, whether of United States or foreign ownership, may be admitted to the United States free of duty, either loaded or unloaded, without limitation as to time. The container owner, prior to using this procedure, must file a security bond with U.S. Customs in the amount of $10,000. This bond covers all containers he owns and is good for an indefinite period of time. No special documentation is required for movement of the container into or through the United States but it must be listed on appropriate customs manifests. Containers admitted on this basis may not engage solely in domestic traffic in their movement. They may, however, carry domestic merchandise in moving from one place to another to discharge their imported cargo. Similarly, they may be used to transport domestic cargo, when such movement is consistent with the movement of export cargo to the port of departure from the United States.

3. *American Goods Returned.* Containers which are the product of the United States, can be serially numbered by the owner and marked with the name and address of the United States manufacturer on a metal plate permanently attached to the container. In lieu of the name and address of the manufacturer, the plate may indicate Item 800.00, Tariff Schedules of the United States, the name of the owner, and the serial number. All markings should be conspicuously shown on the container. For example: 800.00 . . . ZENDA . . . 2469. Containers entered under this provision may be used in domestic traffic without limitation.

4. *Duty-paid Containers of Foreign Origin.* Serially numbered containers of foreign manufacture, which were previously imported and duty paid shall, if duty-free entry is claimed on subsequent importations, be conspicuously marked with Item 808.00, Tariff Schedules of the United States, together with the district and port code of the port of entry, and the year of the entry for which duty was paid on the container. The name of the owner and the serial number assigned by the owner shall also be shown. For example, the markings would appear: 808.00...10-1-366-63...ZENDA... 100314, to indicate the item number of the TSUS, New York district, port

of New York, entry number 366, duty paid in fiscal year 1963, owner ZENDA and serial number 100314.

Markings should all be clearly and conspicuously placed on the outside of the container. When a container is removed from service, all markings shall be permanently obliterated. In the event ownership of the container is transferred, the markings shall be similarly removed, except when the new owner wishes to use the container in the international movement of cargo under this procedure, appropriate changes can be made. Containers entered under this provision may be used in domestic traffic without limitation.

5. *Temporary Importations.* Containers containing merchandise may enter the United States without payment of customs duty as a temporary importation under Item 864.45 of the Tariff Schedules of the United States. To apply this procedure, the importer of the container must obtain a separate bond in an amount equal to double the duties which would have been payable if entered for consumption for every importation of each container. Containers imported under this procedure may be used for the domestic movement of cargo during the 1-year bond period. The bond period may be extended up to 2 years at yearly intervals. Before the expiration of the bond period, the container admitted under the temporary importation procedure must be exported either full or empty.

6. *Customs Treatment of Parts, Accessories, Handling Equipment, etc.* Handling equipment and parts or accessories for containers are dutiable at the time of first importation, unless they are imported and in use with a container moving as an instrument of international trade. If such accessory equipment has entered the United States and duty paid thereon, it can be subsequently exported and reimported without the payment of duty if not advanced in value or condition in a foreign country.

B. Documentation

1. *Exportation.* Reusable containers are not considered to be part of the contents of such conveyance in either value or description. Therefore, at the time of exportation, containers must be manifested and appropriately declared on the shipper's export declaration where applicable. This requirement for export documentation exists for all containers, except vessel equipment which are the object of this text whether of foreign or domestic manufacture.

2. *Transit.* Containers may transit the United States by any means of transport, subject to customs entry requirement. The entry and clearance of a container can be accomplished at the port of first arrival, regardless of whether the merchandise is proceeding in bond to an inland destination with its contents. It must, in all instances, be manifested separately from its contents.

As in the case of the TIR convention for cargo, there also exists a Customs Convention on Containers. The Secretary of State of the United States in his report to President Johnson on June 15, 1966, stated, in part:

"I have the honor to submit to you a copy of the Customs Convention on containers, together with the protocol of signature which forms an integral part thereof, done at Geneva on May 18, 1956. I recommend that the convention be transmitted to the Senate for its advice and consent to accession."

The convention, which was formulated by states' members of the United Nations Economic Commission for Europe (ECE), entered into force on August 4, 1959. It is now in force with respect to 24 countries, including a number of the principal European trading countries such as the United Kingdom, France, the Federal Republic of Germany, and Italy. Though the convention is no longer open for signature, the United States may become a party by accession.

The principal articles of the convention provide for temporary (normally no more than 3 months) duty-free entry of containers, either loaded or empty, having an internal volume of 1 cubic meter or more, such as lift-vans, movable tanks, or other similar structures.

One of the most noteworthy recent developments in international trade is the increase in shipments of goods in containers. This development offers great advantages in reduction of costs of transportation and protection of goods in transit. The development of container traffic is also helping to facilitate the desired expansion of U.S. exports.

The increasing interest of the transportation industry in container traffic gave rise to the recommendation which has been made that the 1959 customs convention on the international transport of goods under cover of TIR (Transport International Routier) carnets be submitted to the Senate. The transportation industry views the present basic container convention as an important companion piece to the TIR carnet convention.

Strong recommendations in favor of U.S. accession to the container convention have been made by the National Facilitation Committee, the Department of Commerce, including the Maritime Administration, and the Treasury Department. The Senate Committee on Commerce, in a formal report (Rept. No. 676) on proposed legislation to amend the *Merchant Marine Act,* recommended that the administration give serious consideration to submitting the container convention for approval by the Senate.

In its investigation of the customs treatment of container traffic, the National Facilitation Committee found that the feeling prevails in European countries that U.S. Customs treatment of container traffic is below the prevailing European standard and below that established by the container convention. While that feeling may result, at least in part, from misunderstanding, the committee believes that it may lead to discrimination by European countries against U.S. shipments in containers. It is believed that U.S. accession to the convention would forestall any adverse action abroad against U.S. exports in containers.

The customs treatment of containers provided for in the convention is, in fact, somewhat more liberal in some respects than U.S. customs law provides. It will be necessary to provide for minor amendment of the tariff schedules of the United States before the convention can be made effective. It is proposed to undertake drafting of needed legislation immediately for submission to the Congress.

As seen, containers at present may be entered duty-free into the United States in two ways. One requires a $10,000 surety bond and permits the container to be kept within the country for an indefinite period. The other requires that a bond be posted equal to twice the container's import duty rate. This bonding must be repeated each passage through customs. Usually, a one-year stay is permitted.

Other signatory countries to the Convention on Containers vary on the details of their duty-free rules, although the convention states the tariff exemption should be for up to three months.

In anticipation of favorable action by the U. S. government, it has been arranged for the Equipment Interchange Association of the American trucking associations to be the issuing agents in the United States for the carnets TIR.

3

Stowage

To properly discuss the stowage of containers used in intermodal transport, we have two factors to consider—the container and the cargo.

The container, unlike any other cargo-carrying unit, is subject to threefold stresses:

1. Stresses from the cargo within.

2. Stresses from movement laterally, while on ship: stop and go while on rail and truck chassis; sudden shift of center of gravity while being lifted on or off different modes.

3. Stresses in relation to other objects and forces, such as containers being loaded on top of one another, and the forces of waves and bad weather hitting the containers loaded on deck.

Container construction has taken almost all of the above into account and proper experience in handling will eventually overcome any remaining possibilities of encountering problems.

CARGO

Packaging of Cargo for Containers (for nonspecialized, nonperishable cargoes). Most cargoes moving overseas by conventional means have traditionally had additional protective packaging added in the form of wood crating or wood casing. Such wooden cases can range to about one-third of the weight of the entire shipment and involve a cost factor which, in addition to the wood and labor, includes U.S. inland transportation, ocean transportation, foreign inland transportation, in most instances duty, and then cost for disposal. It is only natural, therefore, for the tendency toward the other extreme of complete elimination or drastic reduction of packaging when movement takes place by container. However, experience has shown that, although the container can often reduce the need for packaging, it certainly does not eliminate it; and that the same principles of adequate packing, based on detailed study of the product itself and the conditions of climate and handling it is expected to encounter, apply equally well to goods consigned in containers as by other means. The method of packing will sometimes depend on how it is intended to stow the goods in and into the container and vice versa, and both packing and stowage techniques also rely on the type of commodity and on whether it is to be a full through or a consolidated shipment requiring break-bulk.

Progressive exporters are well aware of the importance of shrinking or reducing the size of their export packages. This is done because many articles pay freight on the basis of cubic measurement and, consequently, the smaller the package the lower will be the freight charges. Those who are entering the export field for the first time should bear this in mind, since careful packing in this respect may save them many hundreds of dollars during the course of a year. Reduction of cubic measurement not only will result in reducing freight charges, but will mean saving in storage space, in handling costs, and frequently in the cost of the packaging.

An export package may be reduced in size by the more compact packing of the contents, the disassembly or knocking down of the article being shipped, the redesigning of the carton, or the redesigning of the article being shipped, either to make it smaller in size or to permit its disassembly. It would be difficult to state whether greater savings have been made by redesigning the package or by changing the disposition of the contents in the package, but it is probable that in almost every case something of the two principles enters into every satisfactory export package.

One of the most common errors in packing is the failure to utilize the waste spaces in the carton which are created by the character or form of the contents. This frequently can be rectified by using a smaller carton or by packing additional items. For example, rubber tire manufacturers pack cartons of tubes inside a bundle of tires. Both methods result in more compact packing. Waste of space is frequently occasioned by using cartons which do not fit the contents, and this commonly results from the practice of trying to make a one-sized box do for a number of commodities. A manufacturer who believes he is saving money by using the same size case for his entire line can easily determine whether or not it is so by calculating the freight paid on the extra cubic measurement of a few shipments and comparing the total with the saving that results from the use of a one-size container.

More compact packing can sometimes be achieved by nesting articles that are of suitable shape. This is done with bathtubs, cooking utensils, and many other manufactured articles. Compression may also be used. A shipper of pork products packs hams under very high pressure and thus puts a considerably greater number in a box of the size previously used for a smaller number. Compression is also used in baling clothing and other textile goods, leather, paper stock, tires, cotton waste, etc. The problem differs with each commodity, but every exporter should study his individual product or products to see if more compact packing and consequent reduction of the size of the container or the inclusion of more articles is possible.

Disassembly of articles to be shipped has resulted in the saving of thousands of dollars in freight charges for many companies. One motortruck manufacturer, for example, by shipping certain trucks completely knocked down, has reduced the cubic measurement of the box from 238 to 192 cubic feet, showing a saving of 46 cubic feet. Metal office furniture and wooden and metal household furniture may also be successfully disassembled. Legs may be removed from tables, chairs, and beds, and metal office files of certain types can be almost completely taken apart and shipped in much smaller space than if completely set up.

Redesigning the product itself has been resorted to by many companies to reduce the cubic measurement of their export packages. Projecting parts may be eliminated or be made detachable, or an article may be made of separate parts, easy to disassemble and reassemble, instead of being constructed in a single piece.

Packaging engineers become a very important part of the "containerization train." It is the job of a packaging engineer who will take due consideration of the handling and transit hazards of the total journey, and will make a technical assessment of the strength of the product and its likelihood of spoilage under adverse climatic conditions. He will ensure that any strains imposed by handling and other shocks are transmitted to the strong parts of the item. He may suggest slight changes in the construction of an item (which may increase the production costs fractionally but possibly not at all) whereby the article can be made less prone to damage and savings can be effected in the packaging materials. He may recommend that the design of the item be modified to enable it to be dismantled to form a more useful cube size and so enable a container to hold a larger volume of his goods. Indeed, since freight costs must be reflected in the selling price of a product, the time will come when the size and geometry of these "products" will be based on the efficient utilization of space within the container. If an item is too large, it will be designed so that it breaks down into subassemblies of such a size that will fully utilize the container space.

Even for some fragile items there is a valid argument in favor of no enclosed packing, so that the articles can be seen and the people handling them will see that they are easily damaged and will handle them with respect. The use of the freight container makes this a viable proposition in many instances since the stacking loads are taken by the container. The article is attached to a pallet base and a transparent wrap of polyethylene is used either as a dust cover or as a moisture vapor barrier if heat-sealed, and thus maintains the visible idea.

Goods which can be carried in containers constitute a wide range and vary from powdered chemicals to engines, from whiskey to textiles. Each broad type presents its own particular problems, and is usually suitable for a limited number of packing materials and techniques, which will vary between types. We can, however, make the following generalizations:

1. Subject to weight limitations for purposes of handling, most cargo can be packed in cartons. Such cartons, however, must be of sufficient strength to permit stacking of the cartons in the container. Examples of cartons of insufficient strength are shown in Fig. 8A.

2. Strapping is not necessary, generally doing more damage to the cartons and contents than any protection by its securing feature would afford. Nonmetal fibered strapping as a sealer may be desirable.

3. Where heavier cargoes are involved and wood is necessary for the manipulation of the cargo, domestic crating on skids is usually sufficient.

4. As a protective measure, whenever exposure to moisture may be a possibility, and if such exposure would tend to harm the cargo, a plastic covering, usually polyethylene, should be provided, to cover the cargo, with sufficient packages of silica gel inside. As additional protection, the cargo at the tail end of the container should be covered with a sheet of polyethylene to protect the cargo in the event of seal leakage around the doors. Some carriers even make it a practice of putting additional sealing material around the outside of the container doors after they are locked.

Fig. 8A. Bottom carton creased and bulged from overhead weight. (Assoc. of Amer. R. R.)

5. Wherever possible when the sizes of the cartons or crates or other means of containing the cargo are designed, consideration should be given to alteration of measurements so that, loaded in modules, they will conform as nearly as possible to the inner dimensions of the container, allowing the minimum loss of space. While this may not seem to be of any immediate benefit to the small shipper, he will eventually benefit by lower rates, through more efficient utilization of the container.

6. Because of the use of containers, the cargo in the carton or the crate does not require any less securing than it did before. The cargo must be snug, free from movement, and free from contact with any other cargo in the same carton.

Loading the Containers—Distributing the Load; Securing Methods and Equipment. There are many types of dry cargo containers for general cargo use. These containers will vary as to the kind of adaptations they have in their interiors for the securing of the cargo. The most adaptable containers have a series of vertical, horizontal or combination slotted tracks either built-in, or added, where rigid beams or canvas straps can be attached to securely position cargo. Furthermore, this system allows additional interior decks to be assembled for the most efficient use of the container space.

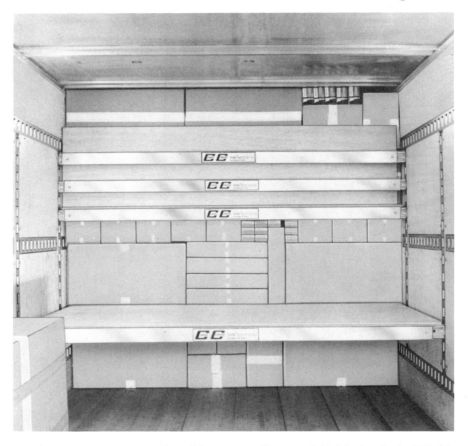

Fig. 9. Cargo in container shored by means of beams slotted in tracks installed in container walls. Beams offer means of multiple decking within container. "Cargo Control System." (Aeroquip Corp.)

Containers not equipped with the foregoing special fittings will usually have tie-down rings to lock the cargo. Alternatively the containers will have interior walls either completely covered with plywood, or with plywood on the upper or lower half of the container. In such instances the ingenuity and experience of the loader are fully called upon. Containers cannot be loaded as cargo arrives. All the cargo should be at the shipping point before loading begins. Then a preliminary loading plan should be made, taking the following into consideration.

1. The heavy cargo must be so planned as to maintain proper balance in the container, taking into consideration the lighter cargo and the weight distribution.

2. The maximum floor-load capacity must be considered and overcome by distribution through judicious use of dunnage. Even if the weight is evenly distributed and properly braced, the knowledge of the maximum load capacity per square inch of the container floor will permit rejection of cargo that exceeds the concentrated limits while not exceeding the load limit of the container.

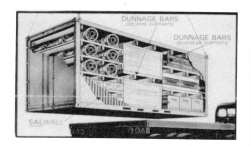

Fig. 10. Heavy gauge pressed metal panels with recessed slots adapted to hold specially bracketed dunnage bars. Can be built into containers during construction or added at a later date. Dunnage bars can be used for flooring as well as bulkheads, "Salwall." (Equipment Mfg. , Inc. , and Metropolitan - Cammell, Ltd.)

3. Highway-weight limitations must be considered and, in most instances, while the container can hold more weight, it may not be able to be legally carried over the highways.

4. Maximum utilization of the cubic capacity of the container must be planned. Considering the above, together with the cargo description including weights, dimensions and densities, a loading pattern is established that will hold the center of gravity and will allow for easy and safe loading, transport, and unloading. At the same time, securing is planned utilizing securing methods such as interior framing to act as bulkheads to separate different cargoes, and strapping and lashing at decisive points.

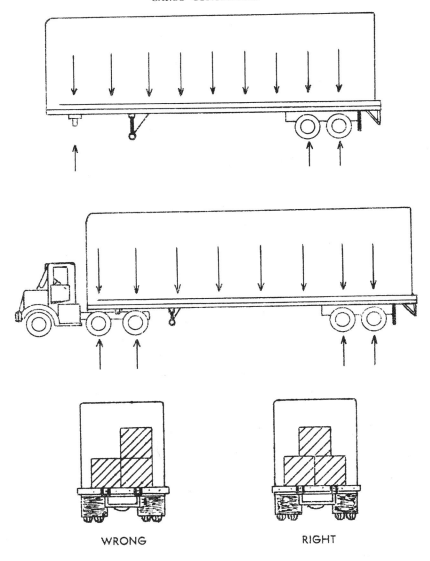

WRONG   RIGHT

Fig. 10A. Weight distribution within trailer. Trailers are designed for uniform load distribution as shown. The payload should be distributed equally between the rear tires and the fifth wheel which transfers its load to the truck tractor. (Assoc. of Amer. R. R.)

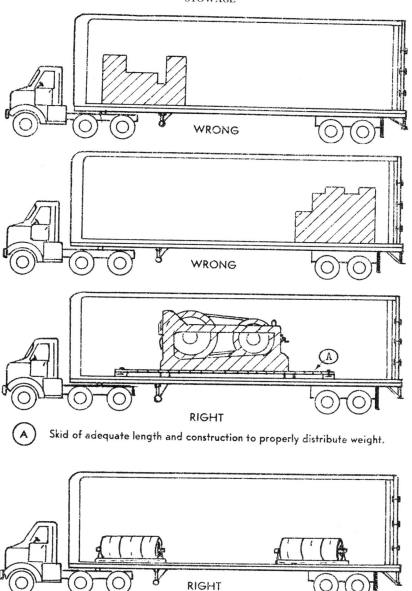

WRONG

WRONG

RIGHT

(A) Skid of adequate length and construction to properly distribute weight.

RIGHT

Fig. 10B. Loading heavy concentrated loads not occupying full trailer floor area. (Assoc. of Amer. R. R.)

In attempting to increase the ease of securing cargo against movement, inflatable cargo dunnage, called "Shor-Kwik" has been devised by Uniroyal, a division of U.S. Rubber, using a series of different sized dunnage sacs of double-wall construction with the outer covering made of neoprene and woven nylon fabric and on the inside a separate sac, like an inner tube, made of butyl. In case of minor damage to the casing while in transit, the inner sac has often been able to maintain the required air pressure to secure the cargo. Repair of the outer casing need only be for retention of mechanical properties, since the patching does not have to retain air pressure as in single-wall containers. If damage to the dunnage is so extensive as to puncture the inner sac, it, too, can be repaired as simply as a bicycle inner tube. Replacement is about 30% of the entire cost. A 4′ x 8′ "Shor-Kwik" inflated with 6 lbs. of air will exert 27,000 pounds of distributed force in a given direction. In use, the sacs are placed in position and then inflated. When the run is completed, the sacs are deflated and rolled up, ready for later use.

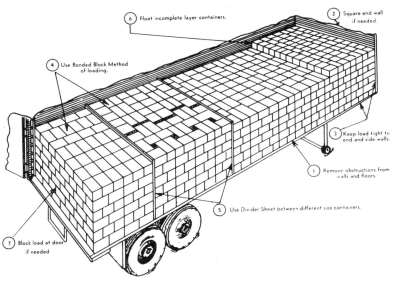

Fig. 10C. Loading fibreboard cartons. (Assoc. of Amer. R.R.)

There are other companies in various parts of the world that have dunnage on the market based on a similar principle.

Manufacturers frequently furnish the carrier with a packing list which gives the mark, number, weight, dimensions and contents of each package. Factory measurements are not always made in accordance with accepted practice, and packing list measurements should not be accepted without verification. The method of measuring usually employed is as follows:

In computing fractions of an inch in cargo measurement, all fractions under a half-inch are to be dropped. Where there is a fraction of one-half inch on three dimensions, two will be included as full inches and the other dropped. When giving and taking fractions where these occur on three dimensions, the ones on the largest and smallest dimensions are to be included and the other dropped. All fractions exceeding one-half inch are to be included as full inches.

In measuring barrels, casks, kegs and drums, the measurements are to be taken on the square of the bilge.

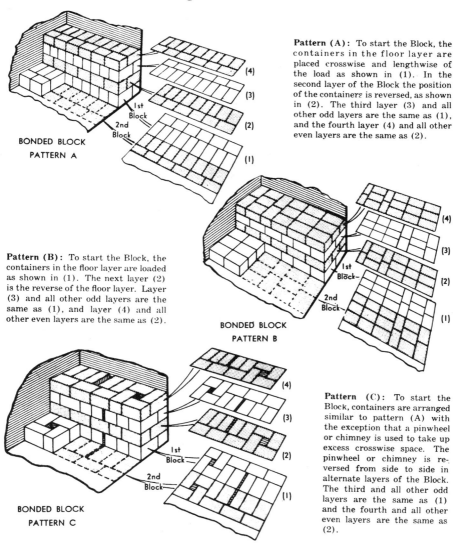

Pattern (A): To start the Block, the containers in the floor layer are placed crosswise and lengthwise of the load as shown in (1). In the second layer of the Block the position of the containers is reversed, as shown in (2). The third layer (3) and all other odd layers are the same as (1), and the fourth layer (4) and all other even layers are the same as (2).

BONDED BLOCK PATTERN A

Pattern (B): To start the Block, the containers in the floor layer are loaded as shown in (1). The next layer (2) is the reverse of the floor layer. Layer (3) and all other odd layers are the same as (1), and layer (4) and all other even layers are the same as (2).

BONDED BLOCK PATTERN B

Pattern (C): To start the Block, containers are arranged similar to pattern (A) with the exception that a pinwheel or chimney is used to take up excess crosswise space. The pinwheel or chimney is reversed from side to side in alternate layers of the Block. The third and all other odd layers are the same as (1) and the fourth and all other even layers are the same as (2).

BONDED BLOCK PATTERN C

Fig. 10D. Bonded block loading method. The three basic patterns of bonded block loading are shown. (Assoc. of Amer. R. R.)

In measuring irregular packages, use the three greatest dimensions to determine cubic.

In computing measurements to determine rate to be applied where weight rate is predicated on measurement per ton, the actual fractions may be used.

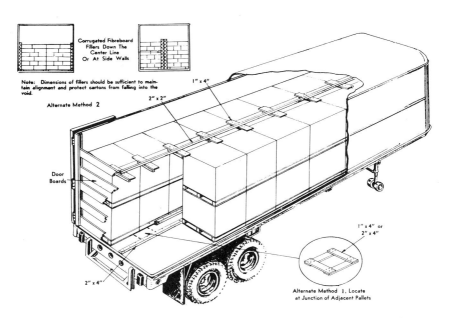

Fig. 10E. Application of cross bracing. (Assoc. of Amer. R. R.)

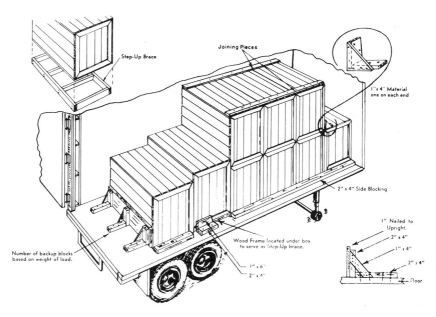

Fig. 10F. Floor blocking large boxes not top-heavy. (Assoc. of Amer. R. R.)

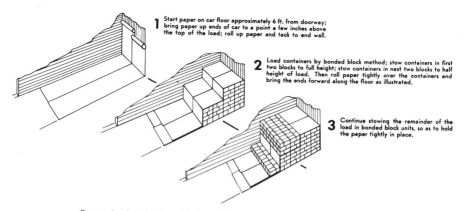

1 Start paper on car floor approximately 6 ft. from doorway; bring paper up ends of car to a point a few inches above the top of the load; roll up paper and tack to end wall.

2 Load containers by bonded block method; stow containers in first two blocks to full height; stow containers in next two blocks to half height of load. Then roll paper tightly over the containers and bring the ends forward along the floor as illustrated.

3 Continue stowing the remainder of the load in bonded block units, so as to hold the paper tightly in place.

By wrapping the end sections of the load with a paper binder, the containers in these sections remain intact during transit and act as floating bulkheads to protect the goods against damage and disorder. The Retaining Paper Method of Unitized Loading has been successfully used in many loads.

Duplex reinforced paper is used for this purpose, such as two sheets of 40 lb. Kraft tough pliable paper with approximately 40 lb. asphaltum lamination impregnated between the sheets to hold in position reinforcing fibres. Rolls of paper are usually available in various convenient widths to fit load requirements.

Fig. 10G. Retaining paper method of unitized loading. (Assoc. of Amer. R. R.)

Fig. 11. Inflatable dunnage, "Shor-Kwik." (U. S. Rubber, "Uniroyal")

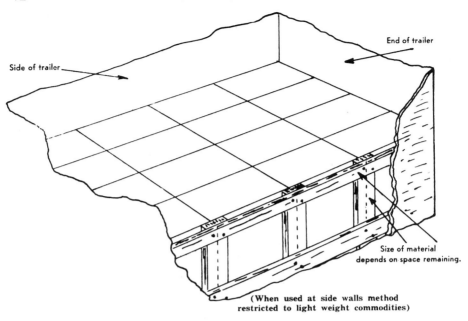

Side of trailer

End of trailer

Size of material
depends on space remaining.

(When used at side walls method
restricted to light weight commodities)

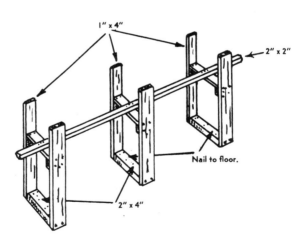

1" x 4"

2" x 2"

Nail to floor.

2" x 4"

Fig. 11A. Fillers for unoccupied floor space. (Assoc. of Amer. R. R.)

As cargo movement in containers is usually rated on its density, both the weights and cubes are necessary so that the volume can be determined.

The spaces that, under usual conditions, are unavoidably "lost" or unoccupied by cargo include: 1) the space between and around packages and containers; 2) the space occupied by dunnage; 3) the space at the sides, ends, and on top of cargo.

These waste spaces, known as "broken stowage," vary with the kind of cargo, ranging from 2% up to as much as 40% of the total cargo space. The average is, perhaps, 10 to 20%, and when handling general cargo, allow about 15% of the container's cubic capacity for this loss in stowage.

While some containers have side doors as well as end doors, most containers have end doors only and loading must take into consideration as well the facilities and equipment that will be available for unloading. Some packing houses, like the 7 Santini Brothers, receive and consolidate containers for shippers, utilizing their packing experiences for good stowage. Some photographs showing container loading and cargo securing at the Santini facility are shown in Figs. 12–17.

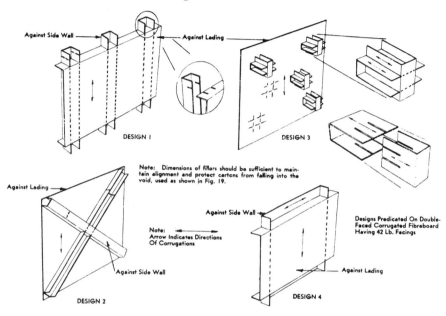

Fig. 11B. Examples of corrugated fillers to occupy unfilled space.
(Assoc. of Amer. R. R.)

The container must be inspected before start of loading to see that it closes properly, that there are no obvious apertures through which water may come (at some container pier installations an overhead mirror is mounted so that the outside of the top of the container can be easily inspected), that the container is clean, undamaged (with floors, roofs, end and side walls and doors sound) and free from protruding nails or other projections which might cause damage to the cargo. At the completion of loading, when the doors are locked, many loaders apply an additional sealer to act as a gasket and to make the container as watertight as possible.

While it is not always possible to choose compatible cargoes for the container, certain general guidelines should be followed.

Red Label Cargo. ICC regulations for over-the-road require, among other things, marking or placard on front, rear and sides of container for: Any quantity of explosives, Class A poison, and radioactive material requiring red label; for 1,000 pounds or more of Class B poison, flammable liquids, flammable solids and oxidizing materials, corrosive liquids; flammable and nonflammable compressed gases. In addition, Coast Guard Regulations for vessel loading require *any* container in which are loaded explosives or other dangerous articles in *any amount* to be marked with the appropriate placard and label.

It is the *shipper's responsibility* to properly describe the shipment by using true shipping name rather than trade name for explosives and dangerous articles as required by the Coast Guard.

It is, therefore, possible to carry such cargo in a container if the above regulations are observed. In this case, the entire container will have to be

Fig. 12. (upper left) Inspection of container before start of loading for proper closure, watertightness, and structural soundness.

Fig. 13. (upper right) Construction of interior bulkhead to secure cargo within the container, protecting cargo from damage and acting as separation between different cargoes within container.

Fig. 14. (lower left) Loading so that maximum cubic space in container is utilized.

Fig. 15. (lower right) Loading so that maximum cubic space in container is utilized.

(7 Santini Bros.)

loaded on deck. In addition, it is not advisable to mix this with other, non "red label" cargo in the same container.

Cargo that is not similarly packed should not be stowed without some form of protective separation. Wet cargoes (liquids, in drums, cans, etc.) should be separated by having such wet cargoes on the bottom of the container, but if this is not possible, then dunnage bulkhead should be built around the wet cargo. Cargo in cases or crates should also be treated like wet cargoes so that other cargoes in the container can be protected against possible movement of this heavier cargo. Very heavy pieces should be fastened to shipper-constructed skids.

Fig. 16. Loading so that maximum cubic space in container is utilized.

(7 Santini Bros.)

For ease of handling and elimination of damage before the cargo is brought to the container and after it leaves the container, side holes for forklift bars should be provided in the skids. Under no circumstances are pieces ever to be bolted through the container, through any surface, for the resultant holes will destroy the watertight character of the container. When heavy pieces are loaded into the container and the floor weight properly distributed, it is important that the piece be braced laterally. It is important that chocking between container walls and cargo not be made with wood wedged or fastened at right angles to the container wall and the cargo. In case of sudden shift of the cargo, the weight would drive the wood through the container wall. It is best to fasten a heavy board of at least 2″ x 6″

Fig. 17. Stowage of specialized equipment in container. (Container Marine Lines)

horizontally along the entire length of the container using the surface of the board to wedge the wood for bracing the cargo. If possible, this should be done on two height levels and on the three stationary surfaces of the containers. The surface toward the doors will have to be protected by a bulkhead as no cargo should come in contact with the doors. Care should be taken in accepting any cargoes that may suffer change because of temperature or humidity conditions.

Stowage of Containers at Plants, at Consolidators, at Piers. Insofar as the responsibilities for proper stowage of cargo in the container rest with the carrier, it is important that the carrier be aware of the consequences of poor stowage. To begin with, he—the carrier, is liable if he issues the bill of lading. While the *Carriage of Goods by Sea Act* was written not to supercede the *Harter Act,* both provide that the carrier cannot absolve himself from the obligation to properly load, stow, care for and deliver cargo.

While it is true that these laws were written primarily for the vessel, their action is transferred to the loading of the container, whether the steamship company or an NVO arranges it. It is possible, under Section 6 of the *Carriage of Goods by Sea Act,* to "enter into any agreement in any terms, as to the rights and immunities of the carrier in respect of such goods—in

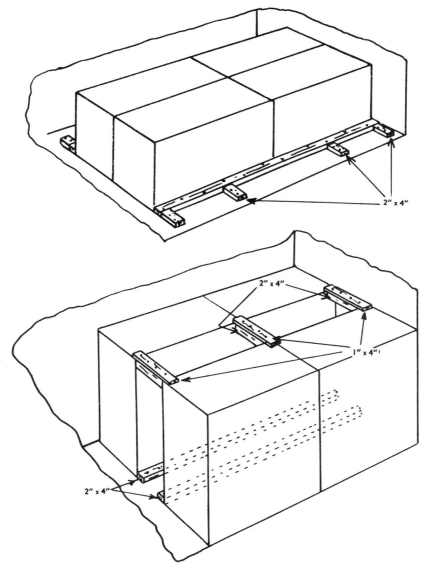

Fig. 17A. (top) Side blocking low center gravity articles; (bottom) Side blocking top-heavy articles. (Assoc. of Amer. R. R.)

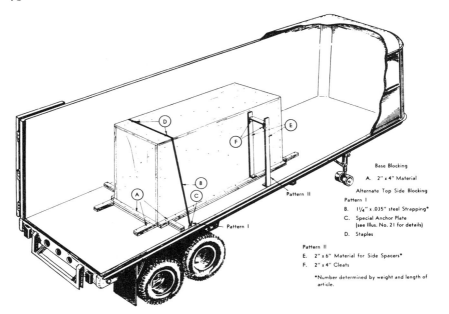

Fig. 17B. Alternate method for side bracing top-heavy articles. (Assoc. of Amer. R.R.)

Base Blocking
A. 2" x 4" Material

Alternate Top Side Blocking

Pattern I
B. 1¼" x .035" steel Strapping*
C. Special Anchor Plate (see Illus. No. 21 for details)
D. Staples

Pattern II
E. 2" x 6" Material for Side Spacers*
F. 2" x 4" Cleats

*Number determined by weight and length of article.

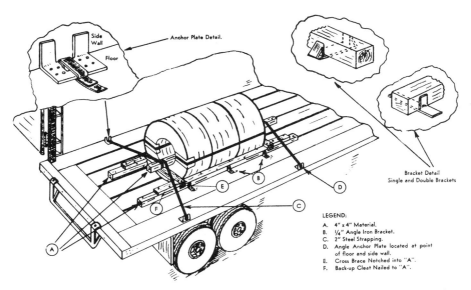

Anchor Plate Detail.

Side Wall
Floor

Bracket Detail
Single and Double Brackets

LEGEND:
A. 4" x 4" Material.
B. ¼" Angle Iron Bracket.
C. 2" Steel Strapping.
D. Angle Anchor Plate located at point of floor and side wall.
E. Cross Brace Notched into "A".
F. Back-up Cleat Nailed to "A".

Fig 17C. Special anchor plate bracing for heavy items. (Assoc. of Amer. R.R.)

the care or diligence of his servants or agents in regard to the loading, handling, stowage, carriage, custody, care, and discharge of the goods carried by sea, Provided, that in this case no bill of lading has been or shall be issued and that the terms agreed shall be embodied in a receipt which shall be a non-negotiable document and shall be marked as such."

In general, the carrier is liable for injury occasioned by defective stowage, in the absence of a contractual or statutory provision to the contrary, and bad stowage may constitute negligence precluding the carrier from relying on exceptions from liability contained in the bill of lading. But although badly stowed, if the carrier can show that damage to the goods must have happened anyhow, so that the bad stowing is not the proximate cause of the

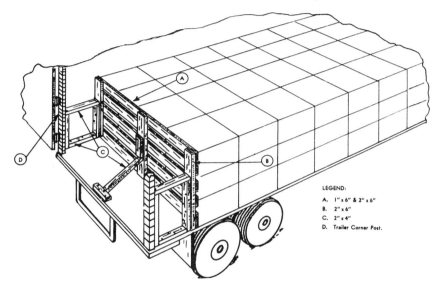

LEGEND:

A. 1" x 6" & 2" x 6"
B. 2" x 6"
C. 2" x 4"
D. Trailer Corner Post.

Fig. 17D. Rear blocking of heavy load not filling trailer. (Assoc. of Amer. R. R.)

loss, he is not liable. It is not sufficient, however, to exonerate the carrier that the loss might have otherwise occurred. The law imposes upon the carrier the duty of using due care to ascertain and consider the nature and characteristics of goods offered for shipment and to exercise the care in their handling, and further, the carrier is charged with notice of any patent defect in the goods. A carrier who accepts goods of a nature which require special care in their stowage must exercise such care. Laboratory compression tests, for example, have shown that a stack of fiberboard cartons loses at least ⅓ of their strength when alignment is not maintained. Therefore, use of fillers provides insurance that cartons will be in good alignment and provide greater strength. An example of what can happen when fillers are not used is illustrated in Fig. 17E.

However, there are two exceptions. One occurs when the cargo is loaded into the container at the shipper's premises by shipper's personnel. While liability for cargo damage is normally assumed by the shipper in such circumstances, liability is not entirely clear if, because of poor stowage by a shipper, the container is damaged or causes damage to property or life. The other type is the degree of packaging that is considered satisfactory for cargo moving in containers. According to a ruling handed down by the United States Circuit Court of Appeals (2nd Judicial Circuit) in March, 1940, Judge Learned Hand stated, "In the carriage of goods the trade must always come to some accommodation between ideal perfection of stowage and entire disregard of the safety of the goods; when it has done so, that

Fig. 17E. Falling apart of stack and start toward possible damage to contents due to lack of dunnage filler. (Assoc. of Amer. R. R.)

becomes the standard for that kind of goods. Ordinarily it will not certainly prevent any damage, and both sides know that the goods will be somewhat exposed; but if the shipper wishes more, he must provide for it particularly." This, then, places a realistic yardstick in determining alleged improper stowage.

There is one other factor which the International Standards Organization is trying to solve and that is the standardization of the inside dimensions of the containers. It is possible at the present time to have variations in inside dimensions by different manufacturers and even by the same manufacturers if the containers were built at different times. If the shipper wants to adjust his packaging to obtain maximum utilization, if the consolidator and the carrier want to preplan their loading, then a standard inside dimension will certainly facilitate matters. Great Britain, as stated earlier, is already moving in this direction.

4

Handling

DESCRIPTION AND USES OF HANDLING EQUIPMENT

In the movement of cargo by container before the advent of the large intermodal container, handling was relatively simple with existing equipment. However the size, weight factor and stress factors of these newer containers have brought into use supplementary equipment which had to be especially designed or adapted. With reference to the loading and unloading of the contents of the container itself, we depend in the final analysis on human planning and labor. We make the task basically easier through the use of conveyors, hand trucks, pallet trucks, various types of forklifts and cranes, among others. Handling packages individually by human chain results in lost manpower and possible damage to the cargo. Moving the cargo to the van or container by avoidance of hand-carrying is the first step toward more efficiency. When the cargo is already on pallets or in unit loads, the use of the pallet truck is simple and practical. From the standpoint of versatility, the forklift trucks are the most desirable. The forklift has the ability to load and unload itself and to elevate its load. It also can discharge or pick up at an elevated level.

Of the power options, these are available for gasoline, diesel and battery electric use. For use on terrain and on graded surfaces, pneumatic or cushion-tired models are made in addition to conventional solid wheels. The principle of the forklift involves carrying the load on the forks, balanced over the weight of the front wheels by the weight of the truck and a counterbalance weight at the rear. Thus a seesaw effect is achieved, with both ends balanced when the truck is loaded, and the rear wheels supporting the counter weight when the forks are empty. In order to provide greater stability while carrying and to facilitate the picking up of a load, most models have masts which tilt forward and backward at approximate angles of 2° and 10°, respectively. Special attachments are available for other than ordinary use, such as extended lengths, rotating forks, coil lifting ram, clamp lift, paper scoop and multi-tine. A hydraulic pusher attachment enables the operator to push the pallet load off the tines of the fork. A combination multi-fork with a pusher attachment enables cargo to be stacked directly on the tines instead of on a pallet. The cargo can then be pushed off the forks with the pusher attachment as a unit load instead of handling individual pieces. This unit was originally designed as part of a stacking system in racks at truck terminals, but lends itself uniquely to load

81

cargoes in the confines of a container. Another attachment, designed for use with a carrier sheet of cardboard, grips the edge of the sheet to draw the load in to the tines of the truck. Various heights of lift are usually available

Fig. 18. (top) "Human chain" loading of container.

Fig. 19. (center) Forklift and pallet loading of container.

Fig. 20. (bottom) Pallet truck.

(Containerisation International, Dec. 1967)

using twin or triple telescopic masts. To facilitate the use of forklifts in containers, adaptations have been made reducing the height of the driver's seat and the height of the mast. In general, the weight lifting capacities of

Fig. 21. A Conveyancer E 22-24 RC battery electric truck operating inside a container stowing palletized goods. This particular truck is fitted with a "full-free lift" mast allowing stacking to container ceiling, S. C. R. control giving accurate "inching" control and a "side shift" carriage to facilitate accurate spotting of the load in the confines of a container. (Conveyance Fork Trucks, Ltd.)

the forklift for container loading use do not exceed 4 to 5 tons and higher capacities are rarely needed for this interior type of work.

There are instances where, because of the nature of the cargo or the facilities of the plant, open-topped containers must be used and the loading and unloading carried out with cranes. In such cases simple overhead cranes would be used, if available at the facility, or a mobile-truck-mounted crane or crawler crane would be used. One adjunct to the forklift as used in loading containers is a British device involving the installation of tracks in the floor of the container. The device is a retractable portable forklift roller and rises above floor level with the action of a simple mechanical bar. In locations where there is no loading dock and the forklift must operate from ground level, this device extends the use of the forklift all the way into the nose of the container.

Fig. 22. A pallet load of cans slides forward easily on the Joloda tracks installed on this trailer. Floor clearance is sufficient to pass a pencil beneath the pallet, but when the "jacks" are withdrawn the pallet rests securely on the trailer floor. (Joloda Transport Equipment, Ltd.)

In preceding paragraphs we have discussed the handling equipment used to put cargo into the container and to take cargo out of the container. We shall now concern ourselves with equipment for handling the container itself.

The container must be "handled" only when the mode of transport changes, i.e., from chassis to pier, or vessel; from train to chassis, etc. It is therefore primarily in marshalling areas of piers or railheads where specialized handling equipment is necessary. During inland moves, as a truck, the container can remain on its chassis or bogie and have no need for being lifted on and off. There are, therefore, the following basic types of equipment for handling the containers.

Fig. 23. Dockside forklift. (L. Lipton, Ltd.)

Fig. 24. Fork truck "top lifts" containers and stacks them three high. (Towmotor)

1. *Forklifts.* This type of truck is used extensively on piers for the lifting on and off of containers up to and including the 20-foot length. The weight limitation of the truck designed for interior use makes it not quite practical for the loads, which, in a 20-footer can, with the container, approach 23 tons. To this end, therefore, trucks have been designed that can lift 30/50 tons. Such trucks are very large and their cost must be measured against other, possibly more versatile equipment. However, the forklift remains as one of the most useful tools for the movement of the containers themselves. Most 20-foot ISO containers built have openings for forklift tines. For the handling of 35-foot or 40-foot containers, the ordinary forklift is not always desirable. Because of the length of the containers, the stresses placed on it by having it just lifted in the middle at the tine width, bowing may occur which would prevent the container from seating properly in the cell of a ship or on a truck chassis.

Fig. 25. Piggy-Packer. (Penn-Central R. R.)

However, the newer forklifts have shovel type adaptive devices to lift and carry 20-foot containers more safely. In addition, some have a top lift cradle which anchors into the corner castings and locks hydraulically. A power side-shifter can move the frame six inches either side of center to facilitate the engagement of the box or its accurate positioning in storage or on a trailer or railcar. These units can handle 40-foot containers.

2. *Piggy-Packer.* This is an adaptation of the forklift truck whereby a gripping mechanism is used to lift a wheeled van on and off railroad cars. Since it does not handle the aforementioned container solely, but has a grabbing "nutcracker type" of action and since the wheeled vans will usually travel on rail and on roll-on/roll-off vessels, the piggy-packer can handle 40-foot equipment this way. While designed primarily for handling the on-lifting and off-lifting of trailers, it can handle containers but, unless great care and skill are exercised, the side rails of the container can be crimped greatly weakening the container, creating danger in handling fur-

ther en route and possible loss of compatibility in interlocking with other ISO equipment.

3. *Straddle Truck*. Another piece of equipment that has been adapted for container handling is the straddle truck. Originally a self-loading device used primarily for carrying lumber and other long lengths, a version of it, known as a straddle carrier is used to carry containers from one place to another. Adapting this still further by having a higher free inside height, straddle carriers were made that could stack containers one on top of the other. By further expanding the inside width, straddle carriers are available that can take a container from a railcar and put it on a truck chassis along-

Fig. 26. **Portal frame straddle carrier.** (Port of Hamburg)

side or vice versa. It can load and unload and move and stack containers within a marshalling area. Most straddle equipment is primarily designed to have top or bottom lifting. The straddle trucks can be generally divided into four types:

A. *The Portal Frame Type.* This vehicle straddles the road lengthwise. Loading and driving are controlled by the driver. It has four-wheel steering, but otherwise is driven like a normal road vehicle. This type usually cannot stack containers, can handle any length container, can approach container from either end, and has fast traveling speeds.

B. *The Open-Top Portal Frame Type.* This vehicle is similar to the portal frame type except that it has an open-top chassis which permits the container to be lifted up through it to allow stacking. The length of the

Fig. 26A. Lifting and stacking portal frame straddle carrier. (United States Lines)

Fig. 26B. Low-level tractor with open-top "U" chassis straddle carrier.   (Mafi
Fahrzeugwerke)

container that can be carried is limited by the chassis aperture, but it can load road vehicles.

Fig. 27. Side-loader.
(Lancer Boss Ltd.)

Fig. 28. Side-loader.
(Belotti Industries Autogru)

Fig. 29. Flexi-Van adapted container. (Container Marine Lines)

C. *The Telescopic Portal Frame Type.* This vehicle, too, is similar to the portal frame type, but it has a telescopic chassis which permits the container to be lifted for stacking, yet has the minimum headroom requirement when transporting. Not limited by length of container, it can approach containers from either end and can traverse stacked containers.

D. *Low-Level Tractor with Open-Top "U" Chassis.* This vehicle is basically similar to the open-top portal frame type but the cab is at ground level. The containers can be approached from one direction only, but they can be stacked. The length of container is limited by the chassis aperture. This unit can load road vehicles but cannot transfer from road to rail.

One interesting design in a straddle truck is the combining of the straddle-carrying feature with the side-loading feature described next.

4. *Side-Loader.* There is equipment designed to side-load and unload containers, as well. One type has a mast which is utilized to operate a spreader mechanism that locks into the corner posts on top of the container, enabling it to be lifted off equipment and stacked up to three high.

Side-loading type of equipment was originally designed for handling containers between railcars and rail-carriage equipment. One of the earliest methods of use was the Flexi-Van railcars used in the United States, which consist of turntables on the railcar. The Flexi-Van container is hooked onto the turntable by the tractor backing the rig up to the edge. The turntable is then swung into place by compressed air activation. The only equipment necessary for this is the ordinary tractor, for both loading and unloading the container. (Further details in *Rail* section.)

The containers used in this system are conventional containers with the addition of the side rails at the base that have holes for locking pins used to lock the container on the turntable, foldaway kingpins and other minor adaptations. Marine containers are loaded together with their chassis which have the side rails for the Flexi-Van cars.

Another method, used by the Canadian Railways and called the Railtainer, involves two tractors, one with the delivery chassis and one with a transfer chassis. These are lined up parallel to the container railcar. The chassis nearest the railcar, a 40-foot side-transfer chassis is equipped with two rear hydraulically-operated levelling rams. The vehicleman ensures that the container slide bolsters on these chassis are lined up with the bolsters on the railcar. He then starts the hydraulic pump and steps out of the truck cab to position himself between the tractor and the front of the transfer chassis. Standing before five hydraulic levers mounted on a pedestal at the front of the transfer chassis, he manipulates the hydraulic controls to effect the side-transfer operation by extending the rear hydraulic backs to level the side-transfer chassis rear bolsters to match the height of the slide bolsters on the container railcar. He then raises the tractor's hydraulic fifth wheel which adjusts the bolsters of the side-transfer chassis in line with the height of the front bolsters on the railcar.

Once both ends of the transfer chassis bolsters are levelled with those on the railcar, he extends the horizontal right push-pull hydraulic side-transfer cylinder with "S"-shaped hook attachment. This hook engages in a tooth rack built into the floor structure under the container. He retracts the transfer cylinder and hook which pulls the container 18″ sideways from the railcar towards the side-transfer chassis and then disengages the hook. Four more horizontal outward and inward movements as the transfer cylinder hook completes the transfer from railcar to container transfer chassis. The operator then levels the side-transfer chassis, if necessary, to match the height of the bolsters with those on the delivery chassis. He extends the left, push-pull side-transfer cylinder equipped with "S"-shaped hook, and pushes the container sideways from the side-transfer chassis, by five transfer cylinder hook motions, to the delivery chassis.

The above transfer of a standard container can be completed by one man in about two minutes. The containers used in this system are also conventional ISO standard containers with the addition of a tooth rack on the base of the container that is used in the side-transfer operation, and square open-leg receptacles for the insertion of removable legs. Mounted on these legs, the container, whether loaded or empty, is left standing for loading, unloading, or storage, while the delivery trailer chassis is freed for other work. The system works best with ISO containers that have the tooth rack built into

Fig. 30. (left) Flat-bottomed ASA/ISO container being transferred between railcar and transfer unit with the Railtainer system; (right, upper) Railtainer push–pull hydraulic transfer cylinder; (right, lower) Transfer cylinder engaged in tooth rack. (Steadman Industries, Ltd.)

the base of the container. However, it has been developed to work with any completely standard ISO container. ISO standard containers are side-transferred by means of a rotating transfer plate accessory on the standard transfer unit. The container designed for the·Railtainer System has square open receptacles next to the lower corner castings for the insertion of re-movable container legs. The container can be stored on these legs.

5. *Miscellaneous Ground Equipment.* There is much equipment still being designed, both ground-mounted, and mobile, which brings even other variations into the handling of the containers. Some of these follow.

A. A truck mounted hydraulic lift that can lift-on and lift-off from chassis to railcar and vice versa, stack containers three high and carry out container loading. This unit can be mounted on flatbed trucks, self-driven chassis, flatbed trailers, semi-trailers, or multiple-purpose chassis; 20- and 40-foot containers can be handled. Major design features are two hydraulic swing mechanisms, each incorporating two slanting rams and a telescoping stabilizer. Support legs are combined with the swing mechanism to secure the load. Sideways and longitudinal adjustments are possible so that the load can be grasped even if the approach of the vehicle is inaccurate. Handling time for an operation is about three minutes.

Container on the rail-car

Transfer loading on a trailer

Setting down on the truck

Lift-on through the Swinglift

Container set down on a trailer

Fig. 31. Swinglift unit removing container from railcar to truck and from truck to trailer. Swinglift can operate from either side of truck.   (Goldhofer Kg. Algauer Fahrzeugwerk)

B. A truck-mounted container crane to handle standard ISO containers from 10 ft. to 40 ft. in length and weighing up to 30 tons. The chassis can be telescopically shortened. The hoisting frames, one at each end of the crane, consist of an articulated parallelogram having a double-acting hydraulic ram set between one diagonal. Extension or retraction of the rams causes the frames to derrick either independently or simultaneously over the left- or right-hand side of the crane. In operation the crane is driven alongside the container and the frames are derricked out to the side. This operation automatically lowers four hydraulically-operated outrigger beams. Twin lifting beams are then lowered by means of four hoist drums located at the top of the frame side members. Hydraulically-actuated latches in the lifting beams engage automatically with corresponding lifting holes at the corners of the containers, and automatic interlocks prevent the load from being raised unless the pins have positively engaged.

C. A container transfer system has been designed to allow a container to be removed from a chassis and deposited in an unloading bay without the

Fig. 32. Coles BL 30 fully mobile container crane.
(Containerisation International, Nov. , 1967)

use of ancillary lifting equipment. The system consists of kingpins installed at both ends of the container on steel plates. A fifth wheel plate is installed on the unloading bay. To unload a container, the truck is backed until the coupling engages with the pin underneath the container. The chassis is then withdrawn after the tractive unit fifth wheel coupling has been released. Supporting legs are fitted at the far end of the container before the tractive unit is driven clear.

D. The container Roadmaster, built by Walter Hunger KG Fahrzeuge und Hydraulikwerk, West Germany, is a self-contained tractor and carrying unit that lifts and lowers and carries the containers from an overhead position, the rear wheel assembly separating for container entry and closing, locking and lifting container after it is in position. The rear unit telescopes to handle sizes between 20 and 40 feet.

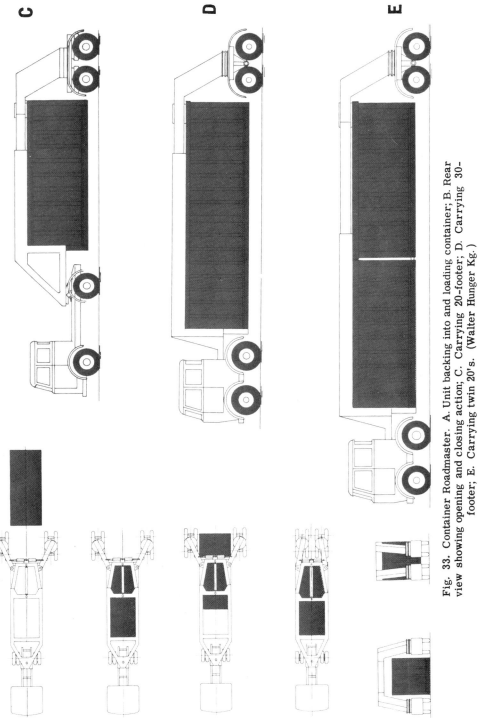

Fig. 33. Container Roadmaster. A. Unit backing into and loading container; B. Rear view showing opening and closing action; C. Carrying 20-footer; D. Carrying 30-footer; E. Carrying twin 20's. (Walter Hunger Kg.)

E. Another container transfer system designed for use without ancillary equipment consists of a specially designed tipping platform and a large torque hydraulic winch. In operation, a container at ground level is raised to the level of the trailer platform, which then tips to the same angle as the raised container. This is then drawn onto the platform by the winch, and the platform is lowered again to a horizontal position for road transport.

6. *Cranes.* Crane equipment for containers as used on piers and railheads consists primarily of ship's gantries and shore gantries. While ship's booms are also used for lifting as are locomotive and crawler and truck-mounted cranes, as well as bridge cranes and floating cranes, these are not

Fig. 34. Shipboard gantries. (American Export Isbrandtsen Lines)

particularly suitable for handling containers and will be discussed in the latter section of this chapter.

The *shipboard gantry* cranes are self-contained diesel or diesel-electric powered or electric driven with current supplied from the ship's electric plant. Gantry cranes are so-called because they are constructed to span an area. Their mechanism is mounted on a gantry, a form of structure that "spans" a given area. The legs are adapted for wheels on rails that permit traveling fore and aft on the vessel.

For securing of the crane while at sea, some gantries can be lifted off the rails by hydraulic jacks and placed on keyplates. In designing ship-mounted gantry cranes, the forces at sea are taken into account. The legs and connections to the main girders, while requiring strength, have this factor balanced against the wish for a minimum weight of the crane so that the carrying capacity will not be reduced. Another consideration is that the height of the crane should not be reduced. Also, the height of the crane should not adversely affect the vessel's stability. There are usually over-hanging jibs in order to obtain overreach over either side of the ship to allow direct loading from the dock. The jibs are designed to fold away when not in use. A motorized trolley, including the hoisting machinery, is mounted between the two bridges. This hoisting unit travels athwartship on wheels. The operator's cabin is also mounted and, in most instances, is attached to the trolley frame.

With few exceptions, containers are lifted by means of a spreader frame whether the lifting is accomplished by a dock or shipboard crane or by the ship's conventional booms or heavy lift. The spreader frame is a structural frame with the same overall dimensions as the container so that it may enter the cell with the container or independently to pick up a container. The frame is generally suspended by four cables (one to each of its corners) from a hook suspended by a boom or to one or more drums on the gantry crane. The spreader frame is equipped with devices at each corner which enter openings in the top corner fittings of the container, lock in place, and join the frame to the container. Both are lifted in unison. To aid in centering the spreader frame on a container, some units are equipped with remotely controlled electric or hydraulic flipper guides that are swung down and act as centering devices. These are folded back before the container enters the cell as they will not clear. When attaching a spreader frame to a container within a cell, the cell guides center the frame upon the container and the flippers are not used. Some units have safety devices provided either in the form of telltale lights or overriding control switches which indicate or pre-vent hoisting when all four corner lifting devices are not fully engaged and locked. The spreader frame is usually equipped with rollers on its corner to reduce friction against cell guide angles. Where a cell guide angle is not continuous and has gaps or interruptions in its length, the spreader frame must be equipped with extensions at each corner which in effect increase the depth of the spreader frame in contact with the guides and span the gaps in them. This feature is necessary when the guide moves through the cell without a container attached.

When containers are handled by a boom with a single fall as opposed to four falls from a gantry crane, the container will tilt in its fore and aft direc-tion if it is not loaded symmetrically. Even if the tilt is not serious enough to cause damage, it will cause difficulty in entering a container into a cell and frequently causes the container to jam when it is in the guides. To cor-rect this, an adjustable spreader frame has been developed containing a hydraulic cylinder which, when activated, shortens the ends of the cables of one end of the container and simultaneously lengthens the others. This action shifts the center of the load under the hook so that the container assumes a level position. This can be done while the load is suspended. This arrangement is excellent while the container is suspended out in the open where it can be seen, but when a container is to be unloaded from within a cell, tilting action cannot be readily seen but will be evident only by its jamming. Marking unsymmetrically loaded containers at the time they are loaded would be a worthwhile measure to correct this disadvantage. However, automatic spreaders have been devised which lift the containers on four falls from a gantry and these, of course, are the safest, fastest and most practical at the present.

Fig. 35. Automatic spreader. (Stothert & Pitt, Ltd.)

Vessels designed for runs that have little or no adequate port handling equipment have multiple gantries for most efficient loading of all cell banks. When multiple gantries are operated simultaneously, dangerous list may occur. This is compensated for by built-in counterbalancing characteristics that reduce any list to a minimum. In setting up the operation of the crane a two-cycle operation can be arranged by unloading one cell with a series of single cycles and then loading containers into this cell, emptying another

Fig. 36. Manual spreader. (German Federal R.R.)

Fig 37. Articulating boom Paceco Shipstainer. Its 31'6" outreach can serve two lanes of traffic on either side of ship to permit simultaneous loading and unloading. When retracted for sailing, it stows flush with the side of the ship entirely above deck level. (Pacific Coast Engrg. Co.)

cell on the return of each cycle. In this way, groups of cells can be loaded and unloaded within the consideration, of course, of number of loading and unloading ports and type of cargo. The actual cycle time for gantry cranes is about one-half that of a conventional boom operation when handling containers.

A ship gantry can handle up to about 20 containers per hour and is adjustable to handle the various sizes of Group I containers. The use of roll-on/roll-off vessels eliminates any crane equipment except when combination

Fig. 38. Shore gantry on rails. (Herbert Morris, Ltd.)

type vessels are constructed. In addition, where vessels are designed for particular runs and where adequate port handling equipment is available, these also will not require any crane equipment. Another design has extra long outrigger girders to permit the servicing of two lanes of traffic on the dock. Servicing two lanes speeds the cycle by eliminating the crane delay while one container is moved away and another is brought under the crane.

A. *Shore Gantry Cranes.* The cargo handling shore cranes take up where the straddle trucks leave off. They are either on tired wheels or on rails. The rails have a distance limiting factor, but are very useful where loading and unloading of railcars is necessary. The wheeled cranes are more versatile because, where needed, they can be moved from area to area. Most gantries of this type are similar to the vessel-mounted units in their operation, lifting containers on and off railcars, on and off truck chassis, bogies, and flatbeds and stacked on the ground.

Handling from these gantry cranes is accomplished with top-lift attachments, such as spreader devices, locking into the ISO corner fittings or with bottom-lifting attachments which envelop the container and lift it from the bottom with grapplers. There is much discussion prevalent relative to

Fig 39. This Diesel-electric tired shore gantry crane spans 60 feet, straddles two roadways and two railroad tracks at a time. Highly maneuverable, this Paceco Transtainer saves turning space when changing work areas by turning all four wheels at right angles and moving sideways. (Pacific Coast Engrg. Co.)

standardizing the contact point pick ups on the 10-foot to 40-foot ISO containers. These lifting centers, when standardized, would be duly marked on each container so that the lifting devices can be easily and accurately placed. Construction would involve reinforced bottom side rails, with lift pads installed at point of lift. Another alternative being suggested is to construct the container so that a 2"–4" indented recess is created between the base corner fittings on the long dimension of the container. Being duly reinforced, this entire bottom rail then acts as a lip for the grappler to hook onto for lifting. These efforts to standardize construction would also affect the handling by straddle carrier and forklift. The dimensions of the forklift pockets would be fixed, as well.

The vessel-loading shore crane is mounted at shoreside and usually on rails so that it can be positioned properly in relation to the vessel. This type of crane has an outreach cantilever section permitting the loading or unloading of containers from any part of the vessel. The outreach can be folded upward, or retracted in use. Some cranes require a backreach for stability, but others do not.

The design of most of the shore gantries permits a loading/unloading cycle of about 2 to 3 minutes per container. The weight design usually allows 30-ton to 45-ton lift capability and adjustments for the various sizes of Group I containers. New designs have combined shore cargo-handling and vessel-loading capability characteristics into one unit so that handlings of the containers can be further reduced. One such design, in Britain, is described as a "merry-go-round" of suspended, self-propelled cars, each carrying its own operator and winch house and lifting beam. The cars run overhead in a track circuit whose range is variable, at will, to command exact positions over the vessel and the stacking area. In operation, a car with its underslung container proceeds along one main rail until it reaches the deflector rail of the forward bridge. It then enters the bridge loop, changing directions through 90° and becoming parallel to the fore and aft arrangement of the ship. The container is lowered into an empty cell and the lifting beam raised. The operator then controls the mobile bridge until it has advanced him to a position commanding the next cell from which a container can be lifted. The carrier now departs, turns off the bridge into the return rail and repeats the bridge-entering routine over the stacking area. The cycle is complete. When a line of the ship's container capacity has been filled, the entire handler on its motorized wheels will be moved along the dock rails, transversely, to position it for the next line of cells. The complement of carriers, usually seven, will maintain a loading/unloading sequence of 30 containers in and out every hour, each working to a 2-minute time cycle to load one container and unload another.

Another unit has been designed to have a wide span so that six parallel rail tracks can be serviced. This gantry is rail-mounted and operates at high speeds over large terminal storage yards. It can rotate a container 360° and will service ground-storage areas, rail tracks and truck roadways. One high-speed gantry with a single operator can replace several pieces of smaller ground-handling equipment, each requiring a labor crew.

The future holds much improvement in this equipment: automation, with radio control from quay and ship and no operator needed in the crane cab, automatic electric-eye-controlled transfer cars for dock to terminal transfer, automated "pigeonhole parking" system for container storage in open steel structures.

7. *Yard Transfer Units.* In most pier storage areas today gasoline-powered towing tractors are used as a supplement to the forklifts to move cargo to and from dockside. In operations involving containers, railcars and trucks can bring the containers directly alongside for handling by the cranes. Containers, however, cannot be stored as simply as general cargo, and in addition to predetermined layout plans, the containers must be stored off the ground. Storage off the ground, on legs, special

Fig. 40. Paceco Portainer shipside gantry. (Pacific Coast Engr. Co.)

tables, or concrete posts (plinths) enables rapid selection and handling. Storing two and three high involves rehandling, which in large yard areas can be kept to an absolute minimum. Storing on chassis involves a large expenditure for a chassis for every container which, in many instances, serves little other purpose than to store the container. To this end supplementary equipment has been devised for use within the storage yard. Such equipment is divided into two parts:

A. *The Dockside Tractor.* This is a specialized tractor that had additional features built in particularly designed to increase versatility in the yard areas. One tractor in use can handle semi-trailers of all types, with varying heights of kingpin, in roll-on/roll-off and container operations. This unit maneuvers trailers in ships and in dock parking areas. It can push as well as pull the unit carried. There is one driver's seat with a reversible backrest and two power-steering wheels. Foot controls are also duplicated. The cab arrangement, because of its unusual mounting, can be moved to any lateral position required within its extreme limits allowing the driver to see down the near- or the off-side of the tractor, when driving backwards. The fifth wheel coupling can be raised or lowered from the cab and the kingpin of articulated semi-trailers can be locked or released from the cab, as well.

B. *The Dockside Semi-Trailer.* Auxiliary equipment, for use in the yard, involves equipment capable of moving containers from one part of the yard to the other and to and from the dock area when the containers are not stored on chassis. Such equipment is used to supplement straddle carriers, side-loaders and piggy-packers. There are two types. One type, an adaptation of the flatbed trailer, is nonadjustable and the containers are placed upon it or removed from it with other equipment. The other type is designed to be used either with or without its own storage table. MAFI makes the unit with the table. The container is stored on or moved about upon this table. (See sketch.)

The other, without its own table unit involves an adaptation of a flatbed truck on which is mounted a hydraulic scissors. This unit is backed under the container resting upon the plinths and lifted for carriage to dockside. In reverse, the container is lowered on the plinths. The scissors device is compressed and the carriage unit driven away. This unit enables containers to be moved about the yard without the need for storage chassis.

In usage at railheads are units called "Commandos," which are adaptations of the dockside tractors that enable them to load and unload containers on and off railcars. (Units further described in *Rail* section.)

Improper Handling—Damage to Container; Damage to Ship; Damage to Cargo. Improper handling causing damage can occur because of lack of operating knowledge of equipment, but more often than not, damage occurs because of lack of proper equipment. To go even further, it is too often the case of adapting existing equipment for container use. This is not to say that the many millions of dollars necessary for properly equipped ships and ports are easily available nor, when they are, that everything can be accomplished overnight. Nonetheless, whether the inept handling is because of lack of education or equipment, whether it is because of a temporary situation, much damage can and does occur.

Fig. 41. Breco system. (Breco Materials Handling)

Fig. 42. Six-track Paceco rail mounted Transtainer with double cantilever shown depositing piggyback trailer in herringbone position. Operator's cab travels with trolley, which rotates 360°. Trailers can be lifted over intervening loaded railcars.
(Pacific Coast Engrg. Co.)

104

Fig 43. "Shipcharger" dockside tractors. Driver can maneuver cab to view either side of unit. (Alexander Stephen & Son, Ltd.)

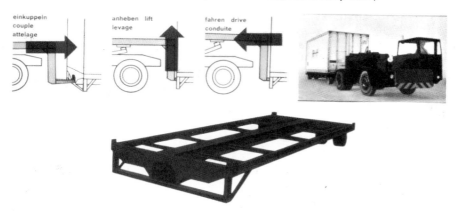

Fig. 44. The Mafi Roll-Trailer transport system. Special gooseneck coupling device is attached to any tractor enabling it to engage Roll-Trailer platform table, lift and transport table with container upon it. (Mafi Fahrzeugwerke)

Fig. 45. Swedish Bollnas Low Loaders. (Containerisation International, Dec.,1967)

In modern terminal installations thorough and frequent inspections are made of the containers. Overhead mirrors are mounted so that the top of the containers can be inspected. A hole, even a small one, can allow much damage to the cargo. Holes can be caused as innocently as by a pebble wedged between two containers being stacked on top of one another. Im-

Fig. 46. Adjust-a-leg equalizing and locking wire rope slings. Designed to handle loads in a perfectly level position regardless of uneven load weight distribution in the container by moving the crane hook over the container center of gravity and lifting the load. The operator adjusts the sling leg lengths for a level lift and locks the slings while the load is on them. (The Caldwell Co., Inc.)

proper use of nails and bolts to fasten heavy equipment within a container can be another cause.

Many ports today are not yet ready with proper lifting equipment for containers. When this condition is coupled with that of vessels without proper lifting equipment and without cellular holds for the containers, then the incidence of damage can be expected to be higher.

The largest container that should be handled by an ordinary forklift is a 20-footer. Any container larger than 20 feet long handled by forklift trucks without spreader adapters has the possibility of having its cross-members spring, especially if loaded to its weight limit.

Fig. 47. Roller Jack. (Joloda Transport Equipment, Ltd.)

The crane equipment at ports is usually of the derrick type capable of lifting containers with one fall. Vessels' derrick booms act in the same manner. Containers are lifted, whether with shore or vessel equipment, with the one fall rigged to the four corners of the container. Manual spreaders are also used, on occasion, and in such instances, the one fall is rigged to the four corners of the spreader. Whether or not the spreader is used, if only one fall controls the lifting, there is danger of cargo improperly stowed in the container shifting and tilting the container dangerously, perhaps causing it to fall and damaging the container and the shore equipment of the vessel. Cargo could break through the container. Wind sway can cause the container to hit against the superstructure of the vessel. The container can tilt in its fore and aft direction if it is not loaded symmetrically. This condition causes difficulty in entering a container into a cell and frequently causes the container to jam when it is in the guides. An equalizing device manufactured by the Caldwell Company, called "Adjust-a-Leg," enables the operator of a manual spreader to find the center of gravity of the container before actual lifting. (See Fig. 46.)

Many vessels attempting to carry containers do have partial conversion to cellular holds. In such case, containers are loaded into the vessel in various ways. To increase the load below decks containers are loaded into the square of the hatches and then manually pulled into the wings by different devices. Some vessels have constructed rollers to assist in this movement. Others use greased dunnage and other makeshift devices. Such handlings do not lend themselves to damage-free movements. Some devices have been put on the market which make the movement of containers into the wings a simpler affair. A "roller jack" has been devised by Joloda Transport Ltd., the same company that has the pallet loader for moving cargo within the container. This equipment is a bi-directional roller beam having a built-in hydraulic jacking arrangement actuated by hand pumps.

There is also a great tendency to load containers on decks of vessels. In addition, containers are stacked on decks so that it is not unusual for some vessels to have deckloads of containers three and four high. There is inherent danger in this practice from many aspects. We have previously discussed the possible damage to the cargo because of heavy weather at sea. Improper lashing or being insufficiently secured can cause damage to the containers and possible loss of the containers at sea. The stacking of the containers on deck can cause loss of vessel stability and decrease maneuverability at critical times when these features are needed at maximum efficiency.

Some vessel construction is already taking the "container-on-deck" factor into consideration. One has its bridge forward, which protects the cargo and gives a clear space over hatches for working shoreside equipment. Another company has built up the sides of their ships to permit a 35 ft. freeboard to the weatherdeck with structures forward to help protect the containers from boarding seas. The containers themselves are stowed up off the weatherdeck. A third, and this one is a roll-on/roll-off vessel, is of an open shelter deck type. The hull is built similarly to that of an aircraft carrier to reduce the hazard of boarding seas and is high-sided.

There still remains quite a long period of time until almost ideal containerization conditions exist but during this interim period valuable experience is being gained so as to constantly reduce the possibilities of damage.

5

Movement

Types of Equipment Inland and at Ports

When moving over the road in the United States, containers move primarily on chassis or bogies. Unfortunately, while there has been standardization of the containers themselves, there has been no attempt to standardize this type of carrying equipment. However, there is an increasing tendency among equipment manufacturers to design equipment that will be able to carry containers other than their own. The following paragraphs describe some of the different types of equipment.

Fig. 48. Single-axle and twin-axle bogies. (Integrated Container Service, Inc.)

Bogies. These are basically trucks on which wheels are mounted. A bogie can be constructed either with a single axle running gear having twin wheels on each side or a tandem axle running gear with eight wheels.

109

Chassis. These are skeletal frames with wheels that are built to carry specific sizes of Group I containers and are fitted with adapters that lock into the ISO corner fittings of the containers. These chassis are either twin- or single-axled. The usual sizes are the 20-foot and 40-foot lengths. To be able to move 20-foot containers over the road economically it is advantageous to be able to move them in pairs. One way to do this is to hitch up two units on two chassis as "double bottoms." The problem with this, however,

Fig. 49. An Adamson Great Dane 20-foot ISO aluminum/steel container. Mounted on an Adamson skeletal semi-trailer capable of carrying up to 30 feet in length. (Containerisation International, Apr., 1968)

is that some states restrict the movement of double bottoms on their highways. Another way, with less problems, is to have the two chassis "couple" or lock together so that, in effect, one 40-foot unit is achieved. For loading and unloading, however, both container units must be broken apart and this does involve extra handling by the trucker. But for the practical combinations of cargo from the same origin for different destinations and also for the carriage of higher weight cargoes, the coupleable chassis can be very useful.

In addition to the foregoing fixed-size non-interchangeable chassis, there is a trend toward the manufacture of chassis that are able to be used for more than one make of container. There are also chassis that can carry more than one size of container, one type having an adjustable length and another having multiple corner fittings that are removable when not in use.

Chassis are designed, however, to accommodate to basic types of construction in containers. Containers are constructed with a flat bottom or a tunnel type of bottom. The flat bottom can be designed to be carried by the varied type of skeletal carriers enumerated later, with tandem wheels or bogies. The tunnel type requires a special type of chassis because the container has been designed with a depression on the outside under surface approximately 6″ high, 30″ wide and 120″ long. This special chassis, sometimes called a gooseneck chassis, is designed so that the nose of the chassis fits into the tunnel, maintaining the necessary kingpin height for the fifth-wheel attachment and allowing the container to sit lower in relation to the road.

Fig. 50. Gooseneck chassis. Note elevated portion near nose. (Dorsey Trailers)

Thus an 8′6″-high container would be within the 12′6″ height limitations placed on many roads when it is a tunnel type on a gooseneck chassis. Advantages of this type of chassis also include better connection between chassis and container because of the additional contact along the surface of the tunnel channel, and easier utilization at loading docks. The usual dock heights for highway trailers is 52″. In using a tunnel type container with a gooseneck chassis this same height is maintained. With the flat-frame chassis and the flat-bottom container the height is increased about 6″, with resultant awkwardness of handling during loading and unloading. However, some flat-frame chassis are designed with an angled height so that the tail end is several inches lower than the nose end.

Some Chassis Types

The Parallel-Frame Chassis. The container carrier which, to date, has enjoyed the greatest prominence and popularity is the straightforward, parallel-frame chassis, having its main frames set in the orthodox manner, similar to the ordinary, straight-framed semi-trailer. Built onto the outsides of this frame are substantial outriggers, designed to accept the corner mounting blocks of one or more modules of containers. There are advantages and disadvantages in operation. Because the frame must accept the container on its outer extremities, the method of passing the load

Fig. 51. Parallel frame skeletal chassis. (Mark 3 Boden, Containerisation International, Apr. , 1968)

Fig. 52. (upper) Perimeter frame platform chassis, M & G Trailers.

Fig. 53. (lower left) Running gear; (lower right) A straight frame semi-trailer carries a 10-foot ISO container at the front and has 23 feet of platform behind for conventional cargo. Hands Trailers, Ltd. (Containerisation International, Apr. , 1968)

stresses back to the main frame and thence to the fifth-wheel plate and running gear, must be via the outriggers. Of necessity these must be substantial and heavy, and because they project outward from the sides of the frame, they are susceptible to accident damage. When under load (this can be as much as 6–8 tons per outrigger), members which are simply attached to the outsides of the frame are placed in bend. The transmission of the load into the main frame sets up considerable high torsional stresses on the actual side members, and even if outriggers are continuous and pass over the whole width of the frame, torsional stresses, though less severe, are always present.

Perimeter Frame Trailer. In this design a main frame is constructed having its main structural members at the extremities of its overall size—hence the name "perimeter." Although thought by many to be a new design, the perimeter, or outside-frame chassis, is, in fact, quite an old concept. It has been used in many designs over the years, and most manufacturers who adopted it have subsequently abandoned the design in favor of the less complicated parallel frame.

There are many reasons for this, the most important being that it is an expensive frame to construct. This is because the main side members need to be reduced in depth at both ends—at the front to accept the fifth-wheel assembly; at the rear to clear the tires. Thus, because the load imposed by the corner mountings has to be borne on these reduced sections, it is essential that the reduction in beam depth is not too great.

There is another important snag with this design: loading height. If more than one size of module of container is to be carried, the need to mount extra locking devices means that the top flanges of the main structure must be cut and, therefore, weakened. The need for heavy cross-members, both at the running gear mountings and at the fifth-wheel mountings, eliminates any savings that might be possible in terms of unladen weight.

In its favor are three important points. This carrier bears its load on its main members; therefore, there are no static torsional stresses. The wide frame allows spring centers to be set at the greatest dimension commensurate with the axle equipment used, creating a high resistance to roll—an inherent drawback with a high load. In addition, due to the frame width, the designer can set the landing legs wide apart and further forward without fouling the tractive unit when turning. This has considerable advantage with regard to stability for loading and off-loading while landed. In other words, there is less chance of the trailer "nosediving" as the load is taken off the rear.

The Cruciform Carrier. A novel kind of skeletal trailer utilizes a cruciform design. One company manufacturing these in the United Kingdom— Hands Trailers Ltd.—has developed this type of container carrier based on the principle that while its frame is set between its wheels and it is not excessive in height, it nevertheless does not suffer from torsional stresses through its structure. The cruciform design ensures that every member of the structure is at all times placed in bend only. This represents quite a step forward in frame design for container carriers. As yet the design accepts only one module of container, but the makers intend to produce an extensible version to cater to all container sizes.

The same company has also experimented with "button-on" type of running gear—a method which has been widely used in the United States for many years but which is not considered to be suitable to the narrow, badly cambered roads found in the United Kingdom and in certain other European countries. The Handsgear utilizes a three-point mounting in place of the American-favored double sliding channel. The advantage of this method is that the single mounting point of the gear is situated at the leading end. Thus, when the container is positioned firmly on the central point, it acts like a drawbar—the running gear following faithfully in perfect alignment at all times.

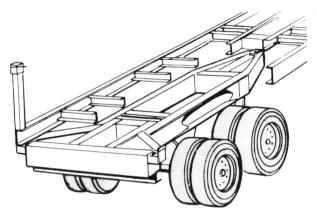

Fig. 54. Cruciform chassis. (Containerisation International, Apr. 1968)

With equipment such as this, one can envisage a fifth-wheel coupling plate manufactured with the same type of mounting described in the previous paragraph. With the two attachments *in situ*, a container or flat would in fact become a semi-trailer, following the lines of the Strick or Fruehauf equipment used so successfully here in the U.S. When the container is lifted off, the running gear could be towed, close-coupled to the tractive unit.

Dual-Purpose Unit. The dual-purpose unit provides over-the-road operators with a combination of container haulage and general cargo carriage. Most manufacturers are producing this kind of semi-trailer, which usually takes the form of skeletals decked over, or flats equipped with container locking devices. Available in lengths of up to 40 ft., several smaller containers can be carried, or a combination load consisting of one or two containers, plus conventional freight.

From the safety angle, all container carriers are now being equipped with twistlocks—rectangular section blocks, with tops shaped like blunt pyramids. The cone is in two pieces, the top half being mounted on a spindle which passes through the bottom half to an operating handle. By movement of this handle, the top half of the cone is turned through 90° when it presents itself as a broad Tee and prevents the snug-fitting hole in the container mounting from passing over.

When mounted on a frame, the four locks provide a positive location for the container. A point to be borne in mind is that a heavy blow of the type often sustained by the outriggers of a semi-trailer can quite easily knock the locks out of alignment—one of the drawbacks of this type of vehicle. Once an outrigger is bent by more than 1.5″, it is impossible to correctly locate a container.

Chassis come equipped with single axle and wheels as well as twin axles and wheels. On the single 40-foot units, these axles are usually fixed in position. On the "marriageable" 20-foot chassis units, there are several methods of coupling and equalizing the loads. With one of these methods, the chassis supports the load completely so that no connecting of the containers is necessary. The system can accommodate any 20-foot containers with ASA-ISO corner castings, and the containers can be of different types. This makes possible lighter, less expensive containers than those built with enough strength to provide support for the coupled assembly. When two chassis are coupled, the front and rear suspension hangers lock automatically into a load equalizing tandem instead of two independent single axles. A sliding frame section at the rear of the chassis can be extended to serve as the lower fifth wheel for coupling. The extended position also enables a single unit to carry extremely heavy loads.

On some running gear in use on trailers, the forward tandem axles can be adjusted forward on a series of lateral pin openings. The trailer is put on an axle scale, the air breaks are set to lock the wheels on the axle to be moved and the entire rig is moved backward. The purpose of this adjustment is to equalize the load on the axles and to comply with the highway regulations covering the maximum weight on each axle. Some manufacturers have adapted this "sliding tandem" so that the axle can be moved on the longitudinal arms of the chassis instead of the pin device.

Different areas, having different problems, have devised solutions which have been useful in solving their problems. We have seen the adaptation of the sliding tandem from trailer usage to chassis usage. In the western United States there are trailers with auxiliary motors mounted underneath that are activated when additional pull is needed in climbing mountains. There are also additional sets of wheels which are available in an "off-ground" position when not needed, but can be locked in a ground contact position for additional traction and load bearing when needed.

In addition to the above fixed-size non-interchangeable chassis, there is a trend toward the manufacture of chassis that may be used for more than the one make of container. There are also chassis that can carry more than one size of container, one type having an adjustable length and another having multiple corner fittings that are removable when not in use.

Flatbed Trucks. Unfortunately there are times when there is a shortage of equipment, temporary or otherwise, and the containers are placed on flatbed trucks, for moving about. This is found in the smaller port areas. The tendency in Europe is toward the "all-purpose" type of carrier. One all-purpose semi-trailer, used in Holland, is 12.20 meters long and is supplied with a floor. A 20-ton tandem axle unit is employed which makes payloads of 24 tons possible. It incorporates locking devices for 20-ft., 35-ft., and 40-ft. containers. For conventional cargo haulage removable sideboards and

Fig. 55. Dorsey break-apart chassis system with single axle suspensions that unite into a load-equalizing tandem called the DX 20-40 coupleable chassis. (Dorsey Trailers)

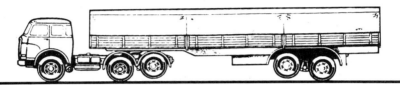

Semi-trailer — trailing axle of traction engine and leading axle of twin-axle semi-trailer unit raised.

Truck and trailer — motor vehicle trailing axle raised.

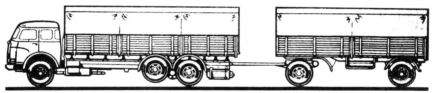

Truck and trailer — front driven axle of motor vehicle raised.

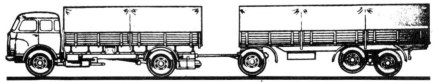

Truck and trailer — leading axle of trailer twin-axle unit raised.

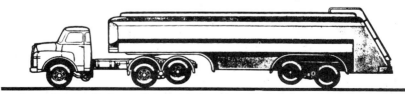

Tractor with tank semi-trailer — trailing axle of traction engine and leading axle of trailer twin-axle unit raised.

Fig. 55A. Axle lifting device showing varied possibilities. Adjustable for loads, empty, partial, or full; axle load control; traction control. (Walter Hunger Kg.)

tarpaulin can be utilized, transforming the chassis into an ordinary semi-trailer with a payload of 22 tons.

There is little adequate handling equipment at truck terminals and so the containers are usually left on their chassis or bogies or flatbeds for the duration of their inland need.

FRONT BOLSTERS ARE DESIGNED TO SECURE
FLAT-BOTTOM AND TUNNEL-TYPE CONTAINERS

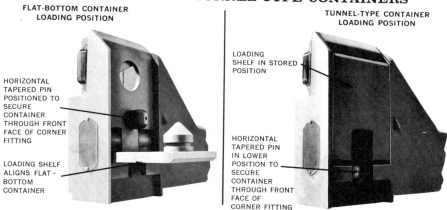

FLAT-BOTTOM CONTAINER
LOADING POSITION

TUNNEL-TYPE CONTAINER
LOADING POSITION

LOADING
SHELF IN STORED
POSITION

HORIZONTAL
TAPERED PIN
POSITIONED TO
SECURE
CONTAINER
THROUGH FRONT
FACE OF CORNER
FITTING

LOADING SHELF
ALIGNS FLAT-
BOTTOM
CONTAINER

HORIZONTAL
TAPERED PIN
IN LOWER
POSITION TO
SECURE
CONTAINER
THROUGH FRONT
FACE OF
CORNER FITTING

Fig. 56. Multi-purpose chassis. Designed to transport 8'6" high tunnel-type as well as 8' high flat-bottom 40-foot containers or coupled 20-foot containers interchangeably and maintaining a 12'6" overall height. Chassis rear bolsters with twist locks and front bolsters with dual pin-locking mechanisms secure all types of corner fittings. (Freuhof Div., Freuhof Corp.)

Consolidation and Break-Bulk Depots

In the United States the great part of the consolidation and break-bulk operation still takes place in or near the port areas. A major breakthrough in changing the thinking away from the port concept was made by United Cargo Corp., a worldwide container consolidator, when they set in motion a program they call "Trailsea Cities." The intent is to move the ports inland to points where the cargo originates or is destined. Since the aim is to reduce the handling of the merchandise to a minimum and so reduce the hazards, such aims must be accomplished at a series of centrally located depots. Here the containers are loaded and sealed, ready for export. Here the import cargo is brought in bond ready for clearance. This inland operation is usually handled at truck terminals, which are the only ready-made facilities fairly well dispersed throughout the U.S. Containers are stored at these locations so that as cargoes flow in to the terminals, consolidations take place and the movement is effected.

However in the usage of truck terminals for the handling of container cargoes, we must look at the facilities available, for it is more than just a building that is necessary, or the number of doors, or the amount of dock space, or any single item. The average truck terminal was not built for container use. The U.S. domestic trucking requirements and, indeed, the overseas trucking requirements are not necessarily synonymous with those of efficient handling. In this connection we are not discussing the use of labor, for this is covered elsewhere. To begin with, we are concerned with the location of the terminal. Accessibility to the new, fast highways is important; the layout of the terminal is important. The dock area designed for a cross-loading operation, with trucks lined up at doors opposite each other separated by a narrow dock area, are not suitable for a container operation. All the container cargo must be received on the dock before even one piece is loaded onto the container. There must be room for this and so a large dock area is necessary with a minimum of at least 50 feet between the cross doors. Such terminals are found when the trucker is already operating that terminal for break-bulk purposes. Truckers will usually have such break-bulk stations at the center of a hub where traffic from feeder terminals converges, or they may use a break-bulk-station as a way point for cargo en route from a more distant terminal, his own or another, to a port city. The handling system for the cargo within the terminal is important. There are four types used primarily:

1. *The Dragline System* is one where carts hooked into a below-floor-level dragline are constantly rotating around the periphery of the dock at slow speed. Carts can easily be added or taken out of the system. This system is an outgrowth of the cross-loading system and is designed primarily for immediate in and out movement of cargo. Cargo is stored in the carts in the center area of the dock.

2. *The Pallet System* is one in which cargo, when received, is placed on pallets and the pallets are kept in assigned spaces on the dock floor. When ready for loading into a truck or container, the pallet is brought over by forklift and the cargo is loaded with or without the pallet. This is but one step above the handling of cargo without any pallets. It is done in very small terminals or with special types of cargo such as skidded items and cartons.

3. *The Stacking System* is one where metal racks are constructed in the terminal so that cargo can be stored several tiers high. Sometimes special multi-tined forklifts with pusher attachments are used. This system has

not proved to be very practical as cargo is not always uniform in size so that much space is wasted, and "fitting" loads within the bins is too time-consuming.

4. *The Pull-Cart System* is similar to the pallet system except that the cargo is put on manual carts. The carts are kept in assigned spaces on the dock floor. When ready for loading, the cart is rolled right into the truck or container. This system has proved to be the most efficient from a time standpoint as well as freedom from damage.

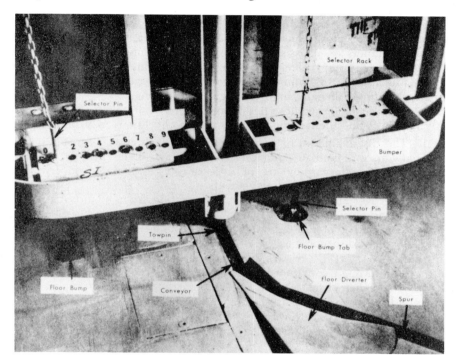

Fig. 57. Automatic shunt system. (S. I. Handling Systems, Inc.)

There are, of course, combinations of systems and adaptations of systems. An interesting adaptation of the dragline system has been installed at the Wilson Freight Terminal in Elizabeth, New Jersey, by the S. I. Handling Systems, called an automatic shunt system. Carts can be programmed to be shunted off the draglines at specific doors. A cart travels the towline path until it reaches its destination spur. A floor diverter opens automatically when selector pins on the cart contact a tab in the floor bump causing a cart to switch into the spur. The floor diverter is closed automatically by the towpin of a cart as the cart moves into the spur—either pushed or conveyed. A full spur control sensing unit prevents carts from switching into a spur when it is full. Extra carts recirculate and switch when room is available in the spur.

We are also concerned with the carrier's route system to the various port areas. Does he have rights direct to the port or does he have to participate in an interchange? With an interchange, the time factor will be affected. Does he have a direct run with perhaps only a change of tractors under a relay system and the van or container continuing on to destination? Does he have to reconsolidate at a station on the way to the port? If so, another delay.

To be considered are the traditional differences in what a trucker considers as desirable cargo and what an ocean carrier considers as desirable cargo, such as volume, applicable rate, etc. In addition, marine containers fall short of the most efficient standards for intercity road equipment. Marine containers on chassis have a higher tare weight and lower cubic capacity than the most efficiently designed semi-trailers of corresponding outside dimensions. Optimum container payload capacities for vessel loading are sometimes illegally high for highway trucking.

So that the meaning of these disadvantages may not be exaggerated, it is important to define their scope. Four major modifications may be noted:

1. The disadvantages of marine containers in highway carriage apply principally to intermediate and long hauls. They affect, principally, line-haul costs. Therefore, they are of little or only moderate concern with respect to local and short-haul movements.

2. The over-all economies of containerization from inland shipper to overseas consignee, involving savings in packaging, loss and damage, insurance, waterfront handling, and marine carriage, in their totality, vastly outweigh such diseconomies as occur in the overland segment of the movement.

3. For through container service, any higher cost for highway movement is somewhat offset by load lightness in the absence of export packaging. When cargo is packaged at an inland point for export by conventional methods, the cost of overland transfer to the waterfront includes hauling the extra weight and bulk of the packaging. Containerization commonly eliminated export packaging. Thus, the extra tare weight and limited cubic capacity of the container is something of a trade-off as against the overland freight saving on export packaging.

4. Finally, the design of containers is subject to continuous improvement. The differentially higher tare weight of the container-on-chassis unit will probably be reduced by progress in lightweight construction, notably through improved application of lightweight materials such as aluminum. This becomes particularly promising as progress continues in product improvement of such materials and in design innovation. In short, the movement of marine containers on the highways imparts a higher importance to lightweight design and construction than previously, and faster progress in this direction may be expected.

In addition to lighter-weight materials, other design improvements may reduce weight. Notable among these is the availability of containers which do not require support of a chassis, such as 20-foot containers which may be directly coupled together.

Subject to these favorable modifications, the tare weight and cubic capacity limitations of marine containers, of prevalent current design, may be described as follows:

Tare Weights: Containers and Trailers

Table I compares tare weights and cubic capacities of marine containers on chassis with conventional equipment. For heavy-density cargo, it will be noted that, within legal gross combination weight limitations, the containers will haul a smaller quantity. For light-density cargo, the cubic capacity of the containers will likewise be less.

Table I

Representative Tare Weights and Cubic Capacities
Marine Containers for Highway Trucking Compared
with Conventional Trucking Equipment

	Conventional 40' Semi-trailer	40' Container	Two 20' Containers
Tare Weight	(Lbs.)	(Lbs.)	(Lbs.)
Tractor	15,000	15,000	15,000
Trailer	10,125	—	—
Containers(s)	—	5,630	7,260
Chassis	—	7,000	7,000
Total Tare Wght.	25,125	27,630	29,260
Container Increment	—	2,505	4,135
Cubic Capacity	(Cu.Ft.)	(Cu.Ft.)	(Cu.Ft.)
Trailer or Container(s)	2,390	2,258	2,220
Container Decrement	—	132	170

Container Capacity and Legal Weight Limitations. In carrying heavy-density cargo, marine carriers have an incentive to open containers at the waterfront to obtain maximum ship loading. This is because allowable weights on the highways are much less than container carrying-capacity aboard ship. Table II illustrates this problem with respect to two 20-foot containers which would constitute a single unit of highway movement. These containers have a carrying capacity of 20 long tons each, or a total for the two of 76,160 lbs. of cargo after allowance for broken stowage. Combined with the tare weight of the vehicle, the full-capacity load would constitute a highway overload of 32,140 lbs. under a GCW limit of 73,280 lbs.

We have previously noted that a complete container service would call for moving a sealed containerload unopened from inland shipper across the ocean to inland consignee. But, when the full cube of a container is not utilized, sometimes the marine carrier will want to open containers at the waterfront to top off the load. On some shipments this will be done. This limitation applies principally to heavy-density cargo and occurs with respect to a minority of shipments. It impairs some of the economies of uninterrupted movement.

Table II

	Pounds
Maximum Load in Two 20-Foot Containers:	
2 Containers @ 20 Long Tons Each	89,600
Allowance for Broken Stowage @ 15%	13,440
Practicable Maximum Two-Container Load	76,160

Add Tare Weight of Highway Vehicles:	Pounds	
Tractor	15,000	
Chassis	7,000	
Containers (2)	7,260	29,260
Gross Combination Weight		105,420
Maximum Allowable GCW		73,280
Overload		32,140

Coupleable 20's. We have noted previously that the higher tare weight and limited cube of containers apply only to current designs and that these are subject to improvement. Such improvement is now well under way. A notable instance, as noted elsewhere, is that of coupleable 20-foot containers. When locked together these containers are self-supporting, eliminating the need for a longitudinal chassis underframe and substantially reducing tare weight.

Referring again to Table I, it will be noted that the tare weight of two 20-foot containers with chassis is shown to exceed that of a conventional 40-foot semi-trailer by 4,135 lbs. Trailer manufacturers report that the use of coupleable 20's substantially reduces this differential.

Steamship carriers also maintain pools of containers at inland cities, at truck depots or railheads but the ocean carriers are generally interested in full loads for their containers. In addition, there are rental container pools at inland locations, but they also are available only to the full container shipper. The consolidation and break-bulk operation as it takes place in port areas started on the piers. The steamship company would discharge the container, on inbound movements, and unstuff the container on the pier leaving the shipper and his agent to take over the shipment at that point. With the increase in container traffic, warehouses in and around the port areas began to set themselves up as container depots, often in conjunction with a trucker, or with a forwarder group. Thus, the container is taken to the depot or container station from the pier and deconsolidated and segregated ready for pick up for final delivery. These depots usually are customs bonded and have provisions for customs clearance right at the depot.

In much of the Near and Far East, in Africa and to a great extent in South

America, the sophistication of inland movement of cargo has not yet taken hold. This is due to many reasons, even over and above the obvious reasons of the supply of containers of two-way movement and of port and vessel facilities. In most of these areas the inland road system is not yet sufficiently developed to permit the movement of 20-foot, much less 40-foot containers. In addition, the customs authorities in many of these areas have not yet unraveled the cumbersome red tape involved in moving a container away from the port area. As a result, when containers are handled, this is done to and from port areas. Countries like Japan have made great strides in container adaptation, but because road strengthening and widening, tunnel widening, bridge raising and other changes take time, the biggest strides are taking place in the updating of ports.

In the United Kingdom and on the European Continent, the greatest progress has been made. In many areas such changes outpace those taking place in the United States. Complete new container ports, like Tilbury, near London, have been built. Others, in Grangemouth and Felixestowe, have been completed and are operating.

The British Road Haulage Association (RHA) and Traders Road Transport Association (TRTA) have arranged agreements with the British Customs and Excise departments, whereby TIR consignments are cleared inland in the U.K. Imported goods under TIR carnets are being allowed to leave ports without examination to proceed to inland clearance depots and approved wharves. Similar arrangements apply to goods being exported. The two associations have given guarantees to cover duties and taxes involved, but these do not cover consignments of spirits and tobacco. In addition, TIR carnets are now accepted in place of bonds as cover for imported wines removed for warehousing at certain designated warehouses.

Action is being taken to free certain classes of international road transport from licensing restrictions. This is another step towards carrying out the policy agreed upon by the Council of European Ministers of Transport in 1965 to liberalize traffic likely to cross and recross frontiers frequently. Now vehicles engaged in international movements of cargoes into England and remain temporarily in England will be exempt from the licensing requirements of the British 1960 *Road Traffic Act*.

On the continent, the northern coast ports have been modernized and updated so that ports like Antwerp, Rotterdam and Hamburg can handle container ships with ease. So far, very little has been done to the port areas in the Mediterranean underbelly of Europe, but some container service will be available to Italian ports of Leghorn and Genoa. Also on the continent, varying consolidation and break-bulk facilities exist, again primarily under the auspices of U.S. NVO's. Cargoes coming into Europe via Northern ports are cleared in transit on TIR documents which permit the inland movement of the cargoes to the public warehouse of the nearest large city in the country of destination. This is the only place that customs entry can be made other than at the entering border of the country of destination. After customs clearance, the cargo is ready for final delivery. For cargo being exported from Europe, customs examination is required, too, and in the case of export of merchandise made with materials not originating in the country of export, the customs requirements become even more complex. Such customs procedures can be made at the consolidation depot and the container may be sealed there and allowed to leave for whatever port of exit it is destined. Some forwarding groups have central collecting depots within a country to which all cargoes are funneled for consolidation. Some, particularly in Germany, still funnel cargoes into the port areas, but inland trucking and rail depots already exist and consolidations take place there, as well.

Highway Limitations; Regulations. The United States is composed of 50 "almost sovereign" states. Each state has different views on economics, safety, and commerce and different groups are espousing these views. Unfortunately the trucking industry and the continued upward development of international trade are hampered. This rein takes the form of varying road limitations in effect in the states. In addition to these limitations, regulations on load factor, weight, double bottoms, etc., vary from area to area and many times from state to state. Weight limitations even vary within one port area, when trucking from one section to another. For example, a truck being loaded to the maximum in the New York port area of Hoboken under New Jersey regulations could not bring his load into a Manhattan or Brooklyn pier, also in the New York port area because he would be overweight by 2280# according to New York law.

About 20 years ago the U.S. Congress passed legislation establishing size and weight limitations for trucks on interstate highways. This legislation was an attempt to standardize the varied limitations in the different states. In 1956, Congress passed the *Federal Aid Highway Act* settling the maximum in effect today, namely, 18,000# as the single-axle limit; 32,000# as the tandem-axle limit, and the gross weight at 73,280#. Unfortunately, the law has poor enforcement provisions so that, while most of the western, southwestern, midwestern and southeastern states have the same limitations as the federal government, most of the eastern states have limitations in excess. There are now 20 states with limitations in excess of the limitations in the federal legislation. The U.S. Congress has presently under consideration a bill which would raise the present single-axle limits from 18,000# to 20,000#; the tandem-axle limit from 32,000# to 36,000#; the width from 96" to 102", and the gross weight would be determined by the number of axles. This bill, if approved and passed, would still leave 14 states with a single-axle limit higher than the 20,000#. It is a good attempt but falls far short of coping with the many problems that exist in present-day trucking. The impetus for the present legislation has been started for the purpose of carrying more freight at a lower cost with resulting lower prices for the American producer and consumer. Another stated purpose is to facilitate the economic growth of the western states. Thus, attractive bulk commodities would be made more attractive because of greater volume. The question of arriving at standards that would be conformed to by all states has not been resolved. The question of standardization of regulations on double- and triple-bottoms has not been resolved. Experimental moves of three 27-footers have already taken place, as have moves of a 40-foot trailer in tandem with a 27-footer. Doubles consisting of two 40-foot trailers and one tractor are already operated on the Ohio Turnpike and the New York Thruway. However, Pennsylvania does not permit doubles, so that even two 20-foot trailers in tandem could not go from Ohio to New York. *The Turnpike or Toll Road Exceptions Are:*

Maximum Allowed Under Special Permit

	Length	Gross Weight	No. of Vehicles in Comb.
Ind. Toll Road	98 ft.	127,400 lbs.	3
Kans. Turnpike	105 ft.	130,000 lbs.	3
Mass. Toll Road	98 ft.	127,400 lbs.	3
N. Y. Thruway	108 ft.	127,400 lbs.	3
Ohio Turnpike	98 ft.	127,400 lbs.	3
Pa. Turnpike	70 ft.	73,280 lbs.*	2

* Pa.—combinations exceeding 73,280 lbs. must have written special hauling permit from Turnpike Commission.

In Europe, economic as well as physical considerations are the factors determining what limitations are placed on trucks. Spain's road system is not yet adequately developed and so, while small containers are permitted, Group I containers are only useable on certain roads. Because of the hazards of mountain driving, Switzerland does not permit 40-footers into the country, except in a few non-mountainous areas. There are two forwarding organizations active in moving cargo from Switzerland, known as Groupement Fer and Groupement Sud, each group being active in a different part of the country. These groups have special arrangements with the Swiss railroads whereby on all movements by rail there are special discounts of up to 17%. In Germany, as in many other countries, licensing of truckers is very rigid and licenses are at a premium. In addition, there are different degrees of licenses with specific distance authorization. The red licenses, which authorize long-distance and international trucking, are owned by about 10,000 firms or individuals. Most operate the traditional platform truck and trailer and, therefore, have no tractive units. These factors, coupled with a strong Bundesbahn organization, which gives special export rates on cargo moving via rail out of German ports, and the views of Transport Minister Georg Leber, who advocates the transport of certain cargoes by rail only plus special taxes on truckers, offer effective limitations on truck transport. Technical limitations, exist such as the legal length limit on German roads being 15 meters, a meter less than a 40-foot trailer and tractor (though no action is taken to prevent movement of 40-foot containers at present). When German truckers pick up a foreign trailer they must sign a special form reporting the condition of the unit. Faulty tires or brakes must be replaced or adjusted, and since these units have many different drivers, they are often in poor condition. In Italy, under Article 32 of the Highway Code, 40-foot containers are not permitted on the roads. But here, too, no action is taken to prevent their movement. However, in Italy, Austria and West Germany, it is against the law to drive a truck on the highways from midnight Saturday to midnight Sunday! The registry regulations are quite involved as well. The chassis, for purposes of TIR documentation, takes on the identity number of the tractor. Therefore, in Italy, for example, only the tractor that brought in the chassis can pull that chassis. So, in Italy an Italian driver with an Italian tractor could not pull a Dutch-registered container and chassis, for example. In the United Kingdom 35-foot containers and 40-foot containers have only temporary permits and the Transport Ministry tends to favor the rail transport over road haulage. In Australia certain state "construction and use regulations" restrict the use of 40-foot containers.

Conditions Existing in Central and South America

Limits of Motor Vehicle Sizes and Weights in Central and Latin America. (See Chapter 6 for weight and size maximums, as well as vehicle descriptions and types.) There are two countries in Latin America where the vehicle limits differ according to class of highway used. Mexico and Panama have designated specific limits for each type of highway.

There is general agreement regarding permissible maximum width for vehicles in Latin America. With the exception of Paraguay, which only allows 2.40 m (7.9 ft.), and Columbia, which sets 2.45 m (8.0 ft.), the remaining countries permit a width of 2.50 m (8.2 ft.). Only Brazil and Venezuela exceed this limit, allowing 2.60 m (8.5 ft.).

According to current regulations, international traffic of containers will be hindered in Paraguay since the 2.40 m (7.9 ft.) limit is less than the standard sizes of the containers. In Paraguay, the Group I container cannot be transported legally. The other countries offer no problem as regards permissible width although the chassis for vehicles for this type of transport should be specially fitted and equipped with ties and special accessories so that the container forms a part of the chassis. This is because the existing limits leave little allowance for placing apparatus or lateral outside support on the container.

This indicates that in all the Latin American countries the Group I container cannot be transported on conventional truck chassis, but will need to be adapted or made to conform to the standard ASA/ISO specifications.

Seven countries of the area set the permissible maximum height at 3.80 m (12.5 ft.) under the strictest regulation. The remaining countries exceed this limit with Mexico having the most liberal regulation, 4.15 m (13.6 ft.) for its Class A highways.

Consequently, in the countries where 3.80 m is in force for the maximum height of vehicles, the upper part of the chassis of any vehicle for container use must have 1.36 m (4.5 ft.) maximum height above ground level.

To study the length of vehicles, the following types must be considered: single truck, truck-tractor and semi-trailer, truck-tractor and trailer, truck-tractor with more than one non-motorized unit.

Two categories of single trucks can be considered: 2-axle (Type 2) trucks, and 3-axle (Type 3) trucks. This distinction is mentioned because some countries establish different limits for each of these categories.

For 2-axle trucks the most common permissible maximum length is 10.0 m (32.8 ft.) which in turn corresponds to the lower limit in the entire area.

The maximum limit is 11.6 m (38.0 ft.) which corresponds to Mexico for Class A highways.

For 3-axle vehicles, Paraguay has stricter limits with a maximum of 10.0 m (32.8 ft.), while the usual regulation is 11.0 m (36.1 ft.) and the most liberal regulation is in Bolivia with a 15.30 m (50.2 ft.) maximum.

Brazil, as well as Mexico, Paraguay and Uruguay do not differentiate between the two types of single trucks for limit in length.

For combination vehicles, truck-tractor and semi-trailer, Colombia and Ecuador have the strictest limit—13.7 m (45 ft.), followed in order by the Central American area and Chile, with 14.0 m (45.9 ft.). Brazil has the most liberal legislation with a limit of 22.0 m (72.7 ft.).

Paraguay and Mexico are the only two countries of the area that specify individual maximum sizes for each of the types of vehicles within this class. For example, Paraguay sets the following maximum limits: 2-S1 has 10.0 m (32.8 ft.), 2-S2 and 3-S1 have 12.0 m (39.4 ft.), and 3-S2 has 20 m (65.6 ft.). Paraguay has the most extreme limits for these types of vehicles. Mexico also has regulations similar to Paraguay but the variation in limits is not as great.

Therefore, if containers of 12.9 m (40 ft.) in length are to be transported in Central America and Chile, it must be remembered that the cab of the truck-trailer will have to be reduced to a maximum length of approximately 1.20 m (4 ft.), a critical factor in the design of such truck-tractors.

Truck-tractors with trailer are only suitable for moving average or small-size containers, 20 or 10 ft. (6 or 3 m) long. This is because the limits are applied independently to the truck-tractor and it is considered as a single unit as well as a combination tractor and trailer. Consequently, all that has been stated regarding trucks is also applicable to truck-tractors for this class of vehicle. Due to the limits of overall length, trailers cannot carry the large-type containers.

The Central American area shows the strictest limit on maximum length for this class of vehicle—14.0 m (45.9 ft.), and Bolivia, with 24.0 m (78.7 ft.), has the most liberal limit.

The truck-tractor with more than one non-motorized unit is a class of vehicle not permitted in every Latin American country, and is allowed only on the highways in Argentina, Panama and Peru. The regulations for length are: Argentina, 20.50 m (67.3 ft.); Panama, 19.85 m (65.2 ft.); Peru, 15.0 m (49.2 ft.).

The Republic of Bolivia has rather broad and liberal rulings for vehicle sizes, but quite strict ones for weights. Bolivia has much lower axle loads and overall weights than the rest of the continent.

The extremes in load regulations for single axles are 8 metric tons (17,637 lbs.) for Central America, and for Peru and Ecuador, 11.0 tons (24,251 lbs.).

As for tandem-axle loads, Paraguay is the country with the greatest restriction, 10.0 tons (22,046 lbs.), while Chile permits the heaviest axle load of 19.0 tons (41,888 lbs.). The most common limit for single-axle loads is 8 tons (17,637 lbs.), while for tandem axles it is 14.50 tons (31,967 lbs.).

With the exception of Bolivia, as we have already seen, Paraguay has the lowest admissible maximum weights for the following vehicle types: 2, 3, 2-3, 3-2 and 3-3.

It should be noted that with the exception of Types No. 2 and No. 3, Ecuador has the highest limits for any type of vehicle.

The road situations seem to add up to almost impossible conditions. Yet the advantages of flexibility of movement, highway improvements, regulation relaxation and standardization, rate negotiation (in Europe), and the gradual improvements in equipment and service forced by the increase in container traffic would seem to offer a bright future, with the passage of time, to the road movement of containers.

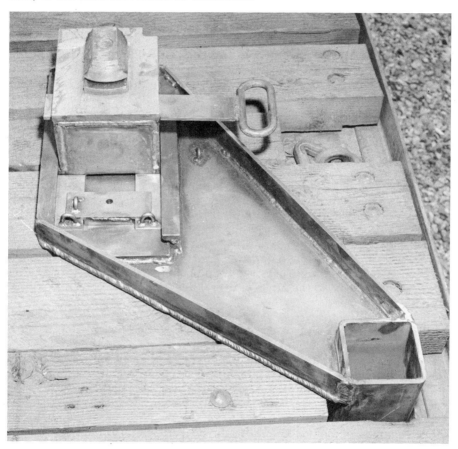

Fig. 57A. Tie-down device for carrying containers on flatcars.(Union Pacific R. R.)

By Rail Here and Abroad

Types of Equipment, Ease of Movement and Interchange

While Group I containers can be carried on ordinary flatcars, this is not at present a practical means of transport, from the standpoint of speed, cost and handling. In the United States there are several means of carrying containers. One of the major means is the Flexi-Van which carries only the container. Another is the piggyback which carries the container on a chassis or a trailer. Existing flatcars are adapted for carriage of containers by adding bolsters. Some of the new equipment has twist-locking devices designed to secure ISO corner fittings. This equipment, when using the adapters, consists of tie-down devices that are inserted in the stake pockets. Each tie-down device set consists of two rigid and two adjustable pieces. The height of the base upon which the twist-lock rests is now about 3½" to 4" between the container bottom and the flatcar deck.

Flexi-Van came into use because of the height of the trailer with wheels. When on railcars this presented problems in clearing some of the tunnels. However, most of these clearances have been corrected. Using piggyback railcars the container on the chassis is driven right onto the railcar and locked to the kingpin. These piggyback cars can handle two 40-foot containers or 2 units of two 20-foot containers, each on a coupled chassis. There are special tariffs with Trailer on Flatcar (TOFC) rates. Such rates fall into different "plan" categories depending on the amount of service the railroad is asked to perform. These plans are:

Plan I: Railroads and Motor Common Carriers. Carriage by railroad of trailers of common carrier truckers at a flat charge per trailer rate—a "substituted" service performed for the trucker, who solicits the business and bills the shipper. The railroad, in effect, works on a subcontract basis for the trucker.

Plan II: A Railroad Operation (Known As a "Retail" Operation). Railroad performs all of the service, including the furnishing of trailer, loading and unloading, and pick up and delivery. Railroad solicits the business at truck competitive rates and bills the shipper.

Plan II¼: Railroad performs pick up service at origin, but consignee must arrange for delivery at destination.

Plan II½: Similar to Plan II except railroad performs ramp-to-ramp service only, does not furnish pick up or delivery. (This plan does not provide for carrier liability as it is on a basis of shipper's load and count.)

Plan II¾: Shipper delivers railroad trailers to origin ramp; railroad performs delivery service at destination.

Plan III: Shipper (Owned or Leased) Trailers. Railroad furnishes flatcar and provides loading and unloading of trailers. Shippers handle pick up and delivery. Ramp-to-ramp rates made for these shipper trailers (owned or leased) based on commodity and quantity moved, or at a flat per-trailer charge.

Plan IV: Shipper Owned or Leased Trailers and Cars. Railroad furnishes only power and rails for shippers, who not only furnish both flatcar and trailers, but perform all loading, unloading and pick up and delivery services. A flat charge per car is made for not exceeding two trailers, loaded or empty.

Plan V: Joint Rates Intermodal Service. Joint rail-truck rates. In effect, such rates extend the territory of each carrier into that served by the other, permitting each to handle shipments originating in or destined to the other's territory. Each may sell for the other.

These trains are specially routed so that a coast-to-coast delivery can take place in six days. Piggyback railcars are of two lengths—one, a 75-foot length that will handle one 40-footer and one 20-footer, three 20-footers, or two 35-footers. The other length is 85 feet and this can handle two 40-footers. Piggyback railheads use "circus-loading" whereby the units are driven onto the railcar through an end ramp, but this method is giving way to the use of gantries and the special lifting unit called a "piggy-packer" and

Fig. 58. (upper left) Piggyback car with trailer hitch, North Bergen Yards.

Fig. 59. (lower left) Manual adjustment of trailer hitch on piggyback car, North Bergen Yards.

Fig. 60. (upper right) Yard Hustler backing unit onto railcar. Bogie has been released, Highbridge Yards.

Fig. 61. (lower right) Yard Hustler on nose wheel moving into position to push unit alongside railcar, Highbridge Yards.

(Penn-Central R. R.)

described in Chapter 4. The piggy-packer can lift a van up to 70,000 pounds on or off the railcar in less than three minutes. The van is held in place on the car by a hitching device attached to the kingpin. On some of the cars this device is manually adjusted to the height of the kingpin and locked. On others the hitch is spring-locked in place. (See Figs. 58 & 59.)

In the Flexi-Van system each railcar has two turntables on which a single 40-footer or two 20's on a coupled chassis can be handled. There are three types of Flexi-Van cars used. The original car, 84.6 feet long, carries two 40-footers. On this car, the container rests on its center on the turntable. There are two turntables on each car which are activated by a hydraulic system. The fluid for the hydraulic system comes from a unit mounted on the railcar, a portable unit independent of the railcar, or the tractor. Power, however, is supplied by the tractor. On this railcar, an adaptation of the

Fig. 62. Yard Hustler with boom in position in socket plate transferring a Freuhof Flexi-Van container onto railcar. The detachable bogie, which acts as the Flexi-Van's tandem rear axle for over-the-road operations remains idle at the side of the railcar. (Freuhof Div. , Freuhof Corp.)

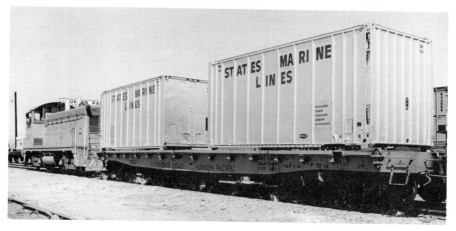

Fig. 62A. Cushioned underframe 53'6" flatcar shown with two 20-foot containers. Note space between containers where doors may be opened while on flatcar. These flatcars are also used for articulation into 107' cars. (Union Pacific R. R.)

original, the container rests on three surfaces. The Mark III railcar, 86.9 feet, and the Mark IV railcar, 87.3 feet, the other two cars, have turntables arranged to hold the container at one end. The turntables are manually operated and do not require hydraulic assistance. The use of refrigerated trailers, which had their units protruding beyond the 40-foot length of the container, necessitated the longer Mark III and Mark IV. The kingpin lock on the Mark III is manually operated. On the Mark IV it is operated by compressed air. In transporting trailers on bogies, only the bogie is left behind.

While Flexi-Van trailers have the side strips built on and have folding landing gear, regular marine containers are handled on rail with the chassis attached, the chassis having the side strips and folding landing gear. In the handling, the yard tug backs the container on its bogie perpendicular to the railcar. The turntable on the railcar is turned into position. A ratchet device pulls back the pins on the turntable. The tug backs the container onto the turntable, the pins on the turntable are released to lock into the holes on the side rails of the chassis. The pins on the bogies are manually released and the tug lifts the fifth wheel so that the container is free of the bogie. The pins that lock the bogie to the trailer or the container chassis are the same ones that lock the turntable to the trailer or the container chassis. The landing gear is folded away and the tug lifts itself up, engaging its nose wheel. It turns on the nose wheel and pushes the container over to almost lateral position with the railcar. The self-contained boom on the tug is released and hooks onto the socket plate on side of container and pushes the container into final position. As an additional precaution, a chain lock is fastened to the container at this end. For removal, compressed air from the tug is hooked up to the kingpin lock on the railcar and the lock is released. The boom engages the socket plate on the container, pulls the unit onto the fifth wheel of the tug and the process is reversed. The tug has fold-away vertical facing forks which are used to lift and move bogies about. These tug units differ from the pier yard tugs or tractor in the two features needed at railheads and not needed at the piers, namely, the boom and the forks. The Flexi-Van service presently operates only between New York, Cleveland, Chicago and St. Louis.

The tariffs for this type of movement, Container on Flatcar (COFC), have rates that are primarily class and commodity rates. The Flexi-Van trains also operate a "through" speed service with special long-distance expresses. There are two other types of cars that are in general use for the carriage of containers. One is an 89' flatcar with raised bolsters that contain locking devices for fastening ISO containers. Two 40-footers or four 20-footers can be hauled. The other is an articulated 107-foot car consisting of two 53½-foot cars. Each of the 53½-foot cars is designed to carry two 20-footers at the extreme ends. A 13-foot space is thus achieved between the containers permitting loading and unloading of the containers without removing them from the railcar.

New rail equipment is constantly being designed. Trailer Train Company is arranging for the manufacture of an all-purpose car. This car will handle trailers and/or containers interchangeably in 30 different combinations (20, 24, 27, 35 and 40 feet). The car will be 89' 4" in length, 9' 6" in width, with hydraulic draft gears, tractor-operated non-cushioned hitches, inside rub rails, and 16 adjustable folding pedestals, which fold flat into the floor when not in use. Trailers can be loaded by use of cranes, yard equipment or by circus type loading.

Trailer Train, comprised of the major U.S. railroads and one domestic freight forwarder, is owned by its participating carriers on an equal basis. Cars are used on a combined per diem and mileage basis. The basic daily rate is $3.30 for an 85-foot, or longer, car plus a mileage rate which decreases as the miles per month increase. If a member railroad holds a car for five consecutive days and no other member calls for it, the car can be shunted back to a trailer train yard without cost to the railroad beyond the five days per diem charge.

A. T. Kearney, a consulting firm, has designed a car which will permit double-tiering of containers without overhead clearance problems. It has a drop-well design and, with two stacked 8 feet high, has a clearance over the rail of 17 feet. The chief advantage of this car is that it would permit the carriage of 31–50% more cubic capacity and up to 2½ times the weight capacity of the present cars. These cars will be able to carry six 20-foot containers, or two 40-foot containers and two 20-footers, or five 24-footers, or three 35-footers, or two 40-foot trailers. (*See* Fig. 64.)

Canadian railroads have also been using piggyback, but are concentrating on the more economical use of the container only on a railcar. The economic advantages of moving only the containers are that the container costs less than the trailer. There are no annual license fees, the container flatcar costs less than the piggyback flatcar, and less power, with its attendant fuel and maintenance costs are required for operation of a container train than is required for a piggyback train. The concentration, however, is not on the Flexi-Van cars, but on railcars whereby the containers can be loaded with the side-loader. Overhead clearances for railroads are not problems in Canada. Canadian Pacific has already built and tested a prototype car that carries eight 20-foot containers double-decked from coast to coast. In the United States, Railway Express Agency utilizes this system with railtainers and is known as Interpool.

In Europe there are several international train services for the movement of containers, both government- and nongovernment-operated. Unlike the United States, all European railroad systems are state-owned.

The transport of transcontainers is a relatively new development in Europe and the French National Railways have taken part in various international schemes to facilitate this traffic. The French National Railways are members of Intercontainer, the association formed by the European Railways for the conveyance of transcontainers across Europe, and they have assisted also in the preparation of the International Transcontainer Tariff which gives an all-inclusive rate, expressed in gold francs, from main ports to destinations in France and other countries. This tariff does not take weight into account, but classifies transcontainers into three categories, according to length (20', 30', and 40').

The close cooperation between the British and French Railways in this field as in others, has resulted in the creation of two direct services for transcontainers. The first, known in Britain as the London-Paris Freightliner and in France as the Transcontainer Service Londres-Paris, offers a "next-morning" delivery service between the two capital cities five days a week. Customs formalities are completed in the two terminals—at Paris, La Chapelle and at London, Stratford—which eliminates difficulties and delays. At Stratford this service is connected with the facilities provided by the internal freightliner services of British Rail. From Paris-La Chapelle good connections are also available to all main French centers and containers arriving there from London one morning will be ready for collection by the receiver the next day in Marseille, Bordeaux or Toulouse, etc.

The second transcontainer service of the French and British Railways is the one between London and Dunkirk, which provides good connections to centers in Northern and Eastern France, as well as to Switzerland and Italy.

British Rail and French Railways are also bringing about great changes to their fleet. From Harwich, main transcontainer port of British Rail, a French Railways container roll-on/roll-off ship will, within a year, provide a direct link to Dunkirk for unit loads, in addition to the B.R. container services to Zeebrugge. The new ship will have a capacity of 185 twenty-foot containers or 60 large road vehicles, or an appropriate combination of both. From Dunkirk fast transit services will be available to any point in France and beyond (Switzerland, Italy, Spain). There are about forty towns in France

LOCATION OF PEDESTAL SLOTS

WILL HANDLE ANY OF THE THIRTY POSSIBLE COMBINATIONS

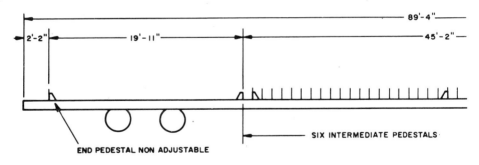

POSSIBLE LOADING COMBINATIONS

4 – 20'	1 – 20', 1 – 24', 1 – 27'	3 – 27'
3 – 20'	1 – 20', 1 – 24', 1 – 30'	2 – 40'
2 – 20', 1 – 24'	1 – 20, 1 – 24', 1 – 35'	2 – 35'
2 – 20', 1 – 27'	3 – 24',	1 – 35', 1 – 40'
2 – 20', 1 – 30'	2 – 24', 1 – 30'	2 – 30'
2 – 20', 1 – 35'	2 – 24', 1 – 27'	1 – 40', 1 – 27'
2 – 20', 1 – 40'	1 – 24', 1 – 27', 1 – 30'	1 – 40', 1 – 30'
1 – 20', 2 – 24'	1 – 20', 1 – 27', 1 – 30'	1 – 40', 1 – 24'
1 – 20', 2 – 27'	1 – 20', 1 – 27', 1 – 35'	1 – 35', 1 – 30'
1 – 20', 2 – 30'	2 – 27', 1 – 24'	1 – 35', 1 – 24', 1 – 24'

Fig. 63. Trailer Train all purpose car. (Trailer Train Co.)

already equipped for the handling of transcontainers and new equipment, similar to that in use in the freightliner terminals of British Rail, is now being installed in fifteen of these.

For the conveyance of transcontainers new types of rolling stock have been developed and these will progressively replace the flat wagons at present in use for this traffic. Two categories of transcontainer wagons have been approved by the International Railways Union (U.I.C.). One wagon without shock absorbers is approved for direct center-to-center traffic, and one with shock absorbers is approved for use on cross-country movements carrying other freight traffic.

The French Railways are also introducing a number of special express container trains linking the main industrial centers overnight. The first of these, known as the Mediterranée Fret Express, has been in service for over a year between Paris and Marseille, and others will soon follow. Even as these new services are inaugurated, it is now possible to send transcontainers overnight from Paris to all the great French cities with the above-mentioned trains.

The containers themselves are of many kinds, depending upon the nature of the traffic for which they are intended. For general cargo the French

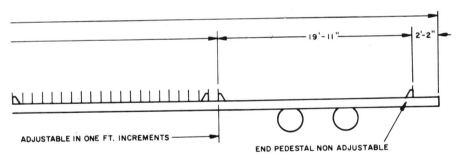

ON TTAX — "ALL PURPOSE CAR"

OF CONTAINERS OF 20', 24', 27', 30', 35', AND 40' LENGTHS

19'-11"   2'-2"

ADJUSTABLE IN ONE FT. INCREMENTS

END PEDESTAL NON ADJUSTABLE

ALL PEDESTALS FOLD INTO DECK WHEN NOT
IN USE OR WHEN LOADING TRAILERS.

TRAILER TRAIN COMPANY

Fig. 63. Trailer train all purpose car. (Trailer Train Co.)

Railways and their associated container company, the C.N.C. (Compagnie Nouvelle de Cadres), are in favor of units equipped with side and end doors. This is to enable firms with their own rail sidings to continue to receive containers on rail at their own premises and to load and unload them without undue difficulties, thus avoiding the need to collect them from a terminal which might be fifty miles away, or be faced with extra road delivery costs. Moreover, in large cities where traffic congestion is severe, the provision of side doors has been found of great assistance for loading and unloading.

Container transport techniques, in both their traditional and new forms, have rendered necessary the provision of an appropriate container tariff. On the French Railways, this is known as "Tariff 106."

The transcontainer section of the tariff has two parts—one for overland movements, involving no sea transport, and the other for containers going overseas or coming from overseas countries. This is the case for the traffic between Britain and France. The rates applicable on transcontainers carried

Fig. 64. Proposed Hi-Cube compatible container car. (A. T. Kearney)

under the conditions of this tariff are expressed as an overall figure for any given mileage, taking into account both the size of the container and the weight. Thus for any one distance there are three possible rates for a 20-ft. container according to the load it contains (up to 8 tons; from 8 to 13 tons; over 13 tons).

There are also three such figures for the 30-ft. containers (up to 10 tons; from 10 to 16 tons; over 16 tons) and three for the 40-ft. units as well (up to 12 tons; from 12 to 20 tons; over 20 tons). Two 20-ft. containers placed together on a wagon would, under the conditions of the tariff, be charged as one of 40 feet with the total load. Inexpensive rates are granted for the empty return of transcontainers when a balanced traffic cannot be achieved.

For transcontainers in transit through France the International Transcontainer Tariff jointly agreed by European railway administrations is applicable.

The ordinary container section of the tariff has three main features:

(a) privately owned containers, of a type approved by a railway company (not necessarily the S.N.C.F.), travel free when loaded;

(b) their return empty, where the traffic is not balanced, is charged at a low rate;

(c) there is a 20% reduction on the tonnage necessary per truck to obtain the lowest transport rate for any commodity; thus, where the lowest rate obtainable for a given kind of goods is—say—5 tons per truck, the same rate, if containers were used, would be obtainable for quantities of 4 tons per truck. In the case of non-hazardous chemicals in solid form, if they were loaded in "tote-bins," the 15-ton rate would apply on quantities of 12 tons per wagon, a loading performance by no means difficult to achieve on a continental truck.

The Compagnie Nouvelle de Cadres (C.N.C.) was created by the French Railways to expand the scope of the possibilities obtainable through container tariff.

Containers are not the only unit load services of the railways. Across the Channel, the joint services of British Rail and French Railways offer four kinds of roll-on/roll-off facilities on three jointly operated routes.

The routes are:

(a) Harwich-Dunkirk, operated by a British Rail ferry boat (this route will be supplemented by the new container ship mentioned previously);

(b) the well-known Dover-Dunkirk train-ferry route, for which a new British Rail ship will soon be built; this service is supplemented by other B.R. services to Calais and Boulogne;

(c) the Newhaven-Dieppe route.

As for the methods of conveyance, apart from containers, previously surveyed, we have: trolleys (known as B.R.U.T.E.) via Dieppe; wagons, by the train ferried via Dunkirk; a special service for road semi-trailers.

Trolleys. These are used on the Newhaven-Dieppe car ferries for the parcel traffic between Great Britain and France and other destinations beyond the French borders. They are rolled into the ship's garage, which ensures safe, speedy and shock-free transport across the Channel. This service is particularly useful to those who have small quantities to send to many destinations on the continent. The conditions of the European part load tariff between Britain and France, applicable for this traffic, are particularly favorable as delivery is included in the rate quoted.

Ferry Wagons. Container traffic will, in some cases, take the place of ferry wagons. The future of the latter is assured, however, for in many instances there is no reason to change to another method: when both sender and consignee are rail connected; when a large capacity vehicle is needed (the new wagons are over 40 feet in length and offer a capacity exceeding 2,350 cu. ft.); when the traffic cannot bear the cost of container hire (for one of the great virtues of the ferry wagon is that its use does not entail hire charges and the user need not worry about depreciation and the cost of repairs). Traffic in ferry wagons was given new scope in Autumn 1967 with the opening of the Harwich-Dunkirk route. This was created especially for the traders of northern Britain, as this route, which avoids the difficulties of a journey through or around London, permits faster transit times to many continental destinations. From Dunkirk, there are fast services for ferry wagons to the large French industrial centers and in transit through France.

For the transport of fruit and vegetables from the continent, when refrigeration is needed, and for traffic with Spain, special ferry wagons render particularly valuable services. The refrigerated ferry trucks are known as Interfrigo wagons, from the name of the international company founded by the European railways to operate them. Trucks known as Transfesa wagons are used to overcome the particular problems caused by the different gauge of the Spanish Railways. From the U.K. they carry back to Spain, without transshipment, an important groupage traffic. Both Transfesa and Interfrigo are represented by an office at the London headquarters of British Rail.

Different private groups have been successful in negotiating operating agreements with the groups of railroads. One of these, originating in France, but already operating in Holland, Belgium and Italy with extensions planned in Germany and Spain, is a piggyback system called the Kangaurou system. Britain presently is a participant in this system, but, because of railway gauge difficulties, the Kangaurou vehicles only operate in Britain as normal, over-the-road semi-trailers.

The Kangaurou conception, while adapted from the intermodal semi-trailers or containers on wheels, can handle only trailers up to 34 feet long and is intended for truck traffic adapted only for rail. Another system, also controlled by Novatrans, the organization controlling Kangaurou system, is the UFT (Union Fet Route). (In Holland, Kangaurou is operated by "Trailstar"; in Belgium by Transport Route Wagons; in France by Novatrans.) Instead of standard road vehicles being adapted to fit special rail wagons, as in the case of the Kangaurou, the order is reversed. Special semi-trailers are loaded onto ordinary railway flatcars which have been modified by the installation of rails on top of the floor. An adaptation of the Flexi-Van has been used by Novatrans for their third system, the MC22. The MC22 is the closest French approach to the handling of containers. This technique does not involve the use of containers at the moment, although the box without wheels is loaded on a Flexi-Van type of rail wagon and it has ISO corner posts and fittings. The box is 38 feet long, 8 feet wide and 8 feet, 6 inches high. These units were designed for traffic within continental Europe and were developed without thought of sending them overseas by ship. However, from the very start the MC22 wagons were devised to carry containers of ISO dimensions, so that it would seem to be only a question of demand before MC22 is used for international movement of intermodal containers. Method of loading and unloading is similar to the Flexi-Van, with the use of a tractor.

The German Federal Railways has placed orders with several container and wagon manufacturers for the building, not only of 20-ft. and 40-ft. transcontainers, but of containers with adjustable roofs, with two side doors, open-topped containers, and containers for loose materials 8 ft. and 4 ft. high; containers for powdered goods with pressure discharge, and 20 ft. and 40 ft. flats. Prototypes have been ordered so that tests can be carried out at the earliest possible moment.

Orders for the building of container rolling stock have also been placed to augment the 230 shock absorbent container carriers already available. These wagons must be biaxial or, as the case may be, have four axles; they must be capable of maximum speeds of 120 kilometers per hour with axle loads of 20 tons, and able to carry units from 10 ft. to 40 ft. in length, as well as smaller containers and exchange boxes on 20-ft. sub-frames. The containers are transferred horizontally and are held down by twist locks.

Individual European railroads have published tariffs for the movement of containers within their borders and the members of U.I.C. also published a container tariff. Such tariff covers movements from terminal to terminal only. The U.I.C. tariff is superseded by the tariff of Intercontainer; this group is covered later in this chapter.

Containers were first used by the Japan National Railroad in 1930 and those were of small box type, made of steel, wood or bamboo, for one-ton loads. However, they were not used extensively, and their use was more or

Fig. 65. Horizontal transfer of container from chassis to cushioned railcar of German Federal Railroad. (German Federal R. R.)

less limited to certain kinds of sundry goods, such as confectionery. At the peak year of 1938, Japan National Railroad had in possession a total of 5,060 containers of all kinds, but ceased to use them in October, 1939 with the aggravation in material supply situation brought about by wartime conditions. The greatest drawback to the prewar containers was that they were loaded on ordinary boxcars and extreme difficulty was involved in loading and unloading because at that time handling machines, such as forklifts, had not been fully developed.

With the rapid progress in motorcar transport after the war, various defects accompanying railway transport became evident, making it necessary to modernize the transport system in some form or other in order to fulfill the mission of the railway as a modern transport medium. As one of the steps, three-ton containers were trial-made in 1956 and used between Tokyo and Osaka. The results were not very satisfactory, however, because only few business dealings were done in three-ton units and there was lack of enthusiasm on the part of private trucking companies. Small-type refrigeration containers for 300 kg. load have been used since 1950 for the transport of small lots of frozen provisions.

After much study on large-size containers, taking past experience as a basis, it was found that five-ton containers were much more advantageous than the three-ton ones, or boxcars, due to measurement and capacity as well as unit of transaction; and it was decided to put them to practical use in 1959. Trial use of these containers was started in June of that year and operation on a commercial basis commenced on November 4 between Tokyo and Osaka—554.5 km, with 330 containers and 55 freight cars specially designed for their conveyance.

The containers owned and operated by the Japanese National Railways are five-ton containers, and their volume is 14 cubic meters (490 cu. ft.). The size of the containers was determined, taking into consideration the unit of transaction and the road condition in Japan. Door-to-door through transport is carried out at the responsibility of the Japanese National Railways. Trucks are used to carry containers between the consignor's and consignee's door and the railway station, and exclusive flat wagons capable of carrying five containers are used for transport between stations. Forklifts are the principal means for transfer between truck and wagon at the station. The containers can thus be handled without inclining them, enabling simpler packing.

The number of containers possessed by the Japanese National Railways has been increasing year after year since 1959. In 1966 the Japanese National Railways owned approximately 10,000 containers and 1,500 wagons exclusively for containers, carrying about 2.5 million tons of goods annually, or about 1.3% of the total freight tonnage of the Japanese National Railways. It is expected that container traffic will increase to 6.5 million tons; and containers to 100,000 in 1971 with container traffic rising to 30 million tons. Containers are handled at over 103 stations. Collection and delivery are made within a range of 20 km. of those stations. The stations to handle containers will be increased to 200 by 1971, though the total number of freight stations is about 3,000 at present.

Commodities making use of containers are mainly such high-class freight as foodstuffs, magazines, household goods, synthetic resins, industrial chemicals and medicines. Container transport service is widely used for shipment of household furniture.

The timetable schedule of container trains is separate from the general freight train schedules. It is directed to speedy delivery and certainty of arrival time. Japanese National Railways fixes 539 sections among the more than 103 container-handling stations and, assigning a certain number of containers to each section for daily transit. It operates flat wagons loaded with the containers by coupling them, with priority, to the previously designated fastest freight train making a through run on the section. Any such train is a limited express, express or yard-passing train. The typical ones are the six high-speed limited express trains linking Northern Kyushu, Osaka, Nagoya and Tokyo at a maximum speed of 100 km. per hour.

It is anticipated that about 50% of the goods moved between Japan and the United States will be containerized by 1970. In line with this tendency, container terminals are being planned or constructed on the coast. By 1971, the ports of Tokyo, Yokohama, Osaka and Kobe will be equipped with necessary terminal facilities. Various quarters concerned are studying how to facilitate feeder service from container terminals of these ports to depots of the principal inland cities or to the end customers.

Feeder service of the international ocean-going containers is proportioned so that 20% will go to car-load and 80% to LCL, or part-load, according to an estimate being made in Japan. But the proportion may eventually become 50/50 when the transport to inland depots is included. The same may be said about part-load shipment. However, the Japanese National Railways' container is about one-half the cubic capacity of the 20' x 8' x 8' marine container, so it is convenient for part-load shipment in particular. The Japanese National Railways will probably have a greater share of wagon-load freight traffic against road transport than the railways in Europe because Japan's roads are too congested to allow traffic of large-sized vehicles, with the exception of some toll roads. The Japanese National Railways operates its network throughout the country, and provides industrial sidings at any such ports, warehouses and factories or other plants that are closely concerned with export or import goods. Thus, Japanese National Railways is in a very favorable position to render door-to-door service for the internationally transported containers.

The 8' x 8' x 20' containers coming across the sea can be carried to any place in Japan by flat wagons owned by the Japanese National Railways. However, they are presently building flat wagons for exclusive use of marine containers so that containers of any size may be carried. The new wagons are equipped with container fastening devices which can be adjusted according to the size of the container, a platform to enable the cargo to be unloaded from the end of the container as it stands intact on the flat wagon, and a high capacity buffer to avoid transmitting the coupling shock to the container. They will be able to run at 85 km. per hour.

Status in Latin America

General description, by countries, of ALAF (Asociación Latinoamericana de Ferrocarriles) (Latin American Association of Railroads) member enterprises. Trackage, gauge, service territories and domestic and international connections.

ARGENTINA

The railways in Argentina are state-owned and their authority is centralized in an agency designated as EFA (Empresa Ferrocarriles Argentinos) which operates them under six management companies.

These railways are:

General Belgrano Railroad operates on 15,320 kilometers of meter-gauge (1.00 meter) main-line track. Its network covers all of Argentina from the latitude through Buenos Aires in the south, to the frontiers with Paraguay and Bolivia in the north, and from the Paraná River in the east to the Chilean border in the west.

It interconnects with the railways of Bolivia and Chile at two points in each case—at La Quiaca-Villazón and Pocitos-Yacuiba with the former and, on the Andean Cordillera, at Socompa and Las Cuevas with the latter country.

These connections afford through rail passage over the following routes: Buenos Aires-La Paz and Buenos Aires-Santa Cruz de la Sierra between Argentina and Bolivia; and Buenos Aires-Mendoza-Santiago-Valparaiso and Buenos Aires-Salta-Socompa-Antofagasta between Argentina and Chile.

General Mitre Railroad operates 6,696 kilometers of wide-gauge (1.676 m) main-line track. Its network serves important Argentina cities, notably Rosario, Santa Fé, Córdoba and Tucumán.

General San Martin Railroad operates 4,658 kilometers of wide-gauge (1.676 m) main-line track serving the area west of Buenos Aires. Its connection with the General Belgrano Railroad makes for the most direct route between Buenos Aires and Santiago–Valparaiso.

General Roca Railroad operates 8,744 kilometers of wide-gauge (1.676 m) main-line track and serves the southern region of the country from east to west, running to important localities in the Andean piedmont, such as San Carlos de Bariloche and Zapala. It does not connect with the south Chilean railways of the same gauge.

General Sarmiento Railroad operates 3,865 kilometers of wide-gauge (1.676 m) main-line track in Buenos Aires Province and adjacent areas.

General Urquiza Railroad operates 3,349 kilometers of standard-gauge (1.435 m) main-line track. Its network serves the entire mesopotamian region of Argentina between the Paraná and Uruguay rivers. At Posadas-Pacú-Cuá it interconnects via ferry with the C.A. López Railroad in Paraguay, of the same gauge, which runs to Asunción.

At Paso de los Libres-Uruguayana it interconnects with the Brazilian meter-gauge network serving Rio Grande do Sul (VIFER). The difference in gauge necessitates transshipment, which can be done on either the Argentine or the Brazilian side, as there are three-rail tracks at both border terminals to carry the rolling stock of both companies despite the gauge difference.

The General Urquiza Railroad could connect with Uruguay at several points along the Uruguay river because the two systems have the same gauge. Although there are no bridges between the two countries across this river, the ferryboats owned by this company could be used. The crossing could be made at Concepción del Uruguay-Paysandú, Concordia-Salto or Monte Caseros-Bella Unión.

In short, 42,632 kilometers of main-line track are operated in Argentina, 36% of it of meter-gauge, 8% standard gauge and the remaining 56% wide gauge.

The Argentine Railways interconnect with Bolivia, Brazil, Chile and Paraguay, and rapid connection could eventually be made with Uruguay. The fact that the connections with Bolivia, Chile and Paraguay involve no change of gauge allows the exchange of rolling stock. The junction with Brazil connects different gauges, but the eventual connection with Uruguay would benefit greatly from the identity of the two gauges.

BOLIVIA

The Empresa Nacional de Ferrocarriles, a government enterprise, operates 97% of the 3,560 kilometers of track in Bolivia. The remaining 3% is owned by the Peruvian Corporation, which operates the 96 kilometers of track between La Paz and Guaqui. The network is entirely of meter-gauge, and is now divided into the western network of 2,337 kilometers and the eastern network of 1,223 kilometers. These two networks are not connected at present, but much effort is being made to complete a connection between Aiquile and Florida.

There are meter-gauge connections with Argentina at two points: Villazón-La Quiaca and Yacuiba-Pocitos, and the Bolivian and Chilean networks also interconnect, with no change in gauge, at Ollogüe for access to the port of Antofagasta and at Charaña to the port of Arica.

The connection with Perú is made by the meter-gauge railroad run by the Peruvian Corporation between La Paz and Guaqui, where the route continues by boat across Lake Titicaca to Puno. From this point standard-gauge (1.43) track runs to the ports of Mollendo and Matarani. At present this rail-boat-rail route can be traversed by transshipment. The transshipment problem would not be entirely solved by introducing ferry service because of the gauge difference between the two railways.

The east Bolivian network connects with the Brazilian of the same guage (1 m) at Puerto Suárez–Corumbá, from which the port of Santos can be reached.

When the laying of track along the Aiquile–Florida section has been completed, the Atlantic and Pacific coasts of South America will have been joined by rail through the heart of the continent.

BRAZIL

The main body of the Brazilian Railways belongs to one or another of two great systems: the RFFSA (Rede Ferroviaria Federal S. A.) and the network operated by the state of São Paulo. There are other, privately operated lines, notably the 570-kilometer-long railroad from Vitória to Minas Gerais State, over which the Cia. Valle do Rio Doce hauls iron ore from the Itabira mines to the port of Vitória for export. Another line, the Madeira-Mamoré Railroad, 366 kms. long and originally built to complement navigation on the two rivers of those names in the territory of Rodônia adjacent to Bolivia, has been transferred to army control as an unprofitable line remotely isolated from the rest of the railway network.

The aforementioned main network contains a total of about 33,100 kms. of main-line trackage, most of it meter or wide-gauge (1.60 m). The latter differs from the wide-gauge track used in Argentina and Chile, which is 1.676 m wide.

The Rede Ferroviaria Federal currently encompasses 13 enterprises operating 25,747 kms. of trackage. These are:

The *Sao Luis-Teresina Railroad*, 452 kms. of meter-gauge track running between those two cities along the course of the Itapecurú River in the state of Maranhão. This line belongs in the so-called isolated group because it is cut off from the main body of the network.

Rede Viacao Cearense-Rede Ferroviaria do Nordeste and Viacao Ferrea Leste Brasileiro, which make up the integrated network serving the Brazilian northeast with 7,078 kms. of meter-gauge trackage. The network covers the states of Ceará, Rio Grande do Norte, Paraiba, Pernambuco, Alagoas, Sergipe and Bahia and serves such important Atlantic coast cities as Fortaleza, Natal, João Pessoa, Recife, Maceió, Aracaju and Salvador.

Viacao Ferrea Centro Oeste-Estrada de Ferro Leopoldina and the *Estrada de Ferro Central do Brasil* constitute the central network serving the states of Goiás, Minas Gerais, Rio de Janeiro, Espiritu Santo and São Paulo. The total trackage of 9,584 kms. is mixed, as the first two lines are of meter-gauge while the third, the Central do Brasil, is wide-gauge (1.60 m) on the São Paulo-Rio de Janeiro-Belo Horizonte route, a distance of 1,139 kms., and meter-gauge everywhere else. It should be noted that part of its trackage is three-railed to allow the passage of rolling stock of both gauges. This central system is interconnected within its own territory and with the northeastern and southern systems by meter-gauge track. The principal cities served by the central network are São Paulo, Rio de Janeiro, Angra dos Reis, Belo Horizonte, Vitória and, soon, Brasilia.

Estrada de Ferro Santos a Jundiai, a wide-gauge (1.60 m) line 139 kms. long, is the axis, anchored on the port of Santos, from which all the other wide-gauge railroad lines fan out into the state of São Paulo and Rio de Janeiro. The line has a special infrastructure for the first 20 kms. out of Santos, where four anchor stations along a steep gradient counterbalance descending against ascending rolling stock to clear a difference in elevation of more than 800 meters between Santos and São Paulo.

Estrada de Ferro Noroeste do Brasil. This meter-gauge line begins at Bauru, where it connects with the Estrada de Ferro Sorocabana, of the same gauge, and with the wide-gauge (1.60 m) Cia. Paulista de Estrada de Ferro. It has 1,636 kms. of main-line trackage running westward through the states of São Paulo and Mato Grosso to interconnect with the Bolivian National Railways at Corumbá-Pto. Suárez (line to Santa Cruz de la Sierra) and to the terminus of Ponta Porã on the Paraguayan frontier.

Rede de Viacao Parana-Santa Catarina-Estrada de Ferro Santa Catarina and *Vivacao Ferrea do Rio Grande do Sul*. These lines consist of 6,586 kms. of meter-gauge trackage covering the southern region.

They serve the states of Paraná, Santa Catarina and Rio Grande do Sul. Except for the 180-km.-long Estrada de Ferro Santa Catarina, which is isolated, this system is connected to the main body of the Brazilian railway network.

The southern system serves important towns and ports, such as Curitiba, Paranaguá, Itajai, Pôrto Alegre and Rio Grande.

The *Vivacao Ferrea do Rio Grande do Sul,* which serves the state of Rio Grande do Sul with 3,400 kilometers of trackage, interconnects the Brazilian network with those of Argentina and Uruguay, with the former over the three-rail track between Uruguaiana and Paso de los Libres, and with the latter country at Livramento-Rivera and Jaguarão-Rio Branco, also over three-rail tracks which give the Brazilian meter-gauge and the Uruguayan standard-gauge (1.435 m) rolling stock access to the opposing border stations.

Estrada de Ferro Dona Teresa Cristina, a line of 272 kms. in the state of Santa Catarina, is cut off from the rest of the network. Its main purpose is to haul coal in bulk.

The railway network operated by the state of São Paulo has 6,414 kms. of trackage and consists of the following companies: Cia. Mogiana, with 1,758 kms. of meter-gauge main lines; the Cia. Paulista, with 2,556 kms. of wide-gauge (1.60 m) track, and the Cia. Sorocabana, with 2,100 kms. of meter-gauge track. The Paulista network influences the economy of the states of São Paulo, Mato Grosso and Goiás, while the Mogiana and Paulista companies serve the northern and northeastern section, with a tendency to extend into the central region of Brazil. The Estrada de Ferro Sorocabana connects southern São Paulo state with the Mato Grosso.

To summarize, there are 33,100 kilometers of railroads in operation in Brazil, of which 4% refers to lines cut off from the main network. Somewhat less than 10% of the remaining 96% is wide-gauge (1.60 m). The rest is meter-gauge trackage, with which Brazil is interconnected with the railways of Bolivia of the same gauge and with the standard-gauge railways of Argentina and Uruguay.

CHILE

The entire network of 6,805 kilometers is controlled by *Empresa de los Ferrocarriles del Estado de Chile.* It runs mainly from north to south, from Puerto Montt to Zapiga. Isolated from this main network is the Arica Railroad, which runs along the frontier with Perú.

Of the aforementioned total trackage, 3,693 kms. belong to the so-called Southern Network, of wide-gauge (1.676 m), which extends from Viña del Mar to the Los Andes Station and goes as far south as Puerto Montt.

The Northern Network, 2,835 kms. of meter-gauge trackage, runs north from Calera Station, where it connects with the Southern Network.

At Socompa the Chilean northern network connects with General Belgrano Railroad, of the same gauge, thereby linking northern Argentina with the port of Antofagasta.

At Ollogüe the northern network of Chile connects with the Bolivian network, also of the same gauge, thereby affording the latter country a rail outlet to the port of Antofagasta.

The meter-gauge Arica Railway, 206 kilometers long, provides a rail connection to the port of Arica from the city of La Paz.

The meter-gauge of Los Andes (trans-Andean) Railroad runs 71 kms. from Los Andes to Las Cuevas, where it connects with the Argentine network (General Belgrano R.R.) of the same gauge.

ECUADOR

The Ecuadorian railway network has a total of 1,150 kms. of trackage in operation. Of these, the *Empresa de Ferrocarriles del Estado* operates 68%, which consists of the following: Guayaquil-Quito R.R., 452 kms. of 1.067 m. gauge; Simbambé-Cuenca R.R., 145.4 kms. of the same gauge as the afore-mentioned line, with which it connects at Simbambé; Puerto Bolivar-pasaje R.R., 25 kms. of 1.067 gauge track and Puerto Bolivar-Piedras R.R., 75 kms. of 0.75 meter-gauge, both of them isolated from the rest of the network; and, lastly, Bahia-Chone R.R., 80 kms. of 0.75 meter-gauge track, also isolated.

Quito-San Lorenzo R.R., one of the most modern facilities in Latin America, consists of 373 kms. of 1.067 meter-gauge track, and it is integrated with the so-called "Nariz del Diablo," which is the infrastructure of Guayaquil-Quito Railroad.

The Ecuadorian railways are not integrated with systems of the same gauge in adjacent countries.

PARAGUAY

The Presidente Carlos A. López Railroad, an agency of the Paraguayan government, operates the 439.4 kilometers of existing standard 1.435 meter-gauge track in the country.

The principal line, 376 kms. of track between Asunción and Pacú-Cuá, is the one that connects with the General Urquiza R.R. in Argentina, of the same gauge. The international connection is made between Pacú-Cuá and Posadas across the Alto Paraná River by train ferry.

It is interesting to note that Paraguayan and Uruguayan railways and the Argentine system between the two rivers are all of the same gauge. Thus, if the two latter systems were to be connected, which could be easily done by train ferries, a multinational system with 6,800 kms. of trackage would be created.

URUGUAY

The Administración de Ferrocarriles del Estado operates all the railway lines in the country, which consist of 2,975 kms. of standard 1.435 gauge track.

The national system resembles a fan that converges on the capital of the country, Montevideo. The lines virtually blanket the country, serving principal cities and ports on the Plate and Uruguay Rivers. They reach the Brazilian frontier at three points: Rio Branco-Jaguarão, Rivera-Livramento and Cuareim-Quarhaim. At the first two points they connect with the meter-gauge Viacão Ferrea do Rio Grande do Sul.

The international tracks between frontier terminals are all provided with three rails to allow the passage of rolling stock of both railroads.

At the last-mentioned point there is no connection with Vifer because this company has removed the track that used to run from Uruguaiana to Quarhaim. Though there is no connection with Argentina, one could be made at any port by means of train ferries across the Uruguay River.

Finally, it may be stressed that the existing rail network and geographical position of the country afford a good land-river route between Argentina and Brazil, which is shorter than any other between the two countries and is coming into increasing use.

General description, by countries, of the South American railroads not yet members of ALAF (Asociación Latinoamericana de Ferrocarriles) (Latin American Association of Railroads)

PERU

In 1967 the country had a network of 2,462 kms. of track of which 1,945 kms. were main lines and the remainder branches and spurs. Of the 1,945 kms. of main lines, 341 kms. are owned by the state and the remaining 1,604 kms. by private companies.

The most important private network is that of the Peruvian Corporation, which has 1,476 kms. of standard 1.435-gauge track—1,311 kms. of it in main lines—in two principal systems, one in southern Peru and the other in central part of the country.

The Southern Railways (Ferrocarriles del Sur) of Peru, with 861 kms. of main lines, provide an outlet for Bolivian products from Puno, on Lake Titicaca, to the port of Mollendo on the Pacific Ocean. The La Joya-Matarani section, 62 kms. of standard-gauge track, connects with the Southern Railroad at Puno. It may be noted that the Peruvian Corporation owns the Guaqui-La Paz line in Bolivia and the shipping company operating on Lake Titicaca from Guaqui to Puno.

The Central Railroad operates 340 kms. of standard-gauge track from the port of Callao through Lima to Huancayo. There is no connection with the Southern R.R.

Service from Huancayo to Huancavelica, a distance of 147 kms., is provided by a government-owned 0.914-gauge line.

The other government-operated lines are: Cuzco-Santa Ana, 131 kms. of 0.914-gauge track, a continuation of the Southern R.R. line from the Cuzco terminal; the Matarani-La Joya line has already been mentioned; lastly, there is the 62-km. standard (1.435 m) gauge international railway from Tacna to Arica, in Chile—it is cut off from the rest of the network.

COLOMBIA

The Empresa de Ferrocarriles Nacionales operates a network of 3,435 kms. of 0.914-gauge track extending from the port of Buenaventura on the Pacific to those of Barranquilla and Santa Marta on the Caribbean Sea, serving the capital, Bogotá, and such important cities as Cali, Popayán, Medellin, etc.

The Colombian network is not integrated with the railway systems of adjacent countries.

VENEZUELA

The Instituto Autónomo de Administración de Ferrocarriles del Estado, a government agency, operates more than 400 kms. of standard (1.435 m) gauge track.

The Gran Ferrocarril de Venezuela connects La Guaira-Caracas with Valencia (176 kms.), and Puerto Cabello with Barquisimeto (174 kms.), and Puerto Cabello with Valencia (55 kms.).

There are other industrial railways in the country connecting iron mines with ports on rivers which flow into the Atlantic.

The Venezuelan network is not integrated with the railway systems of adjacent countries.

Combined Transport Techniques Now in Use; Inter-Carrier Agreements for Combined Transport

ARGENTINA

General Belgrano R.R. has special rates for combined rail-river traffic from Paraná River ports served by its lines to the port of Buenos Aires.

The products to which these special rates apply include: wood and charcoal, in consignments of 230 and 180 tons, respectively, via port of Santa Fé; sugar, in consignments of 350 to 380 tons, via port of Santa Fé, Barranqueras or Formosa; alcohol in drums, baled cotton, sawed wood crossties, logs, beams, posts, salt, tannin, etc., via port of Formosa, Barranqueras or Santa Fé; ingots, billets, sheet and other iron and steel products in consignments of 1,000 tons from General M. N. Savio to the ports of Rosario and Buenos Aires via port of Formosa.

It may be noted that the railroad provides the cars needed to accommodate these minimum consignments. The railroad does not, however, possess special equipment for unitized transport, and cargo is therefore delivered in port to the Flota Fluvial del Estado Argentino (the Argentine National River Fleet), which transships it to lighters and transports it to the port of destination.

The company provides another combined rail-river transport service via the ports of Santa Fé and Bajada Grande to channel merchandise en route from or to the General Belgrano, General Mitre and General San Martin railroads through the General Urquiza R.R. operating on the eastern side of Paraná River. Here, too, freight is transshipped and carried across the river by the River Fleet at special rates.

Relations between the Empresa de Ferrocarriles Argentinos and the River Fleet are maintained through the Secretariat for Transport, to which both enterprises are subject.

Lastly, consideration is being given to providing the aforementioned rail-river service between Santa Fé and Paraná by means of the truck-ferry combination. Only domestic transportation is effected by these means.

The General Mitre R.R. has for some years been using portable containers and piggybacking truck-trailers, although the greater increase has been in the use of the former. This technique is confined in practice to local traffic, although there are isolated instances of combined transportation with the other railroads. Its use in local traffic is due to the fact that, while the first containers were built by the railroad for demonstration purposes, they later came to be made by private firms for their own use; today special cars are assigned for them exclusively—the railroad maintains 17 container and 8 piggyback cars in continuous circulation.

The General Mitre R.R. has not yet carried any international consignments, although its wide (1.676 m) gauge lines have customs connections with international shipping through the various posts.

To activate the use of the door-to-door carriage service introduced at low incentive rates, it is proposed to install portal cranes at the terminals of Buenos Aires and Tucumán, between which two points the greatest volume of traffic takes place.

The General San Martin, General Roca, Sarmiento and General Urquiza Railroads, along with the Mitre R.R., apply uniform regulations to the transport of containers and (by the piggyback method) of full trailers, semi-trailers and tanks loaded with general, perishable and liquid goods.

In regard to rolling stock, in early 1967 there were eight cars suitable for piggyback service and three for the carriage of containers on the General Roca R.R. The piggyback cars that had not been used in 1966 began to serve the Negro River Valley in 1967.

At the beginning of 1967 the General San Martin R.R. had a good number of piggyback flatcars in use in domestic traffic jointly with two large trucking companies.

At the start of 1966 General Urquiza R.R. had two piggyback flatcars to carry the semi-trailers of the Corporación Rodoviaria para América Latina (CORAL), a large Brazilian highway carrier with branches in Argentina.

Transportation is provided from Buenos Aires and Paso de los Libres, and back. At the latter point the trailer unit either is transferred to a piggyback flatcar of the Viacão Ferrea do Rio Grande do Sul or continues with its tractor by road. This is the only case of cargo unitization applied to international traffic. All the Argentine railroads have set differential rates for the carriage of containers and for piggyback service. All these rates are calculated to promote traffic.

BOLIVIA

The Bolivian railroads do not presently use any of the techniques for facilitating the performance of unitized transport service, partly because the continuity of gauge between the railroads of this country and the interconnecting lines in Argentina, Brazil and Chile (they are all meter-gauge) allows the free circulation of rolling stock among them without transshipment.

In the case of the connection with Peru: by rail from La Paz to Guaqui (meter-gauge), transshipment and carriage across Lake Titicaca to Puno for transshipment to a 1.435 gauge railroad leading to the ports of Mollendo and Matarani, it appears highly advisable to introduce containers, and the possibility is now under study.

BRAZIL

The Brazilian railroads make the greatest and most diverse use of existing cargo unitization techniques.

The RFFSA and the São Paulo state railways use:

a. *Servico rodo-ferroviario.* The pickup and delivery of merchandise within a limited radius around urban centers and their environs, to complete the railroad service; performed generally by trucks owned by the railroads themselves.

b. *Rodo-tren.* A rail-truck combination in which owned or leased trucks are used chiefly to pickup and deliver freight outside urban centers. In this type of operation the railroad engages in strictly highway transport, either along routes where there is no rail line or when the combination cannot be made because of lack of cars or nature of the cargo. This service is also provided where the rail line is so short as to make the transshipment operation too expensive.

c. *Auto-tren.* Consists in the hauling of highway vehicles (trucks, trailers, etc.) on railroad flatcars. Contracts are arranged for the leasing of the number of flatcars requested by the private party, who loads, unloads and anchors the vehicles with their drivers, on flatcars the railroad has only to haul. The vehicles are mounted on the cars under their own power by means of ramps.

d. *Piggyback.* Consists in the carriage of highway semi-trailers on specially designed flatcars. After the vehicles are pulled up ramps onto the flatcars by their tractors, the tractors detatched and taken off, because only the semi-trailers are hauled.

e. *Container.* The system was introduced to the country on an experimental basis on the Paraná-Santa Catarina R.R. The type adopted is the box container, which has a base of 2.10 m. x 2.40 m., is 2.30 m. high, and has an inside volume of 8.8m³ and a variable tare ranging from 400 to 1,000 kgs., depending on whether it is made of aluminum or steel. The loaded weight of this container varies between 5.5 and 6 tons.

The auto-tren system is a unitization method created for the door-to-door transportation of merchandise which at one time was carried by rail and had shifted to highway haulage. The system is used chiefly by the railroads of São Paulo state, preferentially between the city of São Paulo and Bauru and Rio de Janeiro. To promote its use, differential rates are applied for carriage inbound to or outbound from São Paulo.

Among the RFFSA enterprises offering unitized transport service are:

Estrada de Ferro Santos-Jundiai operating a rail-truck service with piggyback and auto-tren.

Estrada de Ferro Central do Brasil through its auto-tren service.

Rede de Viacao Paraná-Santa Catarina. The rail-truck and commercial departments of this company developed a joint plan which included:

1. The collection of coffee at production centers in northern Paraná through rodo-tren agencies away from the railway lines, for transport to rail centers such as Maringá, Londrina, Guarapuava, Pôrto União, Santa Cecilia, Lajes and Cacador, where it is shipped by rail to Palmeira, Curitiba and Paranaguá.

2. Direct shipment of bagged goods which cannot be transshipped or when rail accommodations are full, or there is no rail line for the contracted route.

3. Coordinated transport under an agreement signed between the RVPSC and the Carlos A. López R.R. of Paraguay, for rail-truck haulage between Asunción and the free port of Paranaguá. Goods are carried by rail from the

latter port to Guarapuava, where they continue by highway to Asunción, entering Paraguay at Foz do Iguacu. Containers are used on this route to take advantage of the customs privileges that are granted when they are sealed.

4. Coordination of trucking with the railways serving the state of Santa Catarina: Rede de Viacão Paraná-Santa Catarina and the Estrada de Ferro Santa Catarina, which is cut off from the network. In this way, the industrial output of Itajai Valley moves by highway from Blumenau to Curitiba, where it goes on by rail to São Paulo.

5. Acquisition of new containers in order to accumulate 400 of them by the end of 1967. The container in use measures 2.30 x 2.10 x 2.40 meters, with two side doors and top and bottom openings for bulk loads. This container, which weighs 6 tons loaded, affords coordination between meter-gauge flatcars, which carry five containers, and wide-gauge flatcars, which carry six, and with 15–24-ton capacity trucks (four containers), or small 5–6-ton trucks, which can carry one container.

Viacao Ferrea do Rio Grande do Sul utilizes piggyback flatcars in its international service to and from Argentina, with semi-trailers being transshipped at Uruguaiana-Paso de los Libres to the flatcars of the General Urquiza R.R. There are as yet no contracts for combined rail-river carriage.

CHILE

Since 1963 Portland cement has been supplied to Santiago and its environs from the factory 120 kms. away in containers at a rate of 600 tons per day. The containers number about 500 and are owned by the factory. They are hauled in flatcars and at the stations of destination are transferred to trucks for delivery to sites of large civil engineering works.

Pallets are used to transport sheet cellulose from the factory at Laja to the port of embarkation, Lirquén, 90 kms. away. The unitizing equipment is privately owned, and the system first came into use in 1966, handling 10,000 to 12,000 tons per month.

Combined operations—from car to truck—have been attempted on a small scale for door-to-door service, especially in the transporting of cellulose. Efforts are now being made to expand and improve these services; studies are being made of the possibilities of transporting copper concentrates from Los Andes to the foundry on the coast of Valparaíso, and of the profitable carriage of luggage and parcel post via car-truck and vice versa.

In Chile, there is a unitization technique by which the problem of different gauges in the national network has been solved. This is the so-called "dual-gauge" or "bogie-exchange" system. Transshipment is avoided by transferring the freight-carrying superstructure of a railcar from wide to meter-gauge bogies and vice versa.

The Chilean transshipment installations at Calera, Los Andes and Monte Aguila consist basically of elevators with four electromechanical hoists capable of lifting 15 tons each, and bogie-storage sidings.

The bogie-exchange operation is very simple and it takes no longer than 45 minutes to transfer a railcar with 30 tons of freight with the improvised, inexpensive equipment used.

The operation consists in maneuvering the car into position, extracting the brake rods, raising the superstructure with the hoists, withdrawing the bogies to the appropriate siding, the placement of new bogies in position, lowering the superstructure onto them, replacing the brake rods, and hauling the car away. The system, which has been greatly refined in networks on other continents, can become a good solution to the problems of integrating networks of different gauges nationally and internationally.

COLOMBIA

Today Colombian railways have combined their services with those of shipping companies for the carriage of containers inland from the ports of Barranquilla, Buenaventura and Santa Marta, in addition to other inland traffic by that system. The railroads perform this service under special contracts with the Flota Mercante Grancolombiana and the Grace Line.

The Ferrocarriles Nacionales de Colombia have an agreement with the General Customs Administration for clearing these imported goods at their destination points, where there are inland customs stations and special warehousing facilities.

The railroads have also had studies performed on the development of containerized transport, to which end they have contacted all the businesses in the country to ascertain what kinds of merchandise would be carried and the types of containers to use. This initiative has awakened the interest of local industrialists, who are prepared to build their own containers to suit their needs, and some of them have already built them and put them to use on the railways.

ECUADOR

The Empresa de Ferrocarriles del Estado operates a combined land-water-land and rail-oil pipeline-rail service for the transporting of freight and passengers between the port of Guayaquil and the principal inland cities of the country: Quito, Cuenca, Riobamba, Ambato and points between. It does not employ any of the new unitizing techniques in this service.

The combined transportation system is provided mostly with trucks owned by the railroad, although private trucks are leased for service to certain terminals.

One of the obstacles that the Ecuadorian railroad has had to overcome is its access to the principal port, Guayaquil, through which most exports and imports pass. In fact, the railroad terminal is at Durán (on the west bank of the Guayas River), and goods must cross the river to the east bank to continue by land to Puerto Nuevo.

The railroad provides passage over this route by a rail-water-land combination; the third leg by truck. Construction of the Puerto Nuevo-west bank branch line will complete the rail-water-rail combination, with full use of ferrying facilities now being developed.

This combined transportation is provided with equipment owned by the railroad, which has passenger ferries, a 500-ton tugboat, and several barges of 100 to 500 tons displacement with their tugs. The ferrying system will operate with three boarding points, one each at Durán, Guayaquil and the terminus of the branch line to Puerto Nuevo, and will also be equipped with two ferryboats with a bulk-cargo capacity of 500 tons, or of 260 tons with six loaded railcars and one tug. The feasibility of making the ferries self-propelled is being studied.

With the problem of location of the port on the side of the Guayas estuary opposite the railroad terminal solved in the manner described, there remain other deficiencies in the route of the Guayaquil-Quito railroad, which has gradients as steep as 5.5% and which, between the stations of Bucay and Palmira (78.4 kms. apart) rises a total height of 2,944 meters. For the last decade an oil pipeline has been carrying gasoline across this drop at lower cost. The gasoline is conveyed by the rail-oil pipeline-rail combination. It is proposed to expand the system under the program for rehabilitation of the railroad to allow the transportation of a larger number of pure petroleum derivatives.

Thus, modern unitization techniques will come into use as soon as the plan for the entire rehabilitation of the Empresa, now in progress, has been completed.

PARAGUAY

This country uses none of the modern cargo unitization techniques. Integration with the Argentine General Urquiza R.R., of the same gauge, has so far allowed the interchange of rolling stock. This has greatly facilitated the entry and exit of products without need of transshipment.

The difficulty of interconnecting the two networks with frontier stations on either side of the Alto Paraná River has been solved by the use of a train ferry.

Service to Brazil, especially to the Rede Viacão Paraná-Santa Catarina and service between Asunción and the free port of Paranaguá has been described earlier in the section under Brazil.

URUGUAY

The Administración de Ferrocarriles del Estado has a regular river-rail-highway traffic in perishable goods through the country between Argentina and Brazil. Apples are shipped from Buenos Aires to São Paulo and Santos on a 5-day journey that runs by truck ferry from Buenos Aires to Colonia, by rail from Colonia to Jaguarão, and on to São Paulo or Santos by truck. Every vehicle employed in this traffic makes the return trip loaded with bananas. This traffic has been growing as the fruits have gained acceptance on consumer markets; if it becomes permanent, refrigerated containers will be introduced for the shipment of apples out of season.

The railroad generally attaches the freight cars to passenger trains so that they may be hauled rapidly and on a schedule.

In regard to cargo unitization, two piggyback flatcars have been built and are now in use by two large Brazilian highway carriers on the section from Montevideo to Jaguarão and back to carry merchandise for export and import, respectively. Eight other such flatcars are scheduled to be built.

Measures to Encourage the Use of Unitization Techniques in Combined Transport over ALAF Networks

The only noteworthy advances made within the ALAF governments are the regulations issued in the Argentine and Brazilian legislation for the use of containers. In the former case, *Decree N° 5071* of June 18, 1967 directs that this cargo unitizing device must conform to the standards to be issued by IRAM in due course. In the latter case *Law N° 4907* of December 17, 1965 and *Decree N° 59316* of August 29, 1966 provide that these devices shall be built, initially, to the dimensions prescribed by the ISO, and will later be regulated by the Brazilian Technical Standards Association (ABNT).

ALAF has approached the problem in its most general aspect; hence its decision to sign an agreement with COPANT, the parent body for the two standardizing agencies in the region, for the execution of the following high-priority standardization projects (among others):

    a. Technical conditions for the exchange of rolling stock (clearances, height of bumping posts, couplings, etc.) among the countries in the region.

    b. Containers. ALAF also has permanent working teams and auxiliary teams such as the zonal groups. The latter are confined to purely regional aspects and meet to consider or offer solutions to problems of the international integration of networks.

RAIL ROUTES AND LIMITATIONS; REGULATIONS

Container trains of the United States, as mentioned earlier, are primarily limited to the piggyback trains and the Flexi-Van trains. From Chicago eastward only the Penn Central carries this equipment on a full-train basis with the Flexi-Van depot in New York at Highbridge in the Bronx and the piggyback depot in North Bergen, New Jersey. An ancillary service to the rail movement of containers has been started by a company called Interpool. Initial service started in February 1968 between Chicago and New Orleans with plans for servicing other major midwest U.S. points. Steadman Railtainer equipment is used, with a pool of 20 foot, ISO standard containers established at Chicago, side-transfer units, designed to transfer containers laterally between rail flatcars, and highway chassis stationed at the Illinois Central yards in Chicago, Peoria and New Orleans. Interpool arranges all terminal services, including transfer of containers between rail flatcars and road vehicles. Rail, road and ocean carriers are not required to invest in containers and related equipment such as demountable container legs, transfer units or trailer chassis. Interpool provides this equipment as part of the pool cartage service at any domestic point, port or inland, where the customer requires the service. Containers are drawn by members from the pool on a circulating basis at per diem rates which include maintenance, reconditioning, storage at inland points and interchange between carriers. Interpool representatives also arrange cartage, transfer storage and drawing of units from the pool.

In Canada the two railroads, Canadian Pacific and Canadian National, operate piggyback trains. Transition, however, is taking place from the trailer-on-flatcar movements to container-on-flatcar movements. Canadian Pacific operates container cars and trains from Montreal to Toronto, from Montreal and Toronto to Northern Ontario points, and from Montreal to North Bay; and it operates the *M.V. Beaveroak* in the Europe/Canada trade. This vessel has been partly designed to carry 20-foot containers. Special features for this purpose are, twin hatch openings and power and water outlets to accommodate refrigerated containers.

In England, British railroads operate Freightliner all-container trains with depots throughout the United Kingdom.

Fig. 66. Part of British Rail's Freightliner Terminal at Willesden — (London). (British Railways Board)

Lengths of containers are 10', 20', 27' and 30' with heights and widths of 8' x 8'. Freightliner flatcars are 62½' long, feature "spigots" for holding containers and have smaller wheels (2' 8" in diameter) to provide clearance. The international movements converge on Harwich, where, by means of British Rail's own steamship line, Seafreight Line, containers are brought to Zeebruge. There feeder services move the containers to Antwerp and Rotterdam for Freightliner continuation on the continent.

Each of the new British Rail ships is able to carry 148 30' (or other combinations of 20' or 40') high-capacity containers constructed to internationally agreed dimensions, i.e., with a cross-section of 8' x 8'. Of these, 110 can be stacked three-high in the cellular holds and the remaining 38 on the hatch covers. They can include refrigerated containers since there will be provision for the maintenance of controlled temperatures. The payload is 3,000 tons. By handling cargo in unit loads in this way, using fully mechanized equipment for loading and discharge, the ships can be turned around in five hours and will be able to make one round trip every 24 hours.

The Austrian Federal Railways do not utilize Group I containers at the present time. The inventory, as of 1968, consists of: 2,015 units of 1-cubic-meter containers; 1,488 units of 2-cubic-meter containers; 660 units of 3-

Fig 67. British Freightliner handling operation. (British Railways Board)

cubic-meter containers. No Group I containers (transcontainers) are owned by the Railways. The Austrian Railways feel that to obtain economic workability of transcontainer traffic, there must exist large scale industrial and user areas between which the large container trains can be utilized in two-way runs. Traffic is slowly developing in this direction in Austria. The first center evolving in line with these conditions is Vienna. The Austrian Federal Railways have built a terminal at the Vienna-Northwest Railroad Station. All transcontainers up to 35 tons can be handled here with the new crane equipment. A second portal crane will be available in Innsbruck. Additional loading and unloading depots and equipment for servicing later transcontainer traffic centers are planned for the most important Austrian state capitals.

Transcofer is the executive bureau of the pushing group "transcontainer" of U.I.C. The bureau was established February 1, 1967 with the objective

to inform the customers in a better and quicker way on international transport. For this purpose it has compiled the information received from the
individual administrations. Originally intended activity in the field of sales
promotion could not be realized as yet, inasmuch as the installation of container handling facilities on railway premises is still in progress. However,
an international transcontainer tariff has already been established. In the

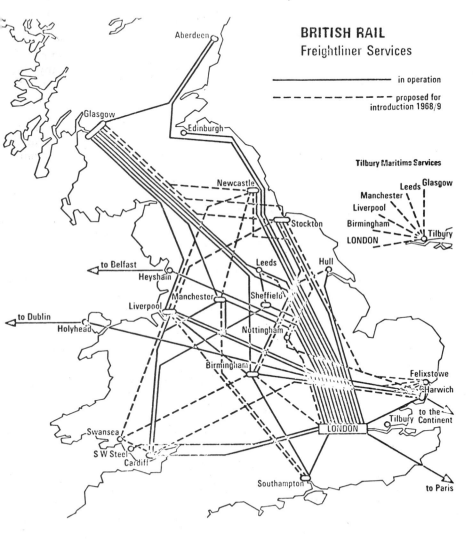

Fig. 68. British Freightliner existing and proposed terminals in the U. K. (British
Railways Board)

meantime it has been found that the possibilities of the railways tending to
influence this sector of transport do not yet suffice to increase the volume of
transcontainer transport by rail.

Since there was no central bureau for the European railroads in connection with operational and rate-making arrangements for containers, fifteen railway administrations and Interfrigo established a private company using the name "Intercontainer." Intercontainer is a cooperative company in accordance with Belgian law and links 160 rail terminals throughout Europe. Headquartered in Basel, the members are Britain, East Germany, Ireland, Belgium, Denmark, France, West Germany, Hungary, Italy, Luxembourg, Holland, Austria, Spain, Sweden and Switzerland, with Norway, Finland, Yugoslavia and Greece joining shortly.

The German Bundesbahn in connection with the Intercontainer Company is setting up an independent subsidiary, National Container Company, to act domestically just as Intercontainer acts internationally. Its purpose will be to operate all the German container terminals and to promote container traffic.

The German Federal Railroad inaugurated the first all-container train *Delphin* in February 1968 to connect the German seaports of Bremen and Hamburg with inland container terminals. In conjunction with the completion of the permanent modern container terminal facilities in Frankfurt (Main), Mannheim and Ludwigsburg near Stuttgart, this event was part of the first step for a massive engagement in intermodal transportation. The express container train *Delphin* departs in both ports, Bremen and Hamburg, at 5:00 P.M. In Hanover both trains will be combined to continue as one to Frankfurt (Main), Mannheim and Ludwigsburg where it arrives in the early morning hours.

If a heavy influx of containers occurs in the ports, four additional express freight trains are available for swift container transportation to the inland terminals. Those trains depart in either port at 3:00 A.M., 1:00 P.M., 4:00 P.M. and 11:00 P.M.

Other permanent container facilities are at present under construction in Basel, Muenchen and Nuernberg and are nearing completion. The domestic container tariff 24 C 1 of the German Federal Railroad, however, was amended recently and contains rate quotations for 20-, 30- and 40-foot containers to be shipped by rail from the seaports of Bremen, Bremerhaven and Hamburg to inland facilities capable to handle TOFC and/or COFC traffic as follows:

Bobingen, Bochum Hbf., Bochum-Langendreer, Dortmund-Eving, Dortmunderfeld, Duesseldorf-Bilk, Duesseldorf-Derendorf, Duisburg Hbf., Duisburg West, Essen-Ruettenscheid, Fischbach-Weierbach, Forchheim (Oberfr.), Frankfurt (Main) Hbf., Frankfurt-Hoeschst, Frankfurt (Main) Ost Gablingen, Gelsenkirchen-Schalke, Giessen, Gundelsheim (Neckar), Haiger, Herzogenrath, Immigrath, Kastl (Oberbay), Koeln-Gereon, Koeln-Messe, Koeln-Nippes, Krefeld-Uerdingen, Leverkusen-Bayerwerk, Ludwigsburg, Mannheim Hgbf., Mannheim Rbf., Mannheim-Waldhof, Marl-Sinsen, Muenchen Hbf., Muenchen Ost, Nuernberg Hgbf., Nuernberg Nordost, Ruesselsheim, Saarbruecken Hgbf., Stolberg (Rheinl.) Hbf., Stuttgart Hbf., Stuttgart-Untertuerkheim, Trier Hbf., Weinheim (Bergstr.), Wengern Ost, Witten Hbf.

The GFR inaugurated TOFC traffic in collaboration with the Swiss Federal Railways and the Italian State Railways between the area of Frankfurt/Darmstadt/Mannheim and the Swiss/Italian border Chiasso/Melide. Later, as the traffic may demand, the service will be expanded to include Duesseldorf and Stuttgart in Germany and Firenze, Genova and Milano in Italy. This TOFC traffic will be moving through the Gotthard-Tunnel in Switzerland.

The Swedish National Railroad operates a daily container train between Stockholm and Gothenburg. It carries 20-foot containers and is one link in a feeder service bringing containers to the few ports now used by container ships interested in rapid turnaround.

Australian railway systems have agreed on the design of a railway wagon to carry containers. This rail wagon is a special 63-foot-long flat-top wagon which will transport 20-foot and 40-foot ISO containers between exporters' rail sidings and the container sea terminals throughout Australia. The wagons will be able to carry three 20-foot containers or one 40-foot and one 20-foot container.

A trans-Siberian container train has started operations between London and Yokohama. Non-intermodal containers traveling on regular rail wagons on regular rail routes go from London to Basle, moving through Germany, Czechoslovakia and Russia to the Russian port of Nachodka, where trans-shipment takes place to Yokohama. Six Dravo type containers can ride on one Russian flatcar.

By Vessel

Specialized Vessels; Roll-on/Roll-off, Lift-on/Lift-off Lash-Conversions. In discussing the effect of the container movement on vessels, we must again differentiate between the relatively small metal and wood containers that have been carried for many years on conventional vessels using conventional methods, and the intermodal Group I containers. In addition, there were the seatrain vessels which, while not in the true sense container ships as we now know them, preceded the container ship by 25 years. Loading was performed by a special fixed shore gantry crane with outriggers extending over ships on both sides of the dock. The freightcar was positioned on, and secured to, a loading cradle and both were then lifted and lowered into a selected cell in the well of the ship and landed on a selected deck. The freightcar was then pulled either fore or aft to its stowed position aboard and secured to the deck. These ships could also carry wheeled or unwheeled containers which were placed on low-profiled wheeled dollies suited for running on railroad tracks. These were then handled aboard as a railroad car would be. There was a great amount of wasted cubic and recently some of these vessel types have been modified to accommodate containers below the deck on a more efficient basis. The 'tween deck has been removed and containers are placed aboard without a cradle and stacked three-high upon powered dollies which run on tracks fore and aft on the tank top. In its stowed position the dolly with its three-high stack is secured for the voyage. The top container in each three-high stack is restrained from athwartship movement by guides projecting down from the underside of the main deck. Those in the lower tiers are restrained by the weight of the containers above them.

Ships that carry containers can be classified into five groups:

1. Full container ships with special features and arrangements for carrying containers in all available spaces and which are generally single-purpose ships.

2. Partial container ships in which only part of the vessel's capacity is especially designed for containers.

3. Convertible container ships in which part or whole of the vessel's capacity may be used for either containers or other cargoes and containing special features which permit the convertibility on a voyage to voyage basis.

4. Ships with limited container carrying ability in which some container handling and securing devices are installed, but the vessel in other respects is of normal construction.

5. Ships without special container-stowing or handling devices in which the container is treated as a larger than usual piece of cargo and secured by conventional means.

Types 1, 2, and 3 in the foregoing may be treated as a group inasmuch as the principles of construction and design to be discussed apply equally to all three.

A distinction must be made between container ships and roll-on/roll-off types sometimes referred to as container ships. Roll-on/roll-off vessels are similar to container ships in that they carry "containerized" cargo in the broad sense of the term. The "container" in this instance is a wheeled vehicle, a trailer, or a truck. The method of loading differs in that the vehicle is driven or towed aboard over ramps through ports in the vessel's shell. This vessel type requires much more cubic for stowing a given number of "containers" than the true container ship. About one-third of the useful cubic is lost to the space under the "container" occupied by the wheels. Additional space is lost to the relatively large clearance between "containers" to permit the necessary access to them and for their maneuvering. Ramps or elevators within the vessel for loading lower decks occupy additional space. Since the vehicles are not stacked, each 'tween deck necessary for stowing the wheeled "containers" and its supporting structure wastes more space as does the necessary clearance between the top of the container and deck beams. A roll-on/roll-off type of vessel requires about 100% more cubic than a true container ship to carry an equal amount of "containerized" cargo. There are the additional arguments that, although there may be a problem with the chassis movement, since a goodly portion may move inland by rail without chassis, there is the additional advantage of vessel space adaptability. Outsized freight would not provide the same type of problem on the trailership as it would on the full container ship. If the trend of building and putting container ships into an increasing number of services continues, there may well be a problem in the availability of space for cargo that cannot be containerized.

Atlantic Container Line has designed its ships to serve all the foregoing purposes. The *Atlantic Span* can carry 405 standard 20-foot or 227 standard 40-foot containers on deck and in cellular holds, plus trailers with wheels, and about 1100 compact cars. Six decks below the weather deck run without bulkheads or other impediments almost the full length of the ship. They run from the after end of the two cellular container holds at the bow of the vessel to the interior vehicular ramp system which starts immediately below the bridge. The trailer decks with individual clearances of 14½ feet are sandwiched between two upper automobile decks and two lower decks, also for autos. All four levels for cars are 5½ feet high, and the one just above the top trailer deck can be removed to provide higher clearances and more stowage area for containers or extraordinarily large roll-on cargo. Twenty traffic lights control movement of vehicles on the ramp system and closed-circuit television affords centralized observation of all cargo areas. Also

Fig. 69. Atlantic Span, first of four Atlantic Container Line ships: Atlantic Span, Atlantic Song, Atlantic Saga, Atlantic Star. Stern ramp view of roll-on/roll-off loading. (Atlantic Container Line)

aboard the *Atlantic Span* is a system of 26 ventilation shafts with heavy-duty fans capable of changing the air throughout the ship every three minutes.

Another vessel type related to the container ship and of particular note is the barge or floating container carrier. These ships carry small barges that are loaded and discharged by special lifting equipment directly from and to the water.

There is still a diversity of opinion as to the future of containerization insofar as ocean voyages are concerned. Some steamship companies have adopted a wait-and-see attitude. These companies attempt to handle

Group I containers with conventional equipment of conventional ships, loading them on the decks or in the square of the hatches, and even, when they feel it is necessary, pulling and pushing them into the wings. Some companies have geared their operations to palletized unit loads and adapted their ships and equipment accordingly. Others, while waiting for new ships to be constructed or while not wanting to go "all the way" in full containerization adapt present ships so that some of the holds are converted for the cellular stacking of containers. One company, the Finnish flag Finnlines, Ltd., wanting the flexibility of both systems is jumboizing some of its vessels, cutting the existing vessels apart and inserting a separately fabricated cellular fitted center section. The ships will be able to carry the equivalent of 168 20-foot containers plus additional general cargo with the addition of the 47-foot midsections. The Swedish flag Swedish Johnson Line is jumboizing five of its ships by adding a 79-foot midsection to each 490-foot existing vessel. These midsections will have a capacity of 90 20-foot containers. In addition, special fittings will allow on-deck carriage of 62 20-foot containers. Lykes Bros. Steamship Co., Inc., is planning to jumboize 9 of the 21 ships of the Lykes Gulf Pride Class by adding a midbody section 97 feet long, thus increasing the cubic cargo capacity of each ship by 40% and the deadweight capacity by 30%. Each of the 9 ships will have the capacity to carry 182 20-foot containers. Sixty of the containers will be stowed underdeck and 122 of them will be carried two tiers high on deck.

The balance of thinking is divided between the lift-on/lift-off vessel, or full container ship—and the roll-on/roll-off vessel, which usually can carry trailers, containers on deck, and conventional cargo if desired.

But there is a good deal of construction of full container ships coming upon the international scene in increasing numbers. Among them are U.S. Lines vessels which will have a capacity of 1,335,000 cubic feet, divided into ten holds (in the case of ships Nos. 1 and 2) and eleven holds (in vessels Nos. 3, 4, 5 and 6). Each vessel will carry a total of 1026 containers loaded below decks in cells plus almost 200 additional ones on deck. The ships will be equipped with plug-in deck facilities for the carriage of a minimum of 90 40-foot refrigerator containers. Stowage area is also provided for liquid tank containers, open top containers and ventilated containers.

In addition, Hapag-Lloyd's four vessels (Hamburg-Amerikanische Packetfahrt Actien-Gesellschaft (Hapag) operating with Norddeutscher Lloyd as Hapag-Lloyd) will carry 616 20-foot units. The container heights have been standardized at 8′6″ and the entire deck is designed to act as cover for the container cells. Some of the new ships designed for service across the English Channel between the U.K. and the continent, such as the *Sea Freight Liner 1*, can carry 148 standard ISO 30-foot containers (or other multiples of 20-footers and 40-footers) in cellular holds and on deck. Turnaround time is restricted to five hours, allowing one round trip per day, therefore the hatch cover design and operation were a prime consideration. These are arranged to give openings of 60′ over the container cells to allow for the carriage of multiples of 20-foot, 30-foot or 40-foot containers. There is an almost continuous level platform when closed; this platform incorporates container securing devices spaced to accept varying container sizes. The main hatch covers comprise six identically-sized panels 31' long x 43' wide (to accept five containers across the width). Of these, three are motorized and can travel fore and aft to act as carriers for the other three, the latter being fitted with a jacking system for raising to enable the motorized panel to become positioned underneath. Both panels can then be moved together as one unit, fore or aft as required. The lifting and traveling movements

Fig. 70. Artist's conception of one roll-on/roll-off vessel in a "second generation" Atlantic Container Line series of six. Five of the six are expected to be in North Atlantic service before the end of 1969. Each ship will be approximately 695 feet long with a speed of 24.5 knots. (Atlantic Container Line)

Fig. 71. American Lancer. First of six full lift-on/lift-off container ships. Note absence of ship gantries or derricks. Full dependence is on shore-based handling equipment. (Port of London Authority)

are controlled by hydraulic operation, the fore and aft hatch covers are designed as hydraulically-operated folding units.

Figure 72 shows the cellular hold of a container ship.

The LASH (Lighter Aboard Ship) system utilizes barges as containers which, in turn, act as the holds of the specially constructed ships. The prime purpose of using lighters as, in effect, "cargo holds" of the ship, is to divorce the time for cargo handling operations from the ship's schedule. From this liberation springs a great increase in transportability of the ship.

Fig. 72. Cellular hold of lift-on/lift-off container ship. (Container Marine Lines)

Port time becomes a small percentage of the vessel's life, and the ship makes a substantially increased number of voyages per year.

The LASH system includes a predetermined number of lighters in ports being loaded, discharged, or standing by. On the day a LASH ship is scheduled to be in a port, any loaded lighters for shipment are towed to a rendezvous area. This may be a pier, or the vessel may anchor in the port, or in a semi-sheltered area. The transfer is completed and the LASH ship returns to sea.

Fig. 73. Artist's impression of LASH ship. (Prudential Lines, Inc.)

To the LASH ship, the term "port congestion" loses much of its fearsome impact upon schedules and profits. On arrival at a congested port, the LASH vessel discharges the lighters for this particular port, picks up such loads as are ready, and heads back to sea. A further advantage occurs from the use of lighters in that they can be handled at small wharves and in shallow waters.

Besides saving turnaround time for the ship, the LASH concept contains potential major savings in cargo handling costs. It is immediately obvious that stevedores need not be worked on an overtime basis, since the low-cost lighter remains in port rather than the high-cost capital investment vessel. Waterside industries could stow the lighters with their products, saving shipment costs to piers for the shipper, and cargo handling costs for the operator.

One of the many advantages that the system offers is its flexibility with respect to the type of cargo that can be carried:

    1. Lighters: palletized, liquid (oil, exotic acid, etc.), containers, bulk (grain, chemicals, etc.), industrial, general break-bulk cargoes, and refrigerated (frozen, chilled)

    2. Liquid in wing tanks, or cargo oil tanks

    3. Containers (over-the-road) in any desired quantity

    4. Bulk cargoes

The LASH system offers the greatest advantages to shippers and carriers when the following system characteristics can be utilized:

    1. A major reduction in port turnaround time

    2. Reduction in fleet investment costs

    3. Express service to all ports served

    4. Virtual immunity to port delays

    5. Bad weather port delays almost eliminated

    6. Economical interchange of cargo between connecting carriers

    7. Reduced pilferage and damage to cargoes

    8. Problems of overstow eliminated

    9. Since lighters only remain in port, stevedoring can be carried out on a normal work shift basis, thus eliminating overtime costs

   10. Industries, on navigable inland waterways, can ship direct to overseas inland marine destinations

   11. Inland ports become sea ports.

One aspect of the LASH system is its compatibility with containerization. This compatibility is true whether the cargo includes one container or hundreds. The LASH vessel can carry containers with the same effectiveness as a highly specialized containership. Using the quick-erecting and quick-dismounting LASH container cell system, a LASH hold can be changed from lighter carriage to container stowage and vice versa overnight and without the use of shipyard equipment.

The holds of the LASH ship are essentially open spaces, with heavy vertical guides at the hatch corners. The guides serve to locate the lighters in vertical alignment, and to keep the lighters from shifting during severe ship motion in heavy weather. They closely resemble the guides provided in a container ship, although of much heavier scantlings. The lighters are supported at four distinct regions of the inner bottom of the vessel, with local structural reinforcement of the vessel's bottom much like that provided by a pillar aboard a conventional cargo ship. Each cargo hold has several levels of walkways (gratings) clear of the lighter's stowage area, enabling shipboard personnel to connect and disconnect cargo ventilation, dehumidification and/or refrigeration ducts and lines to the lighters.

The weather deck hatch covers are single-piece, all-welded steel panels, handled by the LASH crane. These hatch covers are designed for carrying a fully loaded lighter thereon at sea. Further, they are adequate for three-high above deck stowage of ISO containers.

The LASH (Lighter Aboard Ship) system, with company headquarters in New Orleans, was one of the first to receive wide publicity. Another New Orleans Company, Lykes Bros., have announced that they intend to inaugurate a barge/containership service of their own design between U.S. Gulf terminals and Europe early in 1970, using three ships serviced by a total of 266 barges. Called the Seabee Class, the ships can carry either 38 fully-loaded barges or a total of between 1,500 and 1,600 containers of the standard 8′ x 8′ x 20′ size. This is equivalent to approximately 1,500,000 cu. ft. of cargo. The ship can carry special heavy-lift cargo of up to 2,000 tons and can handle vehicles, roll-on/roll-off cargo and unitized loads with equal facility. In addition, each ship can carry about 15,000 tons of liquid cargo in its deep tanks.

Pacific Far East Line, Inc., and Prudential Lines, Inc., are starting a LASH program. The former is planning construction of six vessels, 600 lighters, 1100 dry cargo 20′ x 8′ x 8′ containers and 400 refrigerated 20′ x 8′ x 8′ containers. Each ship will have a capacity of 49 lighters and 356 containers with a bale cubic of 1,208,000 feet. The standard container capacity can be increased as required by the cargo demands of each voyage so that, for example, over 700 containers and 34 lighters could be carried together. A traveling shipboard crane of 500-ton capacity will load and discharge the 61-foot-long lighters. In addition, there will be a 35-ton gantry with both cranes designed so that they can be operated simultaneously.

Yet another solution for the problem of turnaround times is being studied by Hay and Smart (Projects) Liverpool, who are developing a Multipacket Transport (MPT) system. In this system the ship comprises a number of detachable interconnected segments. The rear segment contains the propulsion unit; the front segment forms the bow of the vessel. In between is the cargo-carrying section, which may be subdivided into several segments. At each port of call one or more loaded segments can be left behind for discharging, while the place they occupied in the vessel is taken by other segments already loaded with return cargo. In this manner the time spent in port can be shortened to a minimum. The cost of a shuttle service using three propulsion units and twelve cargo-carrying sections of 8400 tons carrying capacity is estimated at about 10 million dollars.

Land Bridge. A play on words describes man's age-old search for new routes, shortened ways to reach markets and to reduce costs. The term "land bridge," a "bridge" linking two bodies of water, was originated by United Cargo Corp., just after the Arab-Israeli war in June of 1967 when, with the Suez Canal closed, cargo moved from the Far East to the United States West Coast by vessel, across the United States by rail and then across to the United Kingdom and the European continent by another

Fig. 74. Marshalling yard for containers. (Container Marine Lines)

vessel. A voyage that formerly took 40 days could have about 14 days subtracted through the use of the U.S. land bridge. Such movements, involving as they do more than one mode of transportation, with the resultant increase in the amount of handlings, give an additional impetus to the movement of cargo by containers. United States railroads with their highly developed system of routes and services were in a unique position to expedite the movements from coast to coast. Frequent sailings to and from both coasts of the United States added to this gave the element of practicality to the U.S. land bridge, enabling this combination of "know-how" and equipment and service to effectively carry out the undertaking. While much revision in railroad thinking must be made and rate structures will have to be altered downward to make the entire U.S. land bridge program economically feasible, there are no technical limitations. With the new designs in American rail equipment and cooperative arrangements between the U.S. railroads, a new era of usefulness for the American railroads will be created. The Milwaukee Road and Penn Central have already been pioneers in the U.S. land bridge. Subsequently, a Russian land bridge was inaugurated. This route was conceived by the Swiss forwarding firm of M.A.T. Transport A.G. Basle and its British counterpart, M.A.T. Transport Ltd., London. Cargo moves from London to Basle where consolidation takes place with cargoes funnelled in from all of Europe. Cars of the German Federal Railroad move the containers from Basle, via Furth Im Walde in Germany, and Cierna Nad Tisou in Czechoslovakia, to the Czechoslovak-Soviet border at Tschop. There the containers are transferred to the Russian railroad cars and move via the Trans-Siberian Railway to the Russian eastern port of Nachodka, near Vladivostok. At Nachodka the containers are transferred to a Japanese steamship line for movement to Yokohama. The entire movement from London to Yokohama should take about one month. The disadvantages of the Russian route, even though they may only be temporary, are:

1. Delays at Nachodka from one week to one month.
2. Trans-Siberian Railroad, particularly on Eastern portion, heavily overtaxed, with ensuing delays.
3. Change in track gauge at Brest Litovsk
4. Single track road; in case of breakdown—delay.
5. Small size non-ISO containers handled only.
6. Difficulties in obtaining insurance coverage.

Canada has attempted to get into the land bridge picture. Canada has two highly efficient coast-to-coast railroads. One of these railroads controls a link to Chicago from the east, while the other controls a link to Chicago from the west. Thus, Canada has not only two railroad bridges from ocean to ocean, but a bridge from each ocean to the interior of Canada, as well as to the interior of the United States. The west coast port for Canada would be Vancouver and the east coast port, Halifax. Advantages of a Canadian system are overhead clearances which presently permit double-decking of 8-foot-high containers, and more liberal interpretation of the regulations. There is a principle in Canadian law that the railroad must make a profit on whatever transportation function it performs, but also that it is empowered to perform any transportation function on which it can make a reasonable profit. The use of Canada would also shorten the overall time by about 14 days. However, the key to success of any system is economy. While ships coming to the east and west coasts of the United States carry cargoes for the U.S. as well as foreign-to-foreign cargoes, the frequency of such calls could not be substantiated for the foreseeable future for Canadian ports. This would add to the time and cost of the Canadian land bridge. In terms of distance, the present routes between Yokohama and London, using the Panama Canal, is 12,452 miles; the Suez Canal, 11,235 miles; the Cape of Good Hope, 14,788 miles; the U.S. land bridge 11,225 miles; the Canadian land bridge, 10,630 miles; the Russian land bridge, 8,078 miles. But already faster ships have been put into service between U.K./Continent and Japan that are making the voyage in 26 days; so the competition in routing includes the full gamut of alternatives.

Port Facilities; Pier Handling; Marshalling Yards. There are two approaches being used in the adaptation of piers for the handling of containers. In many world areas, entire new ports are being built; in others, the changeover is gradual. An ambitious program of preparing ports in the United States for containerization will be running into billions of dollars. Having much at stake, each port is offering its most attractive facilities and conditions to lure the container traffic. With the ports undergoing a gradual changeover, the handicap of trying to handle a new method of cargo movement with old tools remains. Ideas, too, are slow to change and we find that containers are "stuffed" and "unstuffed" right on the piers—containers are left standing on dock level or on some wood slats. They are stacked on top of one another without being interlocked and they are moved about with forklifts regardless of the weight of cargo they contain or the distribution of such weight or the limitations of the forklift. We shall not discuss the dents and occasional holes made in the side of the container when the forklift operator misses the openings. Containers are loaded and off-loaded with shore or vessel derricks, sometimes with manual spreaders but often without them. It is important that the operating managers, pier superintendents, ships' masters and other personnel directly involved become "container-minded," so that this feeling can be carried down to every single individual handling a container. Even if the cargo in the container must be stowed for the time being in the container while it is on the pier, people must be trained to do it properly, with proper securing and with plans made showing the location of every shipment. Proper areas solely for container use can be set aside on existing piers with containers placed in preplanned, premarked spaces to minimize rehandling and offer a smooth loading sequence. Sufficient turnaround space must be allowed for the forklift. Containers should not be placed on dock level, but raised. Standard height movable wooden or metal corner blocks can be used. Forklift operators must be taught to exercise care in lifting, stacking and handling containers—loaded weight can easily be crayoned on each container to alert the handlers. Containers prepared for lifting by derrick should be carefully inspected, test-lifted, and then securely held with guys at either end to minimize unintended movement. Winch operators, too, should be brought into the educational campaign. In this way, the changeover will take place smoothly with everyone contributing his proper share toward its success.

One of the first steps in the changeover is to remove from the pier the functions of stowing the cargo in the containers. As the full containership operation starts, all general cargo transportation (or as much of it as possible) is taken off the piers. Non-standard characteristics of general cargo transportation constitute the main cause of wasteful use of port resources. The large labor force is poorly utilized because of the large number of cargo units, their variety of sizes and origins and the amount of re-sorting required. This, then, would leave the pier area clear for the marshalling of all containers arriving for ocean departure. In addition, it will serve as a repository for the containers recently off-loaded from a ship and awaiting further inland movement. The quay area of a container installation should be from three to six times the size required for conventional cargoes. Approximately one acre of land is desirable for 100 equivalent 20-foot containers. The area should be designed for maximum efficient utilization of all facilities and allowances made for expansion. Basic facilities are road and rail practically to shipside, receiving and delivery and storage. Supported by proper handling equipment for the containers, we have the tools for pier handling of containers. While each container port terminal will differ in detail from the other, essentially all are the same in fulfilling the function of the rapid, efficient, storage and movement of containers or trailers. A typical terminal for the handling of containerized cargo has been planned in Burchardi, the center of Container Terminal Hamburg in the port of Hamburg. As of March, 1968, the Burchardi terminal consisted of two semi-containership berths, serviced by a conventional crane and a new berth equipped with a Demag 30-ton crane, backed up by 80,000 square meters of paved area and a 7,500-square-meter transit shed. A fourth berth was under construction and a 45-ton Peiner crane was ordered to service it. The capacity of the crane would allow two coupled 20-ton containers to be lifted, and the crane will be equipped with a Peiner telescopic spreader that is adjustable from 20 to 40 feet automatically. The following ground equipment is being used at the terminal: Clark straddle carriers, Peiner van carriers, one Coles mobile container crane and Mafi tractors for moving trailers.

At Felixstowe, the container terminal has a 1,330-foot quay with four sets of railway tracks and a minimum depth of 33 feet of water alongside. There are two 30-ton-capacity Paceco-Vickers Portainer Cranes with 20-, 30- and 40-foot automatic lifting frames, two 30-ton mobile cranes, 10 container transportation and stacking units, 25 acres of open, surfaced storage adjacent to the quay and 160,000 square feet of transit sheds. In addition to the container terminal, there are 2 roll-on/roll-off terminals, rail-connected and equipped with a 32-ton capacity electric traveling crane. There are 13 acres of open storage and 54,000 square feet of transit sheds.

At Antwerp, there are 5 roll-on/roll-off terminals, 12 roll-on/roll-off berths, a container marshalling area of 125 acres and 7 container cranes with a lift capacity between 38 and 53 tons. There is a water depth of 50 feet for the container ships.

In container operation the working of the units on land can play as important a part in deciding the viability of a service as the arrangements for sea transport. Parking of containers in ports and railhead buffers, at land bases, and on the premises of senders and consignees of full-load traffic needs to be on economic lines. Otherwise there is a danger that excessive costs and congestion may erode the financial and operational advantages of containerization.

Containers held in park should occupy the minimum area compatible with efficient operation and the first and running costs of the alternative types of handling equipment available.

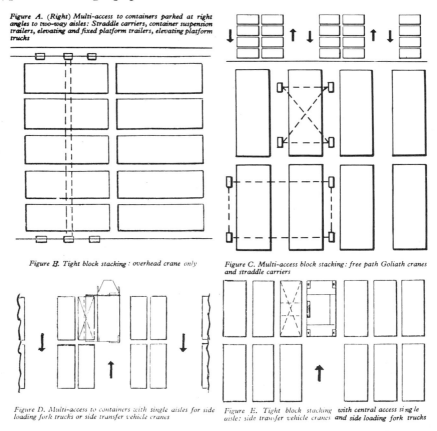

Figure A. (Right) Multi-access to containers parked at right angles to two-way aisles: Straddle carriers, container suspension trailers, elevating and fixed platform trailers, elevating platform trucks

Figure B. Tight block stacking : overhead crane only

Figure C. Multi-access block stacking: free path Goliath cranes and straddle carriers

Figure D. Multi-access to containers with single aisles for side loading fork trucks or side transfer vehicle cranes

Figure E. Tight block stacking with central access single aisle: side transfer vehicle cranes and side loading fork trucks

Fig. 75. Container parking systems in marshalling yard. (Containerisation International, May, 1968)

Speed, efficiency and overall costs of parking will be governed by the layout of the parking area, the detailed arrangement of container stalls and the height of stacking, as well as by the type of handling equipment employed. In turn, these aspects will be influenced by the intensity of loading and discharge operations, the nature of the transshipment gear at a port or railhead and the methods of filling and emptying containers at terminal points. The relative importance of the different factors will probably vary at each location.

At a port the ruling factors usually will be the size of the container ship and the type and speed of handling required of the berth cranes. As efficient operation of the berth cranes is the key to ship turnaround, crane waiting time must be kept to a minimum. Logically this means that the berth cranes should be confined to their essential function of transferring containers between ship and quay, movement between quay and parking areas being performed by ancillary equipment.

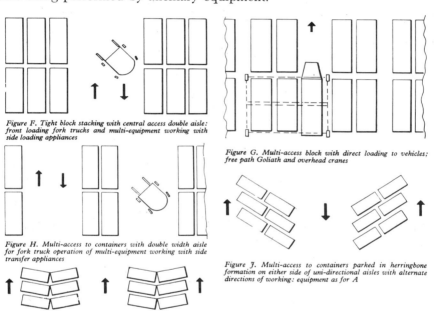

Figure F. Tight block stacking with central access double aisle: front loading fork trucks and multi-equipment working with side loading appliances

Figure G. Multi-access block with direct loading to vehicles; free path Goliath and overhead cranes

Figure H. Multi-access to containers with double width aisle for fork truck operation of multi-equipment working with side transfer appliances

Figure J. Multi-access to containers parked in herringbone formation on either side of uni-directional aisles with alternate directions of working: equipment as for A

Figure K. Multi-access to containers parked in sawtooth formation on either side of uni-directional aisles: equipment as for A

Fig. 75. Container parking systems in marshalling yard. (Containerisation International, May, 1968)

To avoid delays to the cranes, the number of trailers, side-loader fork trucks, straddle carriers, etc., used in support of berth cranes should be in optimal proportion to the number of cranes. This will take into account the horizontal distance involved between berth and parks and the layout of the parking areas.

Some possible layouts suggested for marshalling yards are illustrated in Fig. 75.

Containers in the Steamship Transport System. As it is unlikely that a container off-loaded from a ship would be reloaded in time for the same sailing, the least number of containers that would seem to conceivably suffice for a single ship service is three sets, where a set corresponds to a full load for one ship. This would be made up of one set on the ship and one set at each of the two ports of call. A further set would be required for each additional ship operating in the service. Thus, for a single ship with a capacity of 1,000 containers, the number of containers in the system would be 3,000; for two ships, 4,000 and for three ships, 5,000. This last figure may be compared with the 9,000 containers that would be calculated on the basis of three sets for each and every ship.

In estimating the minimum number of containers that could possibly suffice, allowance must be made for spares while maintenance and repairs are being undertaken. The number of containers on board each ship remains constant—that is, one full set. But the port sets are subject to variables. They will depend upon the time it takes for the containers to move inland, be cleared and then load and return to the port with an export load. This time depends upon the size of the port's hinterland for container traffic, the type of inland distribution set up, and the mode of transport employed.

Fig. 76. Ship deck loaded with containers. (United States Lines)

Of course, if the containers are unloaded coming from one system and then reloaded and moving on another system (such as the land bridge might pose), appropriate arrangements would have to be made for such "leakage." Ideally, the flow of containers should balance, but there will be a certain amount of dead-heading required, due to seasonal factors, sudden changes in market conditions, tariff conditions, strike conditions, or just ordinary characteristics of that trade. This may upset the distribution of inland trip times leading to more trips of short duration and thus reducing the number of containers required. Other relevant factors are the time spent in loading and unloading the ship and the interval between the closing time for acceptance of cargo and the ship's sailing time. These variables make the task of establishing the amount of containers needed for a given system a very difficult one. But the difficulties are minor compared to the possible savings that can be made by avoiding overpurchase of containers.

Certain general conclusions can be made in contrasting port facilities designed for conventional cargo in the past, and the experience of container shipping at present. Some of the more important conclusions follow.

1. A shedded finger pier will have several more years of use for vessels that are combinations of containers and conventional cargo, provided the pier is large enough, has wide aprons and four or five acres of upland area.

2. A quay type of berth with sufficient acreage running perpendicular to the berth can replace more expensive pier facilities, particularly for ships that are 100% containerized.

3. Berths for ships that are containerized require more space for aisles, roads and operations than do berths for conventional general cargo ships, namely, about 73% vs. 64%.

4. The container ships that are now planned for construction require longer berths than conventional ships.

5. Container shipping requires special buildings which need not be located near the berth, for packing and repacking of steamship containers, garage and motor repair facilities, refrigeration yard and administrative offices.

6. The rental cost of containership berths is less per ton of cargo handled than conventional berths by a wide margin.

The experience at Elizabeth Port Authority piers for vessels that are 100% containerized shows that for five berths, each of which is 640 feet long, the containers themselves utilized 58.3 acres, the refrigerated yard utilized 8 acres, and the shed for packing and repacking containers and miscellaneous buildings utilized 13.5 acres, or a total space of about 16 acres for each berth. In this calculation general administrative office space has been omitted because it is believed that most steamship lines will not have their general administrative offices at the seaports.

There exist at the five berths, 2,520 container spots on paved land, each of which can contain one 35-ft.-long container on a chassis, with each slot utilizing 375 square feet. There are, therefore, 22 acres of paved upland area devoted to the placement of containers alone out of a total of 58 acres. This means that 27% of the total paved area where the boxes are stored and handled actually holds the boxes, with 73% devoted to working space, mainly in the form of roadways, and some in the form of small optional buildings.

Experience shows that fully loaded container ships which perform the function of loading and unloading simultaneously require container slots in the paved port area that number 1.5 times the number of actual outbound loads. In other words, it is not practical in actual operation to replace alongside the vessel every box loaded to the vessel with one discharged from it.

Experience further shows that a fully containerized vessel can load and unload about 8,000 tons of cargo in 35-foot-long boxes during a straight-time working day using one set of gear. It is generally recognized that the shore cranes will operate faster than the shipboard cranes.

Experience can be projected into the theoretical conclusion that one container berth working five days a week with 554 container slots could handle as much as 1,800,000 tons of cargo per year after allowing for holidays and lost time due to unforeseen circumstances, such as equipment breakdown.

This is a high tonnage volume and probably would not be realized because most container shipping companies will not have enough vessels operating to keep a berth full every day. Even conventional ships which are in port many more days than a container vessel have not utilized their berths more than about 70% as a maximum in the past, and in some cases the berth occupancy is 50%. It is, therefore, more practical to think in terms of good usage when a container ship is in berth for three days a week rather than five, where the tonnage capacity per berth per year would run a little over 1,000,000 tons.

The probable rental cost for port facilities that accrues against a steamship line for conventional shipping is more than applies to fully loaded container ships of the future. For example, the rental for a quay type berth supported by 12 acres of paved upland area and equipped with cranes and structures should cost about $450,000 per year rent in the New York port area. As a practical matter, this berth could handle 1,000,000 tons of cargo resulting in a tonnage cost of 45 cents for facilities. If the same amount of cargo were handled over conventional facilities rather than the open container berth with the supporting land extending a relatively long distance in a perpendicular direction, the requirement would be at least five berths instead of one, and would involve a rental of about $1.00 per ton.

The difference is not in the amount of space required. The more expensive wharf space is used less by container ships; the less expensive paved land is used more.

The marshalling of containers one-high on chassis on paved areas is not the only system that can be used. Considerable study is being made in the field of land conservation and the use of air space by private engineering firms and transportation agencies.

One company, for example, is pushing a plan whereby containers would be stored in multistoried steel framework structures which could contain as many stories as desired with completely automated systems for putting the containers in and out and to and from vessels. The sponsors of these various plans claim that there are no new and unknown elements in the plans, and that all of the mechanical features have already been used in one form or another.

Vessel Stowage and Handling; Vessel Turnaround; Regulations. The container ship is a vessel equipped to carry, load and discharge containers in its hold and on its weather decks by means of special structural arrangements and devices. The containers are stacked atop each other in vertical cells in the holds and in vertical stacks on deck. Loading is achieved by lowering the container into a cell or stack, vertically, to its stowed position without further shifting in the horizontal plane. In discharging, the reverse procedure is followed. There are exceptions to the rule not requiring a horizontal shift but these are generally less efficient arrangements adopted to suit conditions on converted vessels.

Containers are stacked upon each other within a cell, depending on the design and weight of the container. ASA standard containers may be stacked six-high and ISO containers four-high.

Cells are usually arranged so that the containers are stowed with their length fore and aft in the vessel, but cells may also be arranged for athwartship stowage. In an athwartship stowage arrangement, which is exceptional, the contents of the container must be specially restrained from shifting longitudinally within the container. It is in this direction that the container is loaded and space is frequently left clear in front of the doors. Rolling of the ship may cause shifting and damage to the cargo as well as to the doors. In a fore-and-aft stowage arrangement, several cells are arranged close together side by side to form a cell group containing any number of cells, depending on the vessel's beam. One, two, three, or possibly more cell groups fore and aft of each other may be arranged in a hold, depending on the length of the hold between watertight bulkheads. Heavy web frames or similar structures separate cell groups within a hold to provide support for the cell structure and to provide structural rigidity to the vessel.

Each group of cells is located under a hatch opening slightly larger than the cell group. The hatch opening in the case of a wide cell group may extend athwartships up to about 80–85% of the vessel's beam. Hatch openings may be divided into two or three sections to limit the size of hatch covers and facilitate their handling, and for this purpose the hatch opening may be divided by fore-and-aft girders.

In addition to carrying containers in hold cells, the container ship also carries a considerable number of containers above deck in stacks usually on top of the hatch covers. The container easily lends itself to this scheme because it is a rigid box that gives shelter to its contents and is easily lashed in place. The container in this role extends the carrying ability of the ship beyond the confines of the hull and lessens the cubic losses sustained by the squaring-off of the ship's holds. Containers are carried one, two, three, and as many as four high on deck.

The individual container cell is formed by four vertical guides located at each corner of the container stack running from the hatch coaming down to the tank top. These guides are usually of rolled steel angles but may be of tees or fabricated shapes. They are supported by the bulkhead or structure separating the cell groups to which they are attached by frequently spaced chocks. The guides have three related functions:

a. To guide the containers down to their stowed position, even though the vessel may be listing or the crane not perfectly centered over the cell.

b. To land any container on top of the container below it within prescribed tolerances so that the superimposed loads on the lower containers do not exceed the eccentricities for which they are designed.

c. To hold the containers in their stacked position and absorb the horizontal forces imposed on them by the containers due to the motion of the vessel in the sea.

The guides may be parallel or tapered toward the bottom of the cell. Tapering is used so that the lowest container which has the greatest superimposed load is the least eccentrically loaded. Clearances between the guides and container corners are smaller at the bottom of a tapered cell and may be larger at the top than on a parallel-sided cell. Clearances between containers and the cell guides are dependent on the design of the container.

Guide clearances should not be excessive nor too small. Excessive clearances, the container design permitting, cause jamming of the container in the guide; small clearances cause difficulty in entering a container into a cell and may cause jamming due to unequal thermal expansion or working of the ship. Experience and testing has indicated that with a ½ inch clearance all around, a container will not tend to jam while suspended until a tilt of the container or list of the ship approaches about 5 or 6 deg. Standard ASA containers permit ½″ clearance on each side and ¾″ on each end.

The lower end of the guide terminates on the tank top or deck forming the bottom of the cell to which it is welded. The landing point is reinforced by a doubler or insert of heavy plate to distribute the concentrated load of the cell transmitted by the container corners. In addition to the structural doublers or inserts at each corner, leveling plates conforming to the footprints of the bottom corner castings are provided to give a four-point level landing surface to the bottom container and in the case where there are appendages on the bottom of the container to raise them above the tank top. Where container bottoms are clear of bottom appendages, the structural doubler or insert may be sufficient for leveling purposes. The guides are flared for about a foot at their upper ends in both the transverse and the fore-and-aft directions to give a centering effect for the containers entering the cell. The flared upper ends of the guides may extend above the coaming and beyond the clear opening, in which case the hatch covers must land outside the flared ends and be void of deep members directly over the guides. Or the flared ends may terminate level with the coamings and the coamings provided with fittings which precenter the container before it enters the top centering portion of the guides. These precentering devices on the coaming must be hinged or removable to clear the landing surface of the top of the coaming for seating the hatch covers.

The centering ability of the cell guides or coaming fittings determines in great measure the speed with which containers can be loaded with or without manual aid. The selection of this centering ability in terms of inches is derived from experience and depends on the type cranes used to handle the containers.

In this respect, a shipboard gantry crane offers the greatest accuracy in spotting the container both in the longitudinal and athwartship directions. The ship gantry gives the shortest pendulum length of the suspended load. It can be accurately centered longitudinally with respect to the hatch on the deck rails and need not be moved until an entire group is discharged and loaded. The accuracy of the original spotting is not affected by any surging of the ship. Transverse spotting may be automatic and self-adjusting for list or may be by eye. The former method has not been too successful in service. The latter is facilitated by the installation of a pointer fixed to the athwartship moving bridge which when aligned with a target on the stationary bridge frame spots the container over the selected cell. This is accurate within an inch or so. If the ship has a list, auxiliary target marks can guide the operator in making an adjustment. But if the vessel changes its list after spotting due to the motion of another loaded crane, the alignment will be out by 18.8, 11.3, 7.5, or 3.8 inches with a momentary change in heel angle of 5, 3, 2, or 1 degree, respectively. This is based on a pendulum length of 18 ft., typical of a shipboard gantry with a height suitable for stowing containers two-high on deck.

A shore crane's spotting ability is affected by the longitudinal and transverse shifting or surging as well as by the heeling of the ship. The container is suspended usually on a longer pendulum and spotting ability is lessened by the greater effect of the wind.

The head of a ship's conventional type of boom moves in three dimensions without reference points to aid in spotting, and the container suspended from it is on the end of a long pendulum.

To sum up, centering devices play an important role in handling the container into the ship quickly. Centering is always sensitive to small ship movements, but fore-and-aft centering is not as important when shipboard cranes are used.

Generally, a minimum of about 5″ centering ability is desired longitudinally and transversely on both sides and ends of the container cell. This is easily achieved in the fore and aft directions where there is ample space

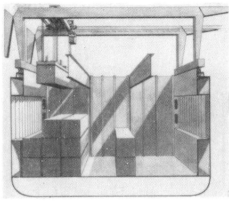

Fig. 77. (left) View looking up from hold in container ship. There are nine container cells in this group subdivided into 3 sections of three cells each. Shipboard container crane is lowering container into a cell.

Fig. 78. (right) Exploded view of a container hatch, pontoon hatch cover, deck-stowed containers, and spreader frame loading container into place. Spreader frame depicted is a self-leveling type, suspended from a single hook. Fixed-type container centering guides are shown on the hatch coaming. Container cell guides, protruding beyond hatch coaming top, self-center the containers and require no precentering fittings on the coamings.

(Society of Naval Architects and Marine Engrs.)

on hatch end coamings. But in the transverse direction, the required 10″ between adjacent cells affects the width of the hatch opening and can result in an increase in the vessel's beam or, in the case of an existing vessel, the loss in one cell per group.

Hatch covers on the weather deck covering a cell group or subdividing group are more frequently of the lift-on/lift-off steel pontoon type. Pontoon covers have the advantage in that they occupy no deck space when open inasmuch as they are stowed upon the hatch covers or upon deck-loaded containers of an adjacent hatch not being worked at the time. Lifting and shifting of pontoon hatch covers is accomplished rapidly by the ship or shore container crane. For this purpose, each cover is fitted with lifting devices duplicating those on the containers.

Where pontoon covers are used, centering devices are fitted on the hatch coamings to center the covers. These may be separate devices or part of the container-centering fittings.

Where rolling and folding covers are employed, the coamings must be clear for the passage of the covers and any container-centering devices above the coaming must be hinged.

Weather-deck hatch covers are designed to meet the normal classification requirements and in addition, when containers are carried upon them, to withstand the concentrated load imposed by the container corners. It is

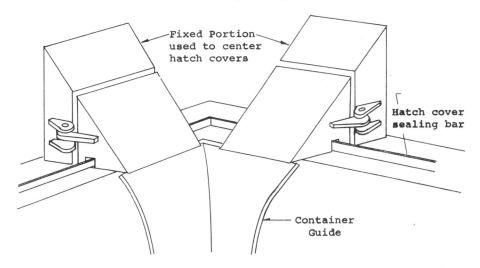

Fig. 79. Hinged precentering fittings on hatch coamings. (Society of Naval Architects and Marine Engineers)

most convenient to arrange deck containers in groups in the same fore-and-aft arrangement as the groups below deck. In this way, the deck-container corners land close to the ends of the hatch covers and the concentrated load is transmitted through the end skirts of the covers to the coaming or structure below without imposing any large bending moment on the covers. For this reason, the end skirts and that portion of the top of the cover upon which the containers land are made of heavier plate, about ⅝″, than the rest of the cover. In addition, a heavy transverse beam is placed at each end in line with the container corners to distribute the load. Where fixed flared cell guides protrude above the top of the coamings, these transverse beams must be set back and heavy shallow header provided under the container corners.

Containers are also frequently carried outboard of the hatch coamings. When fully outboard they are landed on footings on the deck, but when overhanging they are supported by a stanchion the same height at the hatch cover top.

The hatch covers are fitted with container-centering fittings at the landings of each container corner stowed upon them. These fittings can be of several types: fixed corner types, removable corner types, and fixed pyramid points which fit into holes in the bottom corner fittings of the containers lowered upon them. All serve not only to center the container but also to prevent horizontal shifting due to ship's motion. The pyramid points may also be of a rotating type, locking the container down in place, or the corner types may be equipped with locking bars for this same purpose.

Containers may be stacked one, two, three, or even four high upon the deck or hatch covers. The first limiting factor on the number of tiers is the

Fig. 80. Containers loaded outboard of the hatch tops. Note stanchions whereby a constant level is established. (United States Lines)

vessel's stability; and the second, a satisfactory means for securing or lashing the containers in place to prevent toppling and shifting due to ship's motion and to restrain undue racking of the containers, particularly the lowest tier.

Deck containers are stacked and restrained by three different methods. One employs a rigid cell structure built up above the deck. The more conventional method for securing containers on deck are systems of wire rope lashings and tie fittings. The third method is a removable rigid frame and buttress system.

The rigid cell structure above the deck, is constructed similar to below-deck cell structures except that it is self-supporting by means of structural cross-members and struts. All containers must be lifted over this structure, but lashing systems which entail considerable labor are eliminated.

The wire rope lashing systems employ a variety of fittings and hardware. When one tier of containers is carried on deck, the container is restrained from movement in the horizontal plane by the corner centering fittings located on the hatch covers or deck. In addition, the container is tied down to prevent tipping or rotating by diagonal or vertical wire rope lashings or more simply by locking devices included in the corner centering fittings. When two or more tiers are carried, the upper tiers are restrained from horizontal movement by double male corner connecting fittings inserted into the top holes of the lower container's corners and into the bottom holes of the upper container's corners. These may be of the locking type holding the upper tier down to the lower tier, or simple fixed types. The double male fittings of adjacent stacks may also be tied together by adjustable threaded tierods. The cross-connecting fittings tie adjacent stacks into a block. The blocks or individual stacks are then lashed to the deck or hatch coaming by vertical or diagonal wire ropes depending on the degree to which the container can absorb the racking forces imposed by the ship's motion. The wire rope assembly consists of a hook or fitting at the top which ties into one of the holes of the container corner fittings, a tensioning device such as a turnbuckle, and a quick-release device such as a pelican hook, or the tensioning device may be combined with a quick-release mechanism. Tensions in wire rope lashings must be balanced within a block of containers or on a single stack to preclude initial racking forces on any container. Of the single-tier lashing arrangements shown in Fig. 81, (c) and (d) restrain racking but require access to the top of the container and is a very elaborate arrangement. Arrangements (a), (b), and (e) are simpler, but (b) and (e) do not restrain tipping of all containers in the directions indicated by arrows. Should the ship's motions be severe enough, tipping would occur but overturning would be prevented by an adjacent container. However, some damage to the containers or their contents may ensue, and such arrangements should be avoided when there is any chance that dynamic forces are severe enough to cause a tipping action. Similar precautions should be taken with arrangement (g) in two-tier stacking. Three-tier lashing arrangements are similar to two-tier arrangements except that the lashing gear must be either heavier, disposed at a lesser angle to the horizontal or of more parts. The arrangement shown in (h) is used with three and four-tier stacking, but the containers used have superior antiracking strength due to the use of doors which interlock when closed.

Lashing arrangements for three or more high stacking can become complex, heavy, and difficult to handle, and require access up to at least the bottom of the third tier which is a minimum of 16 feet above the deck. These difficulties have led to the development of the third lashing method employing rigid frames and buttresses.

In this system, buttresses or towers extend above the deck at the vessel's side between container groups and are braced inboard between stacked containers. After the first tier of containers is landed on the hatch covers in a conventional manner, a rectangular frame containing pyramid fittings on its top and bottom surfaces is landed on top of the first tier. The pyramid fittings of the frame engage to top corner fittings of the lower tier and large self-centering pins on the outboard corners of the frame engage openings in the buttresses fore and aft of the stack on one side of the vessel. The second tier is then landed on the pyramid fittings of the frame and the process repeated until the third tier is landed. The pyramid fittings on the top of the second frame lock the third tier to the frame. The frames extend to the ship's centerline from both sides of the vessel and are not interconnected. Frames are handled by the container crane.

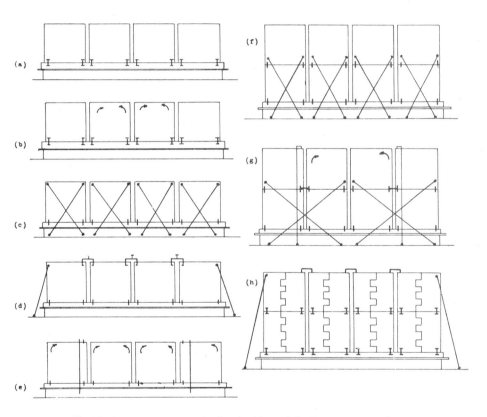

Fig. 81. Various arrangements for lashing of deck containers. (Society of Naval Architects and Marine Engineers)

Vessel speeds are constantly being increased with more efficient propulsion systems so that less time is spent at sea. With the use of containers and modern handling equipment port time, too, is being reduced so that vessel turnaround time is being brought down drastically. Vessel owners can plan on counting port times in hours, instead of days. Even the amount of port calls is being reduced, the vessel falling into the role of ferry.

Enormous investments, and complete design, hardware, and facility changeover are being made. More than 60% of total voyage costs were formerly represented by ship loading and discharge. This becomes a relatively minor portion in a containership operation. As the changeover from a labor-intensive to a capital-intensive operation occurs, the problems of high capital investment and equipment utilization become dominant. For example, assuming, as before, that three containers are required per ship's capacity—a 1,000 container ship would need a logistic backup of more than 3,000 containers—the cost of these units would be at least equal to the

Fig. 82. Three-high tiering and lashing of United States Lines containers on Amer-ican Lancer. (United States Lines)

construction cost of the ship itself. This problem is compounded for U.S. flag subsidized lines by the fact that although the ship construction subsidy may cover more than 50% of construction costs, this subsidy does not apply to containers. However, the Maritime Administration has ruled that Reserve Funds—capital required to be set aside for future construction—may be used.

With the shift in voyage costs, traditional methods of ship operation are being changed. Six and seven days in port at a time, picking up and discharging cargo at several ports on the coast between transocean passages can no longer be economically justified—not with ship amortization costs alone running up to $6.50 a minute. It is absolutely essential that port time be minimized. Prestowage of cargo in containers and the remarkable loading rate achieved by container ships makes this possible. For example, a ship's gang can handle only about 17 long tons of cargo per hour on a conventional cargo ship, but up to 40 tons every three minutes on a container ship. Container ships are regularly unloading and loading full cargoes at the Elizabeth Port Authority Marine Terminal and sailing on the same day as arrived.

Ships are becoming larger and faster in order to achieve suitable utilization. Cargo will come to the ships instead of ships steaming to wherever there is sufficient cargo. Transocean voyages will have one or at most two ports on each end of the sea passage. The number of terminal points will be reduced to a relatively few, very high volume load center ports. These ports in turn will possess the large, highly sophisticated terminals and equipment required for intermodal operation. Port investment in this case can only be economically justified by sufficient volumes. The development of off-pier container consolidation inspection, repair stations, and other ancillary services in these major port districts has already begun.

Nevertheless, in most cases, the heavy initial capital investments required are still among the most serious inhibiting factors. This has been countered by various "share the cost" schemes during the past two years. Best known is the "consortium principle," by which two or more steamship lines share port operating expenses and equipment while maintaining their individual identity. Atlantic Container Line is a classic example: Here six individual lines, Cunard Line, Ltd., flying the flag of Britain; French Line, flying the flag of France; Holland-America Line, flying the flag of Holland; Swedish-American Line, Swedish Transatlantic Line, Wallenius Line, all flying the flag of Sweden, have pooled their material and human resources in what appears to be a successful containership operation. Other consortia formed are the British Associated Container Transportation, Ltd., a service to and from the United Kingdom, the Continent and Australia, and comprise Ben Line, Blue Star Line, Cunard Line, Ellermans Line and Harrison Line; the British Overseas Containers, Ltd., another service to and from the United Kingdom, the Continent and Australia, and comprise the British and Commonwealth Group, Alfred Holt, P. & O. Steamship; the Japanese "K" Line, Mitsui-OSK Line, Yamashita-Shinnehon Line and Japan Lines known as the Interline Consortium and operating between Japan and the United States.

Another method used is equipment interchange agreements between various steamship lines, which is in effect a container pooling compact that occasionally extends to other equipment such as cranes, running gear, and even shore facilities.

Cargo in containers has been just another commodity to the steamship companies and the tariffs have not kept pace with the advances in equipment and movement. Minimum percentage usage of inside space in containers must be guaranteed on some routes, payment based on outside measurement of containers are required on other routes, minimum revenue guaranteed on still other routes. There is little consistency in setting revenue standards—and there is little agreement as to what methods may apply in the future. Some interests feel that system of F.A.K. (Freight All Kinds), or "per container" rates should be introduced similar to Plan 3/Plan 4 Railroad rates. Others feel that it would be difficult to lump all types and all values of cargo together under one rate structure. Still others feel that a series of F.A.K. rates based on different levels of cargo importance may be workable. In the last analysis, it will be the market itself that will determine the turn of rates just as the Overland Common Point (OCP) rates were established by the West Coast steamship lines, in conjunction with the railroads, so that the West Coast could compete for cargo more favorably. If the supply of container ships and containers become plentiful, then competition will force a change in thinking. If the supply of containerable cargo keeps ahead of the available space, then change will come very slowly.

Congress has specifically exempted steamship conferences from prosecution under the anti-trust laws. This was done so as to permit the existence of a dual rate structure. Such dual rates are made to reward the shippers signing agreements to patronize conference lines with a lower "contract" rate. Whether this, too, will become a thing of the past just as whether the conferences themselves will become a thing of the past is being thought about.

It is interesting to note that when volume shipments are involved the United States Court of Appeals for the District of Columbia Circuit in 1967 upheld the Federal Maritime Commissions ruling that "a dual rate contract is not involved where an agreement between shipper and carrier (a) provides a discount for volume from the higher rate applicable without regard to volume and (b) embodies an undertaking by the shipper to tender and by the carrier to transport, the minimum volume that triggers the low rate." This would seem to encourage the NVOCC or the large volume shipper in the direct negotiation of rates with the carrier. If, then, there is competitive bidding, then such bids would lose their efficacy if they were exposed within the conference arrangement so that there is thinking, too, about the continued need or lack of need for conferences. We can see, therefore, the evolution of the regulatory aspects and the rate aspects as well as the physical aspects of the use of containers.

6

Regulations and Sample Forms

As noted in various places elsewhere in this book, there are many organizations both inside as well as outside the United States in addition to international organizations both inside and outside the United Nations, that are concerned with the regulation of transport movement.

A.T.A.	American Trucking Associations
A.A.R.	Association of American Railroads
U.S.A.S.I.	U.S.A. Standards Institute
F.M.C.	Federal Maritime Commission
I.C.C.	Interstate Commerce Commission
C.M.I.	International Maritime Committee
U.I.C.	Unions International Chemin de Fer (International Association of Railways)
I.C.H.C.A.	International Cargo Handling Coordination Association
U.I.T.	Union Internationale des Transports Routiers
I.R.U.	International Road Transport Association
C.I.M.	International Convention concerning the Carriage of Goods by Rail, 1961. Regulated by The Central Office for International Railway Transport
I.S.O.	International Standards Organization

Not all have a direct regulatory influence on the movement of containers, but all of them affect in one way or another, in an advisory or regulatory manner, the international transport of containers.

The American Trucking Association is an organization composed of motor carriers and others interested in the trucking industry. The National Motor Freight Classification Bureau publishes the Motor Freight tariff rates. This Bureau works through regional conferences in which the various motor carriers interested in those regions participate. The regional conferences are: Central & Southern Motor Freight Tariff Association, Central States Motor Freight Bureau, Eastern Central Motor Carriers Association, Middle Atlantic Conference, Middle-West Motor Freight Bureau, New England Motor Rate Bureau, Pacific Inland Tariff Bureau, Rocky Mountain Motor Tariff Bureau, Southern Motor Carriers Rate Conference and Southwestern Motor Freight Bureau. The National Motor Freight Tariff Association in which all motor carriers participate publishes the classification tariff. There are three categories of rates—class rates, class rate exceptions and general commodities.

The Association of American Railroads is an organization composed of the railroads of the United States to further the common interests of the railroads. The association is set up along the same lines as the American Trucking Industry.

The United States of American Standards Institute replaced the American Standards Association on Nov. 24, 1966. Through their Sectional Committee MH-5 designated to develop a modular series of containers that could be carried efficiently in marine, railroad, highway or air transport service, and in conjunction with agreements reached as part of the International Standards Organization of Geneva, promulgated the Standard, known as USASI MH 5.1 which has been temporarily suspended for revision. They have however recommended, for general use, excerpts from:

ISO Recommendation R668 (formerly Draft Recommendation No. 804) Dimensions and Ratings of Freight Containers

ISO Recommendation R830 (formerly Draft Recommendation No. 1055) Terms and Definitions of Freight Containers

Draft ISO Recommendations 1019 Specification of Corner Fittings for Series I Freight Containers (IA, IB, IC, ID)

Draft ISO Recommendation 1496 Specification and Testing of Series I Freight Containers

The Federal Maritime Commission is an agency of the U.S. Government concerned with the execution of the Maritime Policy of the United States as outlined in the *Merchant Marine Act of 1936* which states "To further the development and maintenance of an adequate and well-balanced American merchant marine, to promote the commerce of the United States, to aid in the national defense, to repeal certain former legislation, and for other purposes"—and in the Shipping Act of 1916 which states "To establish a United States Shipping Board for the purpose of encouraging, developing, and creating a naval auxiliary and naval reserve and a merchant marine to meet the requirements of the commerce of the United States with the territories and possessions and with foreign countries; *to regulate carriers by water engaged in the foreign and interstate commerce of the United States,* and for other purposes."

It is this latter function of the Federal Maritime Commission, with which we are concerned. The first of the Regulatory Acts was *The Shipping Act of 1916.*

In 1933 the Intercoastal Shipping Act was passed, primarily to regulate traffic from one part of the United States to another using the Panama Canal. However in 1940, the Transportation Act of 1940 attempted to gather all interstate transportation regulation under the Interstate Commerce Act.

The *Interstate Commerce Commission* is an agency of the United States government concerned with all movements by truck, rail, waterway and air between two U.S. points as well as the U.S. portion of any U.S. to foreign or foreign to U.S. movement. The ICC was set up by the U.S. Congress under the Interstate Commerce Act of Feb. 4, 1897. The original act covered general provisions and railroad and pipe-line carriers. On August 9, 1935, the Motor Carrier section was added becoming known as the Motor

Carrier Act of 1935. On September 18, 1940 the designation, Motor Carrier Act of 1935 was changed to Part II of the Interstate Commerce Act. On September 18, 1940, with the passage of The Transportation Act of 1940, the Water Carrier section was added and was known as Part III. On May 16, 1942 the Freight Forwarder section was added and became known as Part IV. Part V was added August 12, 1958 and covered loans to railroads.

When Part III was added, the following "National Transportation Policy" was inserted into the Act:

"It is hereby declared to be the national transportation policy of the Congress to provide for fair and impartial regulation of all modes of transportation subject to the provisions of this Act, so administered as to recognize and preserve the inherent advantages of each; to promote safe, adequate, economical, and efficient service and foster sound economic conditions in transportation and among the several carriers; to encourage the establishment and maintenance of reasonable charges for transportation services, without unjust discriminations, undue preferences or advantages, or unfair or destructive competitive practices; to cooperate with the several states and the duly authorized officials thereof; and to encourage fair wages and equitable working conditions—all to the end of developing, coordinating, and preserving a national transportation system by water, highway and rail, as well as other means, adequate to meet the needs of the Commerce of the United States, of the Postal Service, and of the national defense. All the provisions of this Act shall be administered and enforced with a view to carrying out the above declaration of policy."

Other U.S. legislation has existed and still exists covering the Carriage of Cargoes.

There are many international conventions of which the United States has become a part. *The American Carriage of Goods by Sea Act of 1936* was the result of U.S. participation in the International Convention for the Unification of Certain Rules Relating to (Ocean) Bills of Lading. This Convention is known as the *Hague Rules of 1924*.

The International Standards Organization is an international nongovernmental organization which enjoys consultative status (Category B) to the United Nations, and maintains a liaison link with the United Nations headquarters, and with its various specialized agencies. It maintains relations with many other international organizations on questions of standardization, etc. and has three official languages—English, French and Russian. Based upon the work of its fore-runners in this field, the International Federation of National Standardizing Associations (ISA) set up in 1926, and the United Nations Standards Coordinating Committee (UNSCC), the ISO was created in 1946 with a membership of some 25 participating countries.

Now, as it comes of age, it has no less than 55 full members and five correspondent members. Together they provide a highly organized forum for the advancement of the art of sharing technical knowledge in a purposeful and workmanlike fashion, freely and without strings. The objects of the organization include action to: coordinate and facilitate the unification of national standards, and to issue the necessary recommendations to member bodies; cooperate with others in related matters; facilitate the international exchange of goods and services and to develop mutual cooperation in the spheres of intellectual, scientific, technological and economic activity.

As presently constituted the ISO consists of a General Assembly, a Council, a President, a Treasurer, a General Secretary. It functions through technical committees and technical divisions. Apart from the permanent and necessary administrative and technical staffs, the burden of the work falls upon the technical committees made up of delegates from member countries who, under arrangements made by a technical secretariat, meet to tender their views and advice and vote upon proposals.

It would be difficult to find a better example of international cooperation across such a wide range of subjects. The reason for the cooperation is not merely to bring about a better-ordered technical system of international exchange of goods and services, but to secure very practical advantages to themselves. For example, greater emphasis on standards has naturally strengthened the tendency to demand that imports as well as national production should comply with the national standard, and countries depending upon their export trade have had to take increasing account of this factor.

In terms of effectiveness, the ISO can give a reasonable account of itself. With some 125 technical committees recorded, its coverage at the present time is enormous. A few examples of its work over the years will suffice. (The country named in parentheses acts as Secretariat.) Starting as it did with its first Technical Committee in 1947 on screw-threads (Sweden), through agricultural machinery (Portugal), refrigeration (United Kingdom), sawn timber (USSR), horology (Switzerland) to nuclear energy (France) and so on, in an almost unending procession of subjects until in recent months a study on "enclosures and conditions for testing" was initiated with France providing the technical secretariat.

Somewhere in between this long list and with direct reference to future requirements of world transportation is Technical Committee 104 (Freight Containers) so called because the word "containers" might have a thousand connotations. What is intended is that the study should refer to containers used for the carriage of freight, the size of which presents a mechanical handling problem or requirement when being transferred between any of the many modes of transport. This work is now well advanced and it is a current example typical of the complexities of international standards problems and their solution.

In the United Kingdom in 1960 a committee of the Engineering Equipment Users Association submitted to the British Standards Institution a proposal concerning the need for interchangeability of freight containers between land, sea and air transport. Recognizing the international significance of this proposal, the BSI not only set up a national committee to consider the question, but simultaneously submitted the matter to ISO for urgent action. Concurrently, the then American Standards Association (not the USASI) took similar action. Thus the first meeting of the international committee TC/104 was held in New York in September 1961, when outline proposals were drafted. Incidentally, the U.S. proposal to introduce the 8' x 8' cross-section external dimension was accepted together with the

modular length dimension of 10 ft. and the outline draft proposal was published in what was then considered record time. The almost unanimous results obtained since that date have been widely implemented.

Each delegation is accepted as representing the concensus of opinion
within its own country, normally obtained through the appropriate national
committee dealing with the subject. In the case of TC/104, detailed evidence was obtained through a series of "working group" meetings, which,
after due consideration of all the national and international factors involved,
advised the technical committee at its plenary sessions held every 12–18
months.

Following upon almost global acceptance of the specification, testing and
external dimensioning proposals, TC/104 is now currently studying what is
regarded as the very heart of the users' problem. What can be loaded into
a standard covered container; how much, and in what fashion? Will it come
out at destination by the same means? These are the questions which
directly concern the seller and the buyer and they must be answered.
Added to these, the experts on the Freight Committee from 15 different
nations, must also take into account the fact that they are dealing with an
international unit of transport equipment which may be required to carry
many and varied commodities during the course of its life. Similarly on any
one journey it must be capable of withstanding the normal climatic hazards,
whether on land or sea, and the diurnal variations of temperature in such
conditions which range from tropical to arctic or antarctic.

How does one allow for all these factors and at the same time fail to
consider the fact that a country's main exports seldom have the same configuration or the same load density as its imports? The answer lies in the
type of shrewd economic appreciation which the Freight Container Committee experts have now become accustomed to making. Always the wider
interests must be taken into account, yet the ultimate solution must be
appropriate and applicable to the trade of the member countries and, more
important, to their future requirements, since it is an accepted duty of the
committee to produce forward-looking solutions rather than standardize an
accommodation of existing practices.

An attempt to answer these and many other relevant questions will be
made when the working groups of Technical Committee 104 meet to consider the question of standardizing the internal dimensions of freight containers together with dimensional specifications for other types such as
"open," "tank," "flat/stackable" and kindred variations.

Whereas the external dimensions have considerable impact on the provision of lifting equipment, size of carrying vehicle and the like, the internal
dimension is of prime importance to the user. As the payer of transport or
freight costs, the user must necessarily have regard to the economics of the
carrying unit from the point of view of (a) protection for his product; (b)
maximum utilization of the carrying power which he is called upon to pay
for. The owner or provider of the container, on the other hand is directly
concerned with the earning power of the unit outward and on its return or
subsequent journey. Given a set of standard internal dimensions, and in the
case of a covered container, a specified degree of protection, (from nil up-

wards) the user is able to plan a complete throughout journey with an accuracy hitherto denied him. Thus he not only controls his costs but is able to give a consistent standard of service to his customer hitherto unthinkable and certainly unattainable before the introduction of standard containers. While guilty on the one hand of taking some of the romance out of the dispatch of goods to "far away places" by obscuring the shape and possibly the aroma of goods, the standard covered container nevertheless conceals only to reveal. It reveals the vast potential for bringing order as well as automated and highly developed electronic systems to bear upon a complex operation which includes not just modern handling methods, but involves a major revolution in documentation, customs procedures and safety considerations. This, in turn, involves banking, insurance, legal and all the associated professional skills encompassed in the need to "facilitate the international exchange of goods." Thus it is that a far-reaching revolution is already under way—one that affects in some degree our very existence.

The part played by the International Organization for Standardization is not inconsiderable. It has established a platform where technicians from all over the world can come and discuss together standardization matters as experts, without any political bias. The efforts required to achieve a satisfactory measure of international agreement, even in a single field such as that of containers, are considerable. They involve the deep understanding of all the industries and trades involved, including a willingness to accept change; they concern the efforts of the national standards organizations and call for a considerable measure of sacrifice of traditional thinking and even the abandonment of traditional techniques.

ICHCA—International Cargo Handling Coordination Assn. is a technical nonprofit, and nonpolitical organization with its membership spread in over 70 countries, having consultive status with the United Nations, the Organization of American States, the International Standards Organization and many other international and national bodies. The Association's goal is that of finding ways and means of improving the efficiency of international cargo movements, covering every aspect of cargo handling, the simplification and expedition. Membership embraces stevedores, ship owners, ship builders, port operators, freight forwarders, insurance underwriters, packing experts, civil engineers, naval architects, manufacturers and government departments. Its principal function is coordinating the views of all interests that need to be brought together to achieve a balanced solution to any given cargo handling problem.

IRU—International Road Transport Association, the only international organization representing road transport, was formed March 23, 1948; it was granted Consultive Status by the Economic and Social Council of the United Nation, February 15, 1949 and was granted similar status by the Council of Europe, March 2, 1959. The IRU's objective is to promote the development of international road transport in the interests of road carriers and the economy as a whole. It has 39 active members (National Road Transport Federations) in 23 countries and 25 associate members (associations, groups and undertakings concerned with international road transport) including five national associations outside Europe: Australia, Brazil, Canada, India and Japan.

International Convention Concerning the Carriage of Goods by Rail, 1961 (CIM). Just as the International Road Union has been instrumental in bringing about order in the international movement of cargoes by road, and in the setting up of the T.I.R. Carnet system for this, so has another organization been successful in accomplishing the simplification of international rail movements. The International Convention concerning the Carriage of Goods by Rail was originally signed at Berne on October 25, 1952 and then revised and concluded as a new convention signed at Berne, February 25, 1961. This convention is known as CIM. The purpose and scope of the convention are covered in Articles 1 to 5 of part I:

Article 1: Railways and traffic to which the Convention applies:

1. This Convention shall apply, subject to the exceptions set forth in the following paragraphs, to the carriage of goods consigned under a through consignment note for carriage over the territories of at least two of the Contracting States and exclusively over lines included in the list complied in accordance with Article 59 of this Convention.

2. Consignments despatched from and destined for stations (the expression "station" includes ports used by shipping services and all road service establishments open to the public in connection with the performance of the contract of carriage) situated in the territory of the same State, which pass through the territory of another State only in transit, shall be governed by the law of the State in which they are despatched:

(a) when the lines over which the consignment is carried in that other State are exclusively operated by a railway of the State in which the consignment is despatched;

(b) when the lines over which the consignment is carried in that other State are not exclusively operated by a railway of the State in which the consignment is despatched, if the railways concerned have concluded agreements under which such carriage is not regarded as international.

3. Consignments between stations in two adjacent States shall, if the lines over which the consignments are carried are exclusively operated by railways of one of those States, be governed by the law of that State, provided that the sender, by his choice of the form of consignment note, elects that the internal regulations relating to those railways shall apply, and provided that such application is not contrary to the laws and regulations of either of the States concerned.

Article 2: Provisions concerning carriage by more than one form of transport:

1. Regular road services or shipping services which are complementary to railway services and on which international traffic is carried may, in addition to railway services, be included in the list referred to in Article 1 of this Convention provided that such services, in so far as they connect at least two Contracting States, may only be included in the list by agreement between those States.

2. The undertakings operating such services shall be subject to all the obligations imposed and enjoy all the rights conferred on railways by this Convention, subject always to such derogations as necessarily result from

the different forms of transport. Such derogations shall not, however, in any way affect the rules as to liability laid down in this Convention.

3. Any State wishing to have a service of the kind referred to in paragraph 1 of this Article included in the list shall take the necessary steps to have the derogations provided for in paragraph 2 of this Article published in the same manner as tariffs.

4. In the case of international traffic making use both of railways and of transport services other than those referred to in paragraph 1 of this Article, the railways, in conjunction with the other transport undertakings concerned, may, so as to take account of the special features of each form of transport, lay down conditions in the tariffs which have a legal effect different from that of this Convention. The railways may, in such a case, prescribe the use of a transport document other than that provided for by this Convention.

Article 3: Articles not to be accepted for carriage:

(a) articles the carriage of which is a monopoly of the postal authorities in any one of the territories concerned;

(b) articles which, by reason of their dimensions, weight or nature or condition are not suitable for the carriage proposed, having regard to the equipment or rolling stock of any one of the railways concerned;

(c) articles the carriage of which is prohibited in any one of the territories concerned;

(d) subject to the exceptions provided for in Article 4 (2) of this Convention, substances and articles which under the provisions of Annex I to this Convention are not to be accepted.

Article 4: Articles accepted for carriage subject to certain conditions:

1. The following articles are accepted for carriage on the following conditions:

(a) the substances and articles set forth in Annex I to this Convention are accepted subject to the conditions laid down therein:

(b) funeral consignments are accepted for carriage subject to the following conditions:

    (i) they shall be carried *grande vitesse* and accompanied by an attendant unless carriage without an attendant is permitted on all the railways concerned;

    (ii) charges shall be paid by the sender;

    (iii) carriage shall be subject to the laws and regulations of each State except in so far as such carriage is governed by special conventions between States:

(c) railway rolling stock running on its own wheels is accepted if a railway verifies that such rolling stock is in running order and so certifies either by marking the rolling stock or by issuing a special certificate; locomotives, tenders, rail motor-coach units and railcars shall in addition be accompanied by a person who is appointed by the sender and shall, in particular, be competent to carry out lubrication;

railway rolling stock running on its own wheels, other locomotives, tenders, rail motor-coach units and railcars, may be accompanied by an attendant who shall, in particular, carry out lubrication. If the sender intends to make use of this facility he shall state the fact in the consignment note;

(d) livestock is accepted subject to the following conditions;

(i) consignments of livestock shall be accompanied by an attendant provided by the sender except in the case of small livestock consigned in cages, crates, baskets, etc. which are properly secure. An attendant shall not however be required in such exceptional cases as are provided for in international tariffs or in agreements between railways. The sender shall state in the consignment note the number of attendants or, if the consignments are unaccompanied, shall insert the words "without attendant;"

(ii) the sender shall comply with the veterinary regulations of the country of departure and destination and of those through which the consignment passes;

(e) articles the carriage of which will give rise to special difficulty by reason of their dimensions, weight, or nature or condition, having regard to the equipment or rolling stock of any of the railways concerned, are only accepted subject to special conditions to be determined by the railway in each case after consultation with the sender; these conditions may derogate from the provisions of this Convention.

2. Two or more Contracting States may arrange, by agreement, that certain substances or articles not acceptable for carriage under the provisions of Annex I to this Convention will be accepted for international carriage between those States subject to certain conditions, or that the substances and articles specified in Annex I to this Convention will be accepted subject to conditions less onerous than those laid down in the said Annex.

Railways may also, by clauses in their tariffs, either accept certain substances or articles not acceptable for carriage under the provisions of Annex I to this Convention, or adopt conditions less onerous than those laid down in Annex I to this Convention for substances and articles accepted under the said Annex.

Such agreements and tariff clauses must be notified to the Central Office for International Railway Transport.

Article 5: Obligation of railways to carry:

1. Every railway shall be bound to undertake the carriage of goods, subject to the terms of this Convention, provided that:

(a) the sender complies with the provisions of the Convention;

(b) carriage can be undertaken by ordinary transport facilities serving the regular traffic requirements;

(c) carriage is not prevented by circumstances which the railway cannot avoid and which it is not in a position to remedy.

2. The railway shall not be obliged to accept articles the loading, transshipment or unloading of which requires the use of special facilities

unless the stations at which these operations are to be carried out have such facilities at their disposal.

3. The railway shall only be obliged to accept consignments the carriage of which can take place without delay; the regulations in force at the forwarding station shall determine the circumstances in which that station is obliged to store temporarily consignments not complying with this condition.

4. When the competent authority decides that:

(a) a service shall be discontinued or suspended totally or partially;

(b) certain consignments shall be refused or accepted only subject to certain conditions; such measures shall, without delay, be brought to the notice of the public and of the railways, which shall be responsible for informing the railways of the other States with a view to their publication.

5. The railways may, by agreement and with the consent of their Governments, decide to limit the carriage of goods between certain places to defined frontier points and transit countries.

These measures shall be communicated to the Central Office which shall notify them to the Governments of the Contracting States. They shall be regarded as having been accepted if within one month of the date of notification they have not been opposed by a Contracting State. In the event of opposition, if the Central Office does not succeed in removing the differences of opinion it shall convene a meeting of the representatives of the Contracting States.

As soon as these measures can be regarded as having been accepted the Central Office shall notify the Contracting States. They shall then be entered in special lists and published in the same manner as provided for the publication of international tariffs.

These measures shall come into force one month after the notification by the Central Office laid down in the third subparagraph above.

6. Any contravention of the provisions of this Article by the railway shall constitute a cause of action to recover compensation for the loss or damage caused thereby.

Dangerous goods are covered in Annex I called International Regulations covering the carriage of Dangerous Goods by Rail (RID) which reads:

"Until a special appendix to Annex I to CIM containing special provisions for the rail-sea carriage of dangerous goods between the Continent and the United Kingdom is agreed and comes into force, dangerous goods carried under CIM to or from the United Kingdom shall comply with the provisions of Annex I and also with the United Kingdom conditions for the carriage of dangerous goods by rail and by sea."

Containers are covered in Annex VIII called International Regulations concerning the carriage of containers (RICO).

CHAPTER I. GENERAL

Article 1: Object and scope of the regulations

1. These regulations shall apply to the carriage of containers belonging

to a railway or belonging to private owners (whether individuals, firms or corporate bodies) and accepted by the railway, which are tendered for carriage under the conditions of the International Convention concerning the Carriage of Goods by Rail (CIM).

2. For the purpose of these regulations, any receptacle (case, crate, tank, etc.) so constructed as to facilitate the door-to-door carriage of goods by rail only or by rail in combination with other means of transport shall be deemed to be a container.

Article 2: General provisions

1. Except as otherwise provided in the tariffs the contents of the container can only be subject to a single contract of carriage.

2. In the absence of special provisions in these regulations the other provisions of the CIM shall apply to the carriage of containers whether empty or loaded.

Article 3: Door-to-door carriage

In the case of consignments to be collected by the railway at the sender's premises or to be delivered by the railway at the consignee's premises, the contract of carriage shall be deemed to be made at the sender's premises or terminated at the consignee's premises, as the case may be.

CHAPTER II. RAILWAY-OWNED CONTAINERS

Article 4: Supply charges

So far as the railway is able to provide containers, it shall make them available to senders. A charge may be made for the use of containers and the amount of such charge shall be fixed by the tariffs or regulations.

Article 5: Particulars in the consignment note

In addition to the particulars required by the CIM, the sender shall enter in the consignment note, in the column headed "Loading tackle—Containers," the category of the container, its marks, its number, its tare in kilogrammes and its capacity in cubic metres or litres.

The tare of containers shall not include the weight of special internal and removable fittings which are for the purpose of packing or securing.

Article 6: Provision, return and handling

The tariffs or regulations shall determine the conditions under which containers will be supplied, the period within which they are to be returned, the charges which will be made for exceeding this period and the conditions under which the operations of loading and unloading are to be carried out.

"Loading" includes placing the container on a wagon and operations ancillary thereto, in particular the securing of the container.

Article 7: Cleaning

The consignee shall be responsible for cleaning containers after unloading. If containers are returned to the railway without having been cleaned, the railway shall be entitled to make a charge the amount of which shall be fixed by the tariffs or regulations.

Article 8: Reconsignment

Containers delivered loaded shall not be reconsigned by consignees on further journeys except with the consent of the railway which has so delivered them.

Article 9: Loss of and damage to containers

1. Any person accepting a container, empty or loaded, from the railway shall check the condition of the container at the time it is placed at his disposal; he shall be liable for all damage found to exist on return of the container to the railway which was not indicated when the container was put at his disposal unless he proves that the damage existed at that time or resulted from circumstances which he could not avoid and the consequences of which he was unable to prevent.

The sender shall be liable for loss of or damage to a container arising during performances of the contract of carriage if it results from his actions or from those of persons acting on his behalf.

3. If the container is not returned within 30 days following the day on which it was put at the sender's disposal, the railway may deem it to be lost and demand payment of its value.

CHAPTER III. PRIVATELY-OWNED CONTAINERS

Article 10: Approval

In order to be accepted for international traffic, privately-owned containers shall be approved by a railway to which the CIM applies, shall be provided by that railway with the distinguishing mark P and shall comply with the conditions laid down for construction and marking.

Article 11: Special equipment

If privately-owned containers are equipped with special apparatus (refrigerating equipment, water tanks, machinery, etc.), the sender shall be responsible for the servicing of such equipment or for arranging for this to be done. This duty shall pass to the consignee as soon as he exercises his rights under Article 16 or 22 of the CIM.

Article 12: Particulars in the consignment note

1. In the case of loaded containers, in addition to the particulars required by the CIM, the sender shall enter in the consignment note, in the column headed "Loading tackle—Containers," the category of the container, its marks, its number, the sign P, the tare of the container in kilogrammes and its capacity in cubic metres or litres.

2. In the case of empty containers, in addition to the particulars required by the CIM, the sender shall enter the following particulars in the consignment note:

    (a) in the column headed "Loading tackle—Containers": the category of the container, its marks, its number and the sign P;

    (b) in the column headed "Description of goods": the tare in kilogrammes and the words "empty container."

Article 13: Return of empty containers or reconsignment

After delivery of the container, and in the absence of special agreements, the railway shall not be bound to take any action to secure the return of the empty container or its reconsignment as a loaded container.

Article 14: "Cash on delivery" charges

Empty containers shall not be consigned subject to "cash on delivery" charges.

Article 15: Liability for exceeding the transit period

In regard to liability for exceeding the transit period railways may, apart from the provisions of the CIM, provide for the payment of special compensation to the owner of the container by special agreement with him.

SHIPPING ACT, 1916

[As amended through the 89th Congress, first session]

(39 Stat. 728, chapter 451, approved Sept. 7, 1916)

[NOTE.—See excerpts from the Transportation Act of 1940, as amended, infra.] [1]

AN ACT

To establish a United States Shipping Board for the purpose of encouraging, developing, and creating a naval auxiliary and naval reserve and a merchant marine to meet the requirements of the commerce of the United States with its Territories and possessions and with foreign countries; to regulate carriers by water engaged in the foreign and interstate commerce of the United States, and for other purposes.

Be it enacted by the Senate and House of Representatives of the United States of America in Congress assembled, That when used in this Act:

The term "common carrier by water in foreign commerce" means a common carrier, except ferryboats running on regular routes, engaged in the transportation by water of passengers or property between the United States or any of its Districts, Territories, or possessions and a foreign country, whether in the import or export trade: *Provided,* That a cargo boat commonly called an ocean tramp shall not be deemed such "common carrier by water in foreign commerce."

46 U.S.C. 801. 39 Stat. 728. 40 Stat. 900. 75 Stat. 522.

Definitions: "Common carrier by water in foreign commerce."

Ocean tramps excepted.

The term "common carrier by water in interstate commerce" means a common carrier engaged in the transportation by water of passengers or property on the high seas or the Great Lakes on regular routes from port to port between one State, Territory, District, or possession of the United States and any other State, Territory, District, or possession of the United States, or between places in the same Territory, District, or possession.

"Common carrier by water in interstate commerce."

The term "common carrier by water" means a common carrier by water in foreign commerce or a common carrier by water in interstate commerce on the high seas or the Great Lakes on regular routes from port to port.

"Common carrier by water."

The term "other person subject to this act" means any person not included in the term "common carrier by water," carrying on the business of forwarding or furnishing wharfage, dock, warehouse, or other terminal facilities in connection with a common carrier by water.

"Other person subject to this act."

---

[1] See section 27(b) of Public Law 85–508 (72 Stat. 351) and section 18(a) of Public Law 86–3 (73 Stat. 12), regarding continuing jurisdiction of the Federal Maritime Board, now the Federal Maritime Commission, after Alaska and Hawaii statehood.

"Person."

The term "person" includes corporations, partnerships, and associations, existing under or authorized by the laws of the United States, or any State, Territory, District, or possession thereof, or of any foreign country.

"Vessel."
46 U.S.C. 801.
40 Stat. 900.

The term "vessel" includes all water craft and other artificial contrivances of whatever description and at whatever stage of construction, whether on the stocks or launched, which are used or are capable of being or are intended to be used as a means of transportation on water.

Shipping Act,
1916, amendment.   Ocean
freight forwarders.
39 Stat. 728.
Definitions.
P.L. 87–254.
75 Stat. 522.

The term "documented under the laws of the United States," means "registered, enrolled, or licensed under the laws of the United States."

The term "carrying on the business of forwarding" means the dispatching of shipments by any person on behalf of others, by oceangoing common carriers in commerce from the United States, its Territories, or possessions to foreign countries, or between the United States and its Territories or possessions, or between such Territories and possessions, and handling the formalities incident to such shipments.

An "independent ocean freight forwarder" is a person carrying on the business of forwarding for a consideration who is not a shipper or consignee or a seller or purchaser of shipments to foreign countries, nor has any beneficial interest therein, nor directly or indirectly controls or is controlled by such shipper or consignee or by any person having such a beneficial interest.

Citizen of
United States.
46 U.S.C. 802,
803.
39 Stat. 729.
40 Stat. 900.
41 Stat. 1008.
73 Stat. 597.
P.L. 86–327.

SEC. 2. (a) That within the meaning of this Act no corporation, partnership, or association shall be deemed a citizen of the United States unless the controlling interest therein is owned by citizens of the United States, and, in the case of a corporation, unless its president or other chief executive officer and the chairman of its board of directors are citizens of the United States and unless no more of its directors than a minority of the number necessary to constitute a quorum are noncitizens and the corporation itself is organized under the laws of the United States or of a State, Territory, District, or possession thereof, but in the case of a corporation, association, or partnership operating any vessel in the coastwise trade the amount of interest required to be owned by citizens of the United States shall be 75 per centum.

Controlling
interest in
corporation.

(b) The controlling interest in a corporation shall not be deemed to be owned by citizens of the United States (a) if the title to a majority of the stock thereof is not vested in such citizens free from any trust or fiduciary obligation in favor of any person not a citizen of the United States; or (b) if the majority of the voting power in such corporation is not vested in citizens of the United States; or (c) if through any contract or understanding it is so arranged that the majority of the voting power may be exercised, directly or indirectly, in behalf of any person who is not a citizen of the United States; or (d) if by any other means whatsoever control of the corporation is conferred upon or permitted to be exercised by any person who is not a citizen of the United States.

(c) Seventy-five per centum of the interest in a corporation shall not be deemed to be owned by citizens of the United States (a) if the title to 75 per centum of its stock is not vested in such citizens free from any trust or fiduciary obligation in favor of any person not a citizen of the United States; or (b) if 75 per centum of the voting power in such corporation is not vested in citizens of the United States; or (c) if, through any contract or understanding it is so arranged that more than 25 per centum of the voting power in such corporation may be exercised, directly or indirectly, in behalf of any person who is not a citizen of the United States; or (d) if by any other means whatsoever control of any interest in the corporation in excess of 25 per centum is conferred upon or permitted to be exercised by any person who is not a citizen of the United States. {.notice 75 per centum of interest in corporation.}

(d) The provisions of this Act shall apply to receivers and trustees of all persons to whom the Act applies and to the successors or assignees of such persons. {.notice Receivers, trustees, successors, and assigns.}

Sections 3–8, inclusive, were repealed by section 903 (a) of the Merchant Marine Act, 1936, supra.

SEC. 9.[2] That any vessel purchased, chartered, or leased from the board, by persons who are citizens of the United States, may be registered or enrolled and licensed, or both registered and enrolled and licensed, as a vessel of the United States and entitled to the benefits and privileges appertaining thereto: *Provided,* That foreign-built vessels admitted to American registry or enrollment and licensed under this Act, and vessels owned by any corporation in which the United States is a stockholder, and vessels sold, leased, or chartered by the board to any person, a citizen of the United States, as provided in this Act, may engage in the coastwise trade of the United States while owned, leased, or chartered by such a person. {.notice Vessels acquired by or from board may engage in coastwise trade. 46 U.S.C. 808. 39 Stat. 730. 40 Stat. 900. 41 Stat. 994. 52 Stat. 964. Vessels acquired from board may be operated under American documentation only.}

Every vessel purchased, chartered, or leased from the board shall, unless otherwise authorized by the board, be operated only under such registry or enrollment and license. Such vessels while employed solely as merchant vessels shall be subject to all laws, regulations, and liabilities governing merchant vessels, whether the United States be interested therein as owner, in whole or in part, or hold any mortgage, lien, or other interest therein. {.notice Subject to all laws and liabilities.}

---

[2] See the Act extending the steamboat inspection laws to Shipping Board vessels (41 Stat. 305 ; 46 U.S.C. 363).

See also section 27A of the Merchant Marine Act, 1920, as amended, supra (Public Law 85–902, 72 Stat. 1736) regarding the term "citizen of the United States".

The fourth paragraph of this section was added by Public Law 89–346 (79 Stat. 1305), p. 282 herein. Section 4 of that act contains provisions with respect to transfers to noncitizens prior to such enactment or within one year thereafter.

Restrictions
upon transfer
of vessels.

Except as provided in section 611 of the Merchant Marine Act, 1936, as amended, it shall be unlawful, without the approval of the United States Maritime Commission, to sell, mortgage, lease, charter, deliver, or in any manner transfer, or agree to sell, mortgage, lease, charter, deliver, or in any manner transfer, to any person not a citizen of the United States, or transfer or place under foreign registry or flag, any vessel or any interest therein owned in whole or in part by a citizen of the United States and documented under the laws of the United States, or the last documentation of which was under the laws of the United States.

The issuance, transfer, or assignment of a bond, note, or other evidence of indebtedness which is secured by a mortgage of a vessel to a trustee or by an assignment to a trustee of the owner's right, title, or interest in a vessel under construction, to a person not a citizen of the United States, without the approval of the Secretary of Commerce, is unlawful unless the trustee or a substitute trustee of such mortgage or assignment is approved by the Secretary of Commerce. The Secretary of Commerce shall grant his approval if such trustee or a substitute trustee is a bank or trust company which (1) is organized as a corporation, and is doing business, under the laws of the United States or any State thereof (2) is authorized under such laws to exercise corporate trust powers, (3) is a citizen of the United States, (4) is subject to supervision or examination by Federal or State authority, and (5) has a combined capital and surplus (as set forth in its most recent published report of condition) of at least $3,000,000. If such trustee or a substitute trustee at any time ceases to meet the foregoing qualifications, the Secretary of Commerce shall disapprove such trustee or substitute trustee, and after such disapproval the transfer or assignment of such bond, note, or other evidence of indebtedness to a person not a citizen of the United States, without the approval of the Secretary of Commerce, shall be unlawful. The trustee or substitute trustee approved by the Secretary of Commerce shall not operate the vessel under the mortgage or assignment without the approval of the Secretary of Commerce. If a bond, note, or other evidence of indebtedness which is secured by a mortgage of a vessel to a trustee or by an assignment to a trustee of the owner's right, title, or interest in a vessel under construction, is issued, transferred, or assigned to a person not a citizen of the United States in violation of this section, the issuance, transfer, or assignment shall be void.

Any such vessel, or any interest therein, chartered, sold, transferred, or mortgaged to a person not a citizen of the United States or placed under a foreign registry or flag, or operated, in violation of any provision of this section shall be forfeited to the United States, and who-

ever violates any provision of this section shall be guilty
of a misdemeanor and subject to a fine of not more than
$5,000, or to imprisonment for not more than five years,
or both.

Sections 10 and 11 were repealed by section 903(a)
of the Merchant Marine Act, 1936, supra.

SEC. 12. That the board shall investigate the relative
cost of building merchant vessels in the United States
and in foreign maritime countries, and the relative cost,
advantages, and disadvantages of operating in the foreign trade vessels under United States registry and
under foreign registry. It shall examine the rules under
which vessels are constructed abroad and in the United
States, and the methods of classifying and rating same,
and it shall examine into the subject of marine insurance, the number of companies in the United States, domestic and foreign, engaging in marine insurance, the
extent of the insurance on hulls and cargoes placed or
written in the United States, and the extent of reinsur-
ance of American maritime risks in foreign companies,
and ascertain what steps may be necessary to develop
an ample marine insurance system as an aid in the development of an American merchant marine. It shall examine the navigation laws of the United States and the
rules and regulations thereunder, and make such recommendations to the Congress as it deems proper for the
amendment, improvement, and revision of such laws, and
for the development of the American merchant marine.
It shall investigate the legal status of mortgage loans on
vessel property, with a view to means of improving the
security of such loans and of encouraging investment in
American shipping.

It shall, on or before the first day of December in each
year, make a report to the Congress, which shall include
its recommendations and the results of its investigations,
a summary of its transactions, and a statement of all
expenditures and receipts under this act, and of the operations of any corporation in which the United States
is a stockholder, and the names and compensation of all
persons employed by the board.

SEC. 13. That for the purpose of carrying out the provisions of sections five and eleven no liability shall be
incurred exceeding a total of $50,000,000 and the Secre-
tary of the Treasury, upon the request of the board,
approved by the President, shall from time to time issue
and sell or use any of the bonds of the United States
now available in the Treasury under the acts of August
fifth, nineteen hundred and nine, February fourth, nineteen hundred and ten, and March second, nineteen hundred and eleven, relating to the issue of bonds for the
construction of the Panama Canal, to a total amount
not to exceed $50,000,000: *Provided,* That any bonds
issued and sold or used under the provisions of this sec-

Investigations
by board.
46 U.S.C. 811.
39 Stat. 732.
Cost of ship-
building.

Rules of con-
struction and
classification.

Marine
insurance.

Navigation
laws.

Vessel
mortgages.

Annual report
and recommen-
dations to
Congress.

Issuance of
$50,000,000 of
Panama Canal
bonds author-
ized.

36 Stat. L., 117,
193, 1013.

Proviso.

To be payable
within 50
years.

Funds of board
permanently
appropriated.

46 U.S.C. 812.
39 Stat. 733.
72 Stat. 574.
P.L. 85–626.
74 Stat. 253.
P.L. 86–542.
No carrier by
water—

To give de-
ferred rebates.

To use "fight-
ing ship."

To retaliate
against any
shipper.

To discrimi-
nate unjustly
or unfairly.

tion may be made payable at such time within fifty years
after issue as the Secretary of the Treasury may fix, instead of fifty years after the date of issue, as prescribed
in the act of August fifth, nineteen hundred and nine.

The proceeds of such bonds and the net proceeds of all
sales, charters, and leases of vessels and of sales of stock
made by the board, and all other moneys received by it
from any source, shall be covered into the Treasury to
the credit of the board, and are hereby permanently
appropriated for the purpose of carrying out the provisions of sections five and eleven.

SEC. 14.[3] That no common carrier by water shall, directly or indirectly, in respect to the transportation by
water of passengers or property between a port of a
State, Territory, District, or possession of the United
States and any other such port or a port of a foreign
country—

First. Pay, or allow, or enter into any combination,
agreement, or understanding, express or implied, to pay
or allow, a deferred rebate to any shipper. The term
"deferred rebate" in this Act means a return of any portion of the freight money by a carrier to any shipper as
a consideration for the giving of all or any portion of
his shipments to the same or any other carrier, or for any
other purpose, the payment of which is deferred beyond
the completion of the service for which it is paid, and is
made only if, during both the period for which computed
and the period of deferment, the shipper has complied
with the terms of the rebate agreement or arrangement.

Second. Use a fighting ship either separately or in
conjunction with any other carrier, through agreement
or otherwise. The term "fighting ship" in this Act
means a vessel used in a particular trade by a carrier or
group of carriers for the purpose of excluding, preventing, or reducing competition by driving another carrier
out of said trade.

Third. Retaliate against any shipper by refusing, or
threatening to refuse, space accommodations when such
are available, or resort to other discriminating or unfair
methods, because such shipper has patronized any other
carrier or has filed a complaint charging unfair treatment, or for any other reason.

Fourth. Make any unfair or unjustly discriminatory
contract with any shipper based on the volume of freight
offered, or unfairly treat or unjustly discriminate against
any shipper in the matter of (a) cargo space accommodations or other facilities, due regard being had for the
proper loading of the vessel and the available tonnage;
(b) the loading and landing of freight in proper condition; or (c) the adjustment and settlement of claims.

[3] See section 3 of Public Law 87–346 (as amended by Public Law
88–5), set out in footnote 6 to section 15 of this Act.

Any carrier who violates any provision of this section shall be guilty of a misdemeanor punishable by a fine of not more than $25,000 for each offense: *Provided*, That nothing in this section or elsewhere in this Act, shall be construed or applied to forbid or make unlawful any dual rate contract arrangement in use by the members of a conference on May 19, 1958, which conference is organized under an agreement approved under section 15 of this Act by the regulatory body administering this Act, unless and until such regulatory body disapproves, cancels, or modifies such arrangement in accordance with the standards set forth in section 15 of this Act. The term "dual rate contract arrangement" as used herein means a practice whereby a conference establishes tariffs of rates at two levels the lower of which will be charged to merchants who agree to ship their cargoes on vessels of members of the conference only and the higher of which shall be charged to merchants who do not so agree.[4]

Penalty.

Dual rate contract agreements.

46 U.S.C. 814.

Sec. 14a. The board upon its own initiative may, or upon complaint shall, after due notice to all parties in interest and hearing, determine whether any person, not a citizen of the United States and engaged in transportation by water of passengers or property—

46 U.S.C. 813.
39 Stat. 733.
41 Stat. 966.

(1) Has violated any provision of section 14, or

(2) Is a party to any combination, agreement, or understanding, express or implied, that involves in respect to transportation of passengers or property between foreign ports, deferred rebates or any other unfair practice designated in section 14, and that excludes from admission upon equal terms with all other parties thereto, a common carrier by water which is a citizen of the United States and which has applied for such admission.

Rebates, fighting ships, retaliation, unfair contracts.
Rebates or unfair practices between foreign ports.

---

[4] This proviso was added by Public Law 85–626, approved August 12, 1958, which reads as follows :

"*Be it enacted by the Senate and House of Representatives of the United States of America in Congress assembled*, That section 14 of the Shipping Act, 1916, is amended by inserting at the end thereof the following : 'Provided, That nothing in this section or elsewhere in this Act, shall be construed or applied to forbid or make unlawful any dual rate contract arrangement in use by the members of a conference on May 19, 1958, which conference is organized under an agreement approved under section 15 of this Act by the regulatory body administering this Act, unless and until such regulatory body disapproves, cancels, or modifies such arrangement in accordance with the standards set forth in section 15 of this Act. The term 'dual rate contract arrangement' as used herein means a practice whereby a conference establishes tariffs of rates at two levels the lower of which will be charged to merchants who agree to ship their cargoes on vessels of members of the conference only and the higher of which shall be charged to merchants who do not so agree.'

"Sec. 2. This Act shall be effective immediately upon enactment and shall cease to be effective on and after June 30, 1960."

The expiration date in section 2 of Public Law 85–626 was changed to June 30, 1961, by Public Law 86–542 and to September 15, 1961, by Public Law 87–75. Public Law 87–252 amended section 2 of Public Law 85–626 to read as follows :

"Sec. 2. This Act shall be effective immediately upon enactment and shall cease to be effective on and after October 15, 1961 : *Provided, however*, That contracts in effect midnight September 14, 1961, shall remain in effect until midnight October 15, 1961, unless such contracts terminate earlier by their own terms, or are rendered illegal under the terms of the first section of this Act."

Dual rate contract agreements.
72 Stat. 574.
74 Stat. 253.
46 U.S.C. 812 note.

Vessels may be denied entry.

If the board determines that any such person has violated any such provision or is a party to any such combination, agreement, or understanding, the board shall thereupon certify such fact to the Secretary of Commerce. The Secretary shall thereafter refuse such person the right of entry for any ship owned or operated by him or by any carrier directly or indirectly controlled by him, into any port of the United States, or any Territory, District, or possession thereof, until the board certifies that the violation has ceased or such combination, agreement, or understanding has been terminated.

Shipping Act, 1916, amendments.
40 Stat. 903.
46 U.S.C. 813a.
Foreign commerce.
Common carriers, dual rate contracts.
P.L. 87–346.
75 Stat. 762.

Sec. 14b.[5] Notwithstanding any other provisions of this Act, on application the Federal Maritime Commission (hereinafter "Commission"), shall, after notice, and hearing, by order, permit the use by any common carrier or conference of such carriers in foreign commerce of any contract, amendment, or modification thereof, which is available to all shippers and consignees on equal terms and conditions, which provides lower rates to a shipper or consignee who agrees to give all or any fixed portion of his patronage to such carrier or conference of carriers unless the Commission finds that the contract, amendment, or modification thereof will be detrimental to the commerce of the United States or contrary to the public interest, or unjustly discriminatory or unfair as between shippers, exporters, importers, or ports, or between exporters from the United States and their foreign competitors, and provided the contract, amendment, or modification thereof, expressly (1) permits prompt release of the contract shipper from the contract with respect to any shipment or shipments for which the contracting carrier or conference of carriers cannot provide as much space as the contract shipper shall require on reasonable notice; (2) provides that whenever a tariff rate for the carriage of goods under the contract becomes effective, insofar as it is under the control of the carrier or conference of carriers, it shall not be increased before a reasonable period, but in no case less than ninety days; (3) covers only those goods of the contract shipper as to the shipment of which he has the legal right at the time of shipment to select the carrier: *Provided, however,* That it shall be deemed a breach of the contract if, before the time of shipment and with the intent to avoid his obligation under the contract, the contract shipper divests himself, or with the same intent permits himself to be divested, of the legal right to select the carrier and the shipment is carried by a carrier which is not a party to the contract; (4) does not require the contract shipper to divert shipment of goods from natural

---

[5] See section 3 of Public Law 87–346 (as amended by Public Law 88–5), set out in footnote 6 to section 15 of this Act.

routings not served by the carrier or conference of carriers where direct carriage is available; (5) limits damages recoverable for breach by either party to actual damages to be determined after breach in accordance with the principles of contract law: *Provided, however,* That the contract may specify that in the case of a breach by a contract shipper the damages may be an amount not exceeding the freight charges computed at the contract rate on the particular shipment, less the cost of handling; (6) permits the contract shipper to terminate at any time without penalty upon ninety days' notice; (7) provides for a spread between ordinary rates and rates charged contract shippers which the Commission finds to be reasonable in all the circumstances but which spread shall in no event be more than 15 per centum of the ordinary rates; (8) excludes cargo of the contract shippers which is loaded and carried in bulk without mark or count except liquid bulk cargoes, other than chemicals, in less than full shipload lots: *Provided, however,* That upon finding that economic factors so warrant, the Commission may exclude from the contract any commodity subject to the foregoing exception; and (9) contains such other provisions not inconsistent herewith as the Commission shall require or permit. The Commission shall withdraw permission which it has granted under the authority contained in this section for the use of any contract if it finds, after notice and hearing, that the use of such contract is detrimental to the commerce of the United States or contrary to the public interest, or is unjustly discriminatory or unfair as between shippers, exporters, importers, or ports, or between exporters from the United States and their foreign competitors. The carrier or conference of carriers may on ninety days' notice terminate without penalty the contract rate system herein authorized, in whole or with respect to any commodity: *Provided, however,* That after such termination the carrier or conference of carriers may not reinstitute such contract rate system or part thereof so terminated without prior permission by the Commission in accordance with the provisions of this section. Any contract, amendment, or modification of any contract not permitted by the Commission shall be unlawful, and contracts, amendments, and modifications shall be lawful only when and as long as permitted by the Commission; before permission is granted or after permission is withdrawn it shall be unlawful to carry out in whole or in part, directly or indirectly, any such contract, amendment, or modification. As used in this section, the term "contract shipper" means a person other than a carrier or conference of carriers who is a party to a contract the use of which may be permitted under this section.

75 Stat. 763.

Notice and hearing.

39 Stat. 733.
46 U.S.C. 814.
Filing of agreements, etc.

P.L. 87–346.

75 Stat. 763.

Discriminatory
agreements, disapproval.

75 Stat. 764

SEC. 15.⁶ That every common carrier by water, or other person subject to this Act, shall file immediately with the Commission a true copy, or, if oral, a true and complete memorandum, of every agreement with another such carrier or other person subject to this Act, or modification or cancellation thereof, to which it may be a party or conform in whole or in part, fixing or regulating transportation rates or fares; giving or receiving special rates, accommodations, or other special privileges or advantages; controlling, regulating, preventing, or destroying competition; pooling or apportioning earnings, losses, or traffic; allotting ports or restricting or otherwise regulating the number and character of sailings between ports; limiting or regulating in any way the volume or character of freight or passenger traffic to be carried; or in any manner providing for an exclusive, preferential, or co-operative working arrangement. The term "agreement" in this section includes understandings, conferences, and other arrangements.

The Commission shall by order, after notice and hearing, disapprove, cancel or modify any agreement, or any modification or cancellation thereof, whether or not previously approved by it, that it finds to be unjustly discriminatory or unfair as between carriers, shippers, exporters, importers, or ports, or between exporters from the United States and their foreign competitors, or to operate to the detriment of the commerce of the United States, or to be contrary to the public interest, or to be in violation of this Act, and shall approve all other agreements, modifications, or cancellations. No such agreement shall be approved, nor shall continued approval be permitted for any agreement (1) between carriers not members of the same conference or conferences of carriers serving different trades that would otherwise be naturally competitive, unless in the case of agreements between carriers, each carrier, or in the case of agreements between conferences, each conference, retains the right of independent action, or (2) in respect to any conference agreement, which fails to provide reasonable and equal terms and conditions for admission and re-

77 Stat. 5
P.L. 88–5.

Existing
agreements,
modifications,
etc.

Dual rate contracts, extension.

---

⁶ Section 14b was added to the Act, and section 15 was amended to read as shown in the foregoing text, by sections 1 and 2 of Public Law 87–346 (except for the proviso of section 15 which was added by Public Law 88–275), approved Oct. 3, 1961. Section 3 of Public Law 87–346 (as amended by Public Law 88–5) reads as follows :
"SEC. 3. Notwithstanding the provisions of sections 14, 14b, and 15, Shipping Act, 1916, as amended by this Act, all existing agreements which are lawful under the Shipping Act, 1916, immediately prior to enactment of this Act, shall remain lawful unless disapproved, canceled, or modified by the Commission pursuant to the provisions of the Shipping Act, 1916, as amended by this Act: *Provided, however,* That all such existing agreements which are rendered unlawful by the provisions of such Act as hereby amended must be amended to comply with the provisions of such Act as hereby amended, and if such amendments are filed for approval within six months after the enactment of this Act, such agreements so amended shall be lawful for a further period but not beyond April 3, 1964. Within such period the Commission shall approve, disapprove, cancel or modify all such agreements and amendments in accordance with the provisions of this Act.

admission to conference membership of other qualified carriers in the trade, or fails to provide that any member may withdraw from membership upon reasonable notice without penalty for such withdrawal.

The Commission shall disapprove any such agreement, after notice and hearing, on a finding of inadequate policing of the obligations under it, or of failure or refusal to adopt and maintain reasonable procedures for promptly and fairly hearing and considering shippers' requests and complaints.

Any agreement and any modification or cancellation of any agreement not approved, or disapproved, by the Commission shall be unlawful, and agreements, modifications, and cancellations shall be lawful only when and as long as approved by the Commission; before approval or after disapproval it shall be unlawful to carry out in whole or in part, directly or indirectly, any such agreement, modification, or cancellation; except that tariff rates, fares, and charges, and classifications, rules, and regulations explanatory thereof (including changes in special rates and charges covered by section 14b of this Act which do not involve a change in the spread between such rates and charges and the rates and charges applicable to noncontract shippers) agreed upon by approved conferences, and changes and amendments thereto, if otherwise in accordance with law, shall be permitted to take effect without prior approval upon compliance with the publication and filing requirements of section 18(b) hereof and with the provisions of any regulations the Commission may adopt. 75 Stat. 762.

Infra.

Every agreement, modification, or cancellation lawful under this section, or permitted under section 14b, shall be excepted from the provisions of the Act approved July 2, 1890, entitled "An Act to protect trade and commerce against unlawful restraints and monopolies," and amendments and Acts supplementary thereto, and the provisions of sections 73 to 77, both inclusive, of the Act approved August 27, 1894, entitled "An Act to reduce taxation, to provide revenue for the Government, and for other purposes," and amendments and Acts supplementary thereto. 26 Stat. 209.
15 U.S.C. 1–7.

28 Stat. 570.
15 U.S.C. 8–11.

Whoever violates any provision of this section or of section 14b shall be liable to a penalty of not more than $1,000 for each day such violation continues, to be recovered by the United States in a civil action: *Provided, however,* That the penalty provisions of this section shall not apply to leases, licenses, assignments, or other agreements of similar character for the use of terminal property or facilities which were entered into before the date of enactment of this Act, and, if continued in effect beyond said date, submitted to the Federal Maritime Commission for approval prior to or within ninety days after the enactment of this Act, unless such leases, Shipping Act,
amendment.
Terminal leases
75 Stat. 763.
78 Stat. 148.
P.L. 88–275.

licenses, assignments, or other agreements for the use of terminal facilities are disapproved, modified, or canceled by the Commission and are continued in operation without regard to the Commission's action thereon. The Commission shall promptly approve, disapprove, cancel, or modify each such agreement in accordance with the provisions of this section.

46 U.S.C. 815.
39 Stat. 734.
False billing.

SEC. 16. That it shall be unlawful for any shipper consignor, consignee, forwarder, broker, or other person, or any officer, agent, or employee thereof, knowingly and wilfully, directly or indirectly, by means of false billing, false classification, false weighing, false report of weight, or by any other unjust or unfair device or means to obtain or attempt to obtain transportation by water for property at less than the rates or charges which would otherwise be applicable.

That it shall be unlawful for any common carrier by water, or other person subject to this Act, either alone or in conjunction with any other person, directly or indirectly:

Undue preference or advantage.

First. To make or give any undue or unreasonable preference or advantage to any particular person, locality, or description of traffic in any respect whatsoever, or to subject any particular person, locality, or description of traffic to any undue or unreasonable prejudice or disadvantage in any respect whatsoever: *Provided*, That within thirty days after enactment of this Act, or within thirty days after the effective date or the filing with the Commission, whichever is later, of any conference freight rate, rule, or regulation in the foreign commerce of the United States, the Governor of any State, Commonwealth, or possession of the United States may file a protest with the Commission upon the ground that the rate, rule, or regulation unjustly discriminates against that State, Commonwealth, or possession of the United States, in which case the Commission shall issue an order to the conference to show cause why the rate, rule, or regulation should not be set aside. Within one hundred and eighty days from the date of the issuance of such order, the Commission shall determine whether or not such rate, rule, or regulation is unjustly discriminatory and issue a final order either dismissing the protest, or setting aside the rate, rule, or regulation.

Filing
of protests.
49 Stat. 1518.
75 Stat. 766.
P.L. 87–346.

Unfair means
to obtain lower
rates.

Second. To allow any person to obtain transportation for property at less than the regular rates or charges then established and enforced on the line of such carrier by means of false billing, false classification, false weighing, false report of weight, or by any other unjust or unfair device or means.

To influence
insurance companies to discriminate.

Third. To induce, persuade, or otherwise influence any marine insurance company or underwriter, or agent thereof, not to give a competing carrier by water as favorable a rate of insurance on vessel or cargo, having due regard to the class of vessel or cargo, as is granted to such carrier or other person subject to this Act.

Whoever violates any provision of this section shall be guilty of a misdemeanor punishable by a fine of not more than $5,000 for each offense.

SEC. 17. That no common carrier by water in foreign commerce shall demand, charge, or collect any rate, fare, or charge which is unjustly discriminatory between shippers or ports, or unjustly prejudicial to exporters of the United States as compared with their foreign competitors. Whenever the board finds that any such rate, fare, or charge is demanded, charged, or collected it may alter the same to the extent necessary to correct such unjust discrimination or prejudice and make an order that the carrier shall discontinue demanding, charging, or collecting any such unjustly discriminatory or prejudicial rate, fare, or charge.

46 U.S.C. 816.
39 Stat. 734.

Common carriers in foreign commerce.

Not to charge discriminatory rates or rates prejudicial to American exporters.

Every such carrier and every other person subject to this act shall establish, observe, and enforce just and reasonable regulations and practices relating to or connected with the receiving, handling, storing, or delivering of property. Whenever the board finds that any such regulation or practice is unjust or unreasonable it may determine, prescribe, and order enforced a just and reasonable regulation or practice.

SEC. 18(a) That every common carrier by water in interstate commerce shall establish, observe, and enforce just and reasonable rates, fares, charges, classifications, and tariffs, and just and reasonable regulations and practices relating thereto and to the issuance, form, and substance of tickets, receipts, and bills of lading, the manner and method of presenting, marking, packing, and delivering property for transportation, the carrying of personal, sample, and excess baggage, the facilities for transportation, and all other matters relating to or connected with the receiving, handling, transporting, storing, or delivering of property.

46 U.S.C. 817.
39 Stat. 735.

Common carriers in interstate commerce:
Must observe just and reasonable classifications, rates, fares, practices, etc.

Every such carrier shall file with the board and keep open to public inspection, in the form and manner and within the time prescribed by the board, the maximum rates, fares, and charges for or in connection with transportation between points on its own route; and if a through route has been established, the maximum rates, fares, and charges for or in connection with transportation between points on its own route and points on the route of any other carrier by water.

No such carrier shall demand, charge, or collect a greater compensation for such transportation than the rates, fares, and charges filed in compliance with this section, except with the approval of the board and after ten days' public notice in the form and manner prescribed by the board, stating the increase proposed to be made;

but the board for good cause shown may waive such notice.

Whenever the board finds that any rate, fare, charge, classification, tariff, regulation, or practice, demanded, charged, collected, or observed by such carriers is unjust or unreasonable, it may determine, prescribe, and order enforced a just and reasonable maximum rate, fare, or charge, or a just and reasonable classification, tariff, regulation, or practice.

(b)(1) From and after ninety days following enactment hereof every common carrier by water in foreign commerce and every conference of such carriers shall file with the Commission and keep open to public inspection tariffs showing all the rates and charges of such carrier or conference of carriers for transportation to and from United States ports and foreign ports between all points on its own route and on any through route which has been established. Such tariffs shall plainly show the places between which freight will be carried, and shall contain the classification of freight in force, and shall also state separately such terminal or other charge, privilege, or facility under the control of the carrier or conference of carriers which is granted or allowed, and any rules or regulations which in anywise change, affect, or determine any part or the aggregate of such aforesaid rates, or charges, and shall include specimens of any bill of lading, contract of affreightment, or other document evidencing the transportation agreement. Copies of such tariffs shall be made available to any person and a reasonable charge may be made therefor. The requirements of this section shall not be applicable to cargo loaded and carried in bulk without mark or count, or to cargo which is softwood lumber. As used in this paragraph, the term "softwood lumber" means softwood lumber not further manufactured than passing lengthwise through a standard planing machine and crosscut to length, logs, poles, piling, and ties, including such articles preservatively treated, or bored, or framed, but not including plywood or finished articles knocked down or set up.

(2) No change shall be made in rates, charges, classifications, rules or regulations, which results in an increase in cost to the shipper, nor shall any new or initial rate of any common carrier by water in foreign commerce or conference of such carriers be instituted, except by the publication, and filing, as aforesaid, of a new tariff or tariffs which shall become effective not earlier than thirty days after the date of publication and filing thereof with the Commission, and each such tariff or tariffs shall plainly show the changes proposed to be made in the tariff or tariffs then in force and the time when the rates, charges, classifications, rules or regulations as changed are to become effective: *Provided, however*, That the Commission may, in its discretion and for good cause,

---

Marginal notes:

Board may prescribe reasonable maximum rates, etc.

Filing of carrier rates, etc.

75 Stat. 764.

77 Stat. 129.

Lumber. Tariff filing requirements, exclusion.

77 Stat. 129.
P.L. 88–103.
79 Stat. 1124.
P.L. 89–303.
Rate changes.

P.L. 87–346.

allow such changes and such new or initial rates to become effective upon less than the period of thirty days herein specified. Any change in the rates, charges, or classifications, rules or regulations which results in a decreased cost to the shipper may become effective upon the publication and filing with the Commission. The term "tariff" as used in this paragraph shall include any amendment, supplement or reissue.

"Tariff."

(3) No common carrier by water in foreign commerce or conference of such carriers shall charge or demand or collect or receive a greater or less or different compensation for the transportation of property or for any service in connection therewith than the rates and charges which are specified in its tariffs on file with the Commission and duly published and in effect at the time; nor shall any such carrier rebate, refund, or remit in any manner or by any device any portion of the rates or charges so specified, nor extend or deny to any person any privilege or facility, except in accordance with such tariffs.

Collection of specified rates only.

(4) The Commission shall by regulations prescribe the form and manner in which the tariffs required by this section shall be published and filed; and the Commission is authorized to reject any tariff filed with it which is not in conformity with this section and with such regulations. Upon rejection by the Commission, a tariff shall be void and its use unlawful.

(5) The Commission shall disapprove any rate or charge filed by a common carrier by water in the foreign commerce of the United States or conference of carriers which, after hearing, it finds to be so unreasonably high or low as to be detrimental to the commerce of the United States.

(6) Whoever violates any provision of this section shall be liable to a penalty of not more than $1,000 for each day such violation continues, to be recovered by the United States in a civil action.

Penalty.

SEC. 19. That whenever a common carrier by water in interstate commerce reduces its rates on the carriage of any species of freight to or from competitive points below a fair and remunerative basis with the intent of driving out or otherwise injuring a competitive carrier by water, it shall not increase such rates unless after hearing the board finds that such proposed increase rests upon changed conditions other than the elimination of said competition.

46 U.S.C. 818. 39 Stat. 735.

Restriction on increase of rates reduced to drive out competitor.

SEC. 20. That it shall be unlawful for any common carrier by water or other person subject to this Act, or any officer, receiver, trustee, lessee, agent, or employee of such carrier or person, or for any other person authorized by such carrier or person to receive information knowingly to disclose to or permit to be acquired by any person other than the shipper or consignee, without

46 U.S.C. 819. 39 Stat. 735.

the consent of such shipper or consignee, any information concerning the nature, kind, quantity, destination, consignee, or routing of any property tendered or delivered to such common carrier or other person subject to this act for transportation in interstate or foreign commerce, which information may be used to the detriment or prejudice of such shipper or consignee, or which may improperly disclose his business transactions to a competitor, or which may be used to the detriment or prejudice of any carrier; and it shall also be unlawful for any person to solicit or knowingly receive any such information which may be so used.

Nothing in this Act shall be construed to prevent the giving of such information in response to any legal process issued under the authority of any court, or to any officer or agent of the Government of the United States, or of any State, Territory, District, or possession thereof, in the exercise of his powers, or to any officer or other duly authorized person seeking such information for the prosecution of persons charged with or suspected of crime, or to another carrier, or its duly authorized agent, for the purpose of adjusting mutual traffic accounts in the ordinary course of business of such carriers; or to prevent any common carrier by water which is a party to a conference agreement approved pursuant to section 15 of this Act, or any other person subject to this Act, or any receiver, trustee, lessee, agent, or employee of such carrier or person, or any other person authorized by such carrier to receive information, from giving information to the conference or any person, firm, corporation, or agency designated by the conference, or to prevent the conference or its designee from soliciting or receiving information for the purpose of determining whether a shipper or consignee has breached an agreement with the conference or its member lines or of determining whether a member of the conference has breached the conference agreement, or for the purpose of compiling statistics of cargo movement, but the use of such information for any other purpose prohibited by this Act or any other Act shall be unlawful.

SEC. 21. That the board may require any common carrier by water, or other person subject to this Act, or any officer, receiver, trustee, lessee, agent, or employee thereof, to file with it any periodical or special report, or any account, record, rate, or charge, or any memorandum of any facts and transactions appertaining to the business of such carrier or other person subject to this Act. Such report, account, record, rate, charge, or memorandum shall be under oath whenever the board so requires, and shall be furnished in the form and within the time prescribed by the board. Whoever fails to file any report, account, record, rate, charge, or memorandum as re-

---

Marginal notes:

Information detrimental to shipper or consignee not to be disclosed, solicited, etc.

Giving of information on legal process, etc.

In adjusting traffic accounts, etc. Filing of reports, records, rates, etc., may be required of any person subject to the act.

75 Stat. 765.
P.L. 87-346.

46 U.S.C. 820.
39 Stat. 736.

Penalty for failure to file.

quired by this section shall forfeit to the United States the sum of $100 for each day of such default.

Whoever willfully falsifies, destroys, mutilates, or alters any such report, account, record, rate, charge, or memorandum, or willfully files a false report, account, record, rate, charge, or memorandum shall be guilty of a misdemeanor, and subject upon conviction to a fine of not more than $1,000, or imprisonment for not more than one year, or to both such fine and imprisonment. *[Penalty for filing false reports, etc.]*

SEC. 22. That any person may file with the board a sworn complaint setting forth any violation of this Act by a common carrier by water, or other person subject to this Act, and asking reparation for the injury, if any, caused thereby. The board shall furnish a copy of the complaint to such carrier or other person, who shall within a reasonable time specified by the board satisfy the complaint or answer it in writing. If the complaint is not satisfied the board shall, except as otherwise provided in this Act, investigate it in such manner and by such means, and make such order as it deems proper. The board, if the complaint is filed within two years after the cause of action accrued, may direct the payment, on or before a day named, of full reparation to the complainant for the injury caused by such violation. *[46 U.S.C. 821. 39 Stat. 736. Complaint of any violation may be filed by any person.] [Remedy for violations.]*

The board, upon its own motion, may in like manner and, except as to orders for the payment of money, with the same powers, investigate any violation of this Act. *[Investigations by board on own motion.]*

SEC. 23. Orders of the board relating to any violation of this Act shall be made only after full hearing, and upon a sworn complaint or in proceedings instituted of its own motion. *[46 U.S.C. 822. 39 Stat. 736.]*

All orders of the United States Maritime Commission, other than for the payment of money, made under this Act, as amended or supplemented, shall continue in force until its further order, or for a specified period of time, as shall be prescribed in the order, unless the same shall be suspended, or modified, or set aside by the Commission, or be suspended or set aside by a court of competent jurisdiction. *[53 Stat. 1182. Orders of Commission.]*

SEC. 24. That the board shall enter of record a written report of every investigation made under this Act in which a hearing has been held, stating its conclusions, decision, and order, and, if reparation is awarded, the findings of fact on which the award is made, and shall furnish a copy of such report to all parties to the investigation. *[46 U.S.C. 823. 39 Stat. 736. Written reports of hearings to be kept.]*

The board may publish such reports in the form best adapted for public information and use, and such authorized publications shall, without further proof or authentication, be competent evidence of such reports in all courts of the United States and of the States, Territories, Districts, and possessions thereof. *[Such reports as evidence.]*

46 U.S.C. 824.
39 Stat. 736.

Board may reverse, modify, etc., orders and grant rehearings.

SEC. 25. That the board may reverse, suspend, or modify, upon such notice and in such manner as it deems proper, any order made by it. Upon application of any party to a decision or order it may grant a rehearing of the same or any matter determined therein, but no such application for or allowance of a rehearing shall, except by special order of the board, operate as a stay of such order.

46 U.S.C. 825.
39 Stat. 737.

Investigation of discriminations by foreign Governments against American vessels.

SEC. 26. The board shall have power, and it shall be its duty whenever complaint shall be made to it, to investigate the action of any foreign Government with respect to the privileges afforded and burdens imposed upon vessels of the United States engaged in foreign trade, whenever it shall appear that the laws, regulations, or practices of any foreign Government operate in such a manner that vessels of the United States are not accorded equal privileges in foreign trade with vessel of such foreign countries or vessels of other foreign countries, either in trade to or from the ports of such foreign country or in respect of the passage or transportation through such foreign country of passengers or goods intended for shipment or transportation in such vessels of the United States, either to or from ports of such foreign country or to or from ports of other foreign countries. It shall be the duty of the board to report

Board to report on such to President.

the results of its investigation to the President with its recommendations and the President is hereby authorized and empowered to secure by diplomatic action equal privileges for vessels of the United States engaged in such foreign trade. And if by such diplomatic action the President shall be unable to secure such equal privileges, then the President shall advise Congress as to the facts and his conclusions by a special message, if deemed

President to secure equal privileges for American vessels.
46 U.S.C. 826.
39 Stat. 737.

important in the public interest, in order that proper action may be taken thereon.

Power to subpoena witnesses, etc.

SEC. 27. That for the purpose of investigating alleged violations of this Act, the board may by subpoena compel the attendance of witnesses and the production of books, papers, documents, and other evidence from any place in the United States at any designated place of hearing. Subpoenas may be signed by any commissioner, and oaths or affirmations may be administered, witnesses examined, and evidence received by any commissioner or examiner, or, under the direction of the board, by any person authorized under the laws of the United States or of any State, Territory, District, or possession thereof to administer oaths. Persons so acting under the direction of the

Fees of witnesses.

board and witnesses shall, unless employees of the board, be entitled to the same fees and mileage as in the courts of the United States. Obedience to any such subpoena

Enforcement of subpoenas.

shall, on application by the board, be enforced as are orders of the board other than for the payment of money.

SEC. 28. That no person shall be excused on the ground that it may tend to incriminate him or subject him to a penalty or forfeiture, from attending and testifying, or producing books, papers, documents, and other evidence, in obedience to the subpœna of the board or of any court in any proceeding based upon or growing out of any alleged violation of this Act; but no natural person shall be prosecuted or subjected to any penalty or forfeiture for or on account of any transaction, matter, or thing as to which, in obedience to a subpœna and under oath, he may so testify or produce evidence, except that no person shall be exempt from prosecution and punishment for perjury committed in so testifying.

46 U.S.C. 827.
39 Stat. 737.

Giving of self-incriminating evidence by witnesses may be enforced.

But no witness to be prosecuted on such evidence.

SEC. 29. That in case of violation of any order of the board, other than an order for the payment of money, the board, or any party injured by such violation, or the Attorney General, may apply to a district court having jurisdiction of the parties; and if, after hearing, the court determines that the order was regularly made and duly issued, it shall enforce obedience thereto by a writ of injunction or other proper process, mandatory or otherwise.

46 U.S.C. 828.
39 Stat. 737.
Enforcement of orders.

Where suits to be filed.
(a) Other than for payment of money.

SEC. 30. That in case of violation of any order of the board for the payment of money the person to whom such award was made may file in the district court for the district in which such person resides, or in which is located any office of the carrier or other person to whom the order was directed, or in which is located any point of call on a regular route operated by the carrier, or in any court of general jurisdiction of a State, Territory, District, or possession of the United States having jurisdiction of the parties, a petition or suit setting forth briefly the causes for which he claims damages and the order of the board in the premises.

(b) For payment of money.

46 U.S.C. 829.
39 Stat. 737.

In the district court the findings and order of the board shall be prima facie evidence of the facts therein stated, and the petitioner shall not be liable for costs, nor shall he be liable for costs of any subsequent stage of the proceedings unless they accrue upon his appeal. If a petitioner in a district court finally prevails, he shall be allowed a reasonable attorney's fee, to be taxed and collected as part of the costs of the suit.

Findings and order of board as evidence.

Costs and attorney's fees.

All parties in whose favor the Board has made an award of reparation by a single order may be joined as plaintiffs, and all other parties to such order may be joined as defendants, in a single suit in any district in which any one such plaintiff could maintain a suit against any one such defendant. Service of process against any such defendant and not found in that district may be made in any district in which is located any office of, or point of call on a regular route operated by, such defendant. Judgment may be entered in favor

Joinder of parties permitted.

Service of process.

of any plaintiff against the defendant liable to that plaintiff.

Time for filing of suits.

No petition or suit for the enforcement of an order for the payment of moneys shall be maintained unless filed within one year from the date of the order.

46 U.S.C. 880.
39 Stat. 738.

Venue and procedure.

SEC. 31. That the venue and procedure in the courts of the United States in suits brought to enforce, suspend, or set aside, in whole or in part, any order of the board shall, except as herein otherwise provided, be the same as in similar suits in regard to orders of the Interstate Commerce Commission, but such suits may also be maintained in any district court having jurisdiction of the parties.

46 U.S.C. 881.
39 Stat. 738.

General penalty provision.

SEC. 32. That whoever violates any provision of this Act, except where a different penalty is provided, shall be guilty of a misdemeanor, punishable by fine not to exceed $5,000.

46 U.S.C. 882.
39 Stat. 788.

Jurisdiction of Interstate Commerce Commission not to be encroached upon.

SEC. 33. That this Act shall not be construed to affect the power or jurisdiction of the Interstate Commerce Commission, nor to confer upon the board concurrent power or jurisdiction over any matter within the power or jurisdiction of such commission; nor shall this Act be construed to apply to intrastate commerce.

46 U.S.C. 883.
39 Stat. 738.

Effect of decisions on constitutionality.

SEC. 34. That if any provision of this Act, or the application of such provision to certain circumstances, is held unconstitutional, the remainder of the Act, and the application of such provision to circumstances other than those as to which it is held unconstitutional, shall not be affected thereby.

Section 35 was repealed by section 903(a) of the Merchant Marine Act, 1936, supra.

46 U.S.C. 834.
39 Stat. 738.

Clearance may be denied vessel for refusal to take cargo offered by citizen of United States.

SEC. 36. The Secretary of the Treasury is authorized to refuse a clearance to any vessel or other vehicle laden with merchandise destined for a foreign or domestic port whenever he shall have satisfactory reason to believe that the master, owner, or other officer of such vessel or other vehicle refuses or declines to accept or receive freight or cargo in good condition tendered for such port of destination or for some intermediate port of call, together with the proper freight or transportation charges therefor, by any citizen of the United States, unless the same is fully laden and has no space accommodations for the freight or cargo so tendered, due regard being had for the proper loading of such vessel or vehicle or unless such freight or cargo consists of merchandise for which such vessel or vehicle is not adaptable.

SEC. 37.[7] That when the United States is at war or
during any national emergency, the existence of which is
declared by proclamation of the President, it shall be
unlawful, without first obtaining the approval of the
board.

46 U.S.C. 835.
40 Stat. 901.

During war or
emergency.

(a) To transfer to or place under any foreign registry or flag any vessel owned in whole or in part by any
person a citizen of the United States or by corporation
organized under the laws of the United States, or of any
State, Territory, District, or possession thereof; or

No vessel to be
transferred to
foreign
registry.

(b) To sell, mortgage, lease, charter deliver, or in any
manner transfer, or agree to sell, mortgage, lease, charter, deliver or in any manner transfer to any person not
a citizen of the United States, (1) any such vessel or any
interest therein, or (2) any vessel documented under the
laws of the United States, or any interest therein, or (3)
any shipyard dry dock, ship-building, or ship-repairing
plant or facilities, or any interest therein; or

No vessel, ship-
yard, etc., to
be sold mort-
gaged, to
foreigner.
No contract to
construct for
foreign
account.

(c) To issue, transfer, or assign a bond, note, or other
evidence of indebtedness which is secured by a mortgage
of a vessel to a trustee or by an assignment to a trustee of
the owner's right, title, or interest in a vessel under construction, or by a mortgage to a trustee on a shipyard,
drydock, or ship-building or ship-repairing plant or
facilities, to a person not a citizen of the United States,
unless the trustee or a substitute trustee of such mortgage
or assignment is approved by the Secretary of Commerce:
Provided, however, That the Secretary of Commerce
shall grant his approval if such trustee or a substitute
trustee is a bank or trust company which (1) is organized
as a corporation, and is doing business, under the laws of
the United States or any State thereof, (2) is authorized
under such laws to exercise corporate trust powers, (3) is
a citizen of the United States, (4) is subject to supervision or examination by Federal or State authority, and
(5) has a combined capital and surplus (as set forth in
its most recent published report of condition) of at least
$3,000,000; or for the trustee or substitute trustee ap-
proved by the Secretary of Commerce to operate said
vessel under the mortgage or assignment: *Provided
further*, That if such trustee or a substitute trustee at any
time ceases to meet the foregoing qualifications, the Secretary of Commerce shall disapprove such trustee or sub-

P.L. 89–346.
79 Stat. 1305.

---

[7] The proclamation by the President of a national emergency in World
War I as provided herein was made Aug. 7, 1918.
 The state of war with respect to World War II and the national emergencies proclaimed by the President on Sept. 8, 1939 and May 27, 1941
were terminated July 25, 1947 for the purposes of this section by the Act
of July 25, 1947 (Public Law 239, 80th Congress; 61 Stat. 449).
 The President on December 16, 1950, issued a proclamation (No. 2914)
declaring the existence of a national emergency (15 F.R. 9029).
 See also section 27A of the Merchant Marine Act, 1920, as amended,
supra (Public Law 85–902, 72 Stat. 1736) regarding the term "citizen
of the United States".
 Subsection (c) of the first paragraph hereof and the third paragraph
hereof were added by Public Law 89–346 (79 Stat. 1305), p. 282 herein.
Section 4 of that Act contains provisions with respect to transfers to
noncitizens prior to such enactment or within one year thereafter.

stitute trustee, and after such disapproval the transfer or assignment of such bond, note, or other evidence of indebtedness to a person not a citizen of the United States, without the approval of the Secretary of Commerce, shall be unlawful; or

(d) To enter into any contract, agreement, or understanding to construct a vessel within the United States for or to be delivered to any person not a citizen of the United States, without expressly stipulating that such construction shall not begin until after the war or emergency proclaimed by the President has ended; or

(e) To make any agreement or effect any understanding whereby there is vested in or for the benefit of any person not a citizen of the United States, the controlling interest or a majority of the voting power in a corporation which is organized under the laws of the United States, or of any State, Territory, District, or possession thereof, and which owns any vessel, shipyard, drydock, or ship-building or ship-repairing plant or facilities; or

No agreement to vest control in foreigners.

(f) To cause or procure any vessel constructed in whole or in part within the United States, which has never cleared for any foreign port, to depart from a port of the United States before it has been documented under the laws of the United States.

No American-built vessels to depart until documented.

Whoever violates, or attempts or conspires to violate, any of the provisions of this section shall be guilty of a misdemeanor, punishable by a fine of not more than $5,000 or by imprisonment for not more than five years, or both.

Penalty.

If a bond, note, or other evidence of indebtedness which is secured by a mortgage of a vessel to a trustee or by an assignment to a trustee of the owner's right, title, or interest in a vessel under construction, or by a mortgage to a trustee on a shipyard, drydock or ship-building or ship-repairing plant or facilities, is issued, transferred, or assigned to a person not a citizen of the United States in violation of subsection c of this section, the issuance, transfer or assignment shall be void.

P.L. 89–346. 79 Stat. 1305.

Any vessel, shipyard, drydock, ship-building or ship-repairing plant or facilities, or interest therein, sold, mortgaged, leased, chartered, delivered, transferred, or documented, or agreed to be sold, mortgaged, leased, chartered, delivered, transferred, or documented, in violation of any of the provisions of this section, and any stocks, bonds, or other securities sold or transferred, or agreed to be sold or transferred, in violation of any of such provisions, or any vessel departing in violation of the provisions of subdivision (e), shall be forfeited to the United States.

Any such sale, mortgage, lease, charter, delivery, transfer, documentation, or agreement therefor shall be void, whether made within or without the United States, and any consideration paid therefor or deposited in connection therewith shall be recoverable at the suit of the person who has paid or deposited the same, or of his successors or assigns, after the tender of such vessel, shipyard, drydock, ship-building or ship-repairing plant or facilities, or interest therein, or of such stocks, bonds, or other securities, to the person entitled thereto, or after forfeiture thereof to the United States, unless the person to whom the consideration was paid, or in whose interest it was deposited, entered into the transaction in the honest belief that the person who paid or deposited such consideration was a citizen of the United States.

SEC. 38. That all forfeitures incurred under the provisions of this Act may be prosecuted in the same court, and may be disposed of in the same manner, as forfeitures incurred for offenses against the law relating to the collection of duties.

SEC. 39. That in any action or proceeding under the provisions of this Act to enforce a forfeiture the conviction in a court of criminal jurisdiction of any person for a violation thereof with respect to the subject of the forfeiture shall constitute prima facie evidence of such violation against the person so convicted.

SEC. 40. That whenever any bill of sale, mortgage, hypothecation, or conveyance of any vessel, or part thereof, or interest therein, is presented to any collector of the customs to be recorded, the vendee, mortgagee, or transferee shall file therewith a written declaration in such form as the board may by regulation prescribe, setting forth the facts relating to his citizenship, and such other facts as the board requires, showing that the transaction does not involve a violation of any of the provisions of section nine or thirty-seven. Unless the board, before such presentation, has failed to prescribe such form, no such bill of sale, mortgage, hypothecation, or conveyance shall be valid against any person whatsoever until such declaration has been filed. Any declaration filed by or in behalf of a corporation shall be signed by the president, secretary, or treasurer thereof, or any other official thereof duly authorized by such corporation to execute any such declaration.

Whoever knowingly makes any false statement of a material fact in any such declaration shall be guilty of a misdemeanor and subject to a fine of not more than $5,000, or to imprisonment for not more than five years, or both.

Marginal notes:

Forfeiture. Contracts and agreements in violation of section void. Consideration may be recovered.

46 U.S.C. 836. 40 Stat. 902.

Prosecutions of forfeitures. 19 U.S.C. 1618.

46 U.S.C. 837. 40 Stat. 902.

Prima facie evidence.

46 U.S.C. 838. 40 Stat. 902.

Transferee must file declaration as to citizenship, etc., with collector of customs.

Penalty for false statement.

46 U.S.C. 839.
40 Stat. 902.

Board may
approve trans-
actions condi-
tionally.

Penalty for
breach of
conditions.

SEC. 41. That whenever by said section nine or thirty-seven the approval of the board is required to render any act or transaction lawful, such approval may be accorded either absolutely or upon such conditions as the board prescribes. Whenever the approval of the board is accorded upon any condition a statement of such condition shall be entered upon its records and incorporated in the same document or paper which notifies the applicant of such approval. A violation of such condition so incorporated shall constitute a misdemeanor and shall be punishable by fine and imprisonment in the same manner, and shall subject the vessel, stocks, bonds, or other subject matter of the application conditionally approved to forfeiture in the same manner, as though the Act conditionally approved had been done without the approval of the board, but the offense shall be deemed to have been committed at the time of the violation of the condition.

Penalty for
making false
statement of
fact to secure
board's ap-
proval.

Whenever by this Act the approval of the board is required to render any act or transaction lawful, whoever knowingly makes any false statement of a material fact to the board, or to any member thereof, or to any officer, attorney, or agent thereof, for the purpose of securing such approval, shall be guilty of a misdemeanor and subject to a fine of not more than $5,000, or to imprisonment for not more than five years, or both.

46 U.S.C. 840.
40 Stat. 903.

Vessels to be
considered
documented
until registry,
enrollment, or
license is can-
celed by board.

SEC. 42. That any vessel registered, enrolled, or licensed under the laws of the United States shall be deemed to continue to be documented under the laws of the United States within the meaning of subdivision (b) of section thirty-seven, until such registry, enrollment, or license is surrendered with the approval of the board, the provisions of any other Act of Congress to the contrary notwithstanding.

Rules and
regulations.
P.L. 87–346.
75 Stat. 766.
46 U.S.C. 841a.

SEC. 43. The Commission shall make such rules and regulations as may be necessary to carry out the provisions of this Act.

SEC. 44. (a) No person shall engage in carrying on the business of forwarding as defined in this Act unless such person holds a license issued by the Federal Maritime Commission to engage in such business: *Provided, however*, That a person whose primary business is the sale of merchandise may dispatch shipments of such merchandise without a license.

46 U.S.C. 841b.

Issuance.

(b) A forwarder's license shall be issued to any qualified applicant therefor if it is found by the Commission that the applicant is, or will be, an independent ocean freight forwarder as defined in this Act and is fit, willing, and able properly to carry on the business of forwarding and to conform to the provisions of this Act and the requirements, rules, and regulations of the Commission issued thereunder, and that the proposed forwarding business is, or will be, consistent with the national maritime policies declared in the Merchant Marine Act, 1936;

otherwise such application shall be denied. Any inde- 49 Stat. 1985.
pendent ocean freight forwarder who, on the effective 46 U.S.C. 1245.
date of this Act, is carrying on the business of forwarding under a registration number issued by the Commission may continue such business for a period of one hundred and twenty days thereafter without a license, and if
application for such license is made within such period,
such forwarder may, under such regulations as the Commission shall prescribe, continue such business until
otherwise ordered by the Commission.

(c) The Commission shall prescribe reasonable rules Conditions.
and regulations to be observed by independent ocean
freight forwarders and no such license shall be issued or
remain in force unless such forwarder shall have furnished a bond or other security approved by the Commission in such form and amount as in the opinion of the
Commission will insure financial responsibility and the
supply of the services in accordance with contracts, agreements, or arrangements therefor.

(d) Licenses shall be effective from the date specified
therein, and shall remain in effect until suspended or
terminated as herein provided. Any such license may,
upon application of the holder thereof, in the discretion
of the Commission, be amended or revoked, in whole or in
part, or may upon complaint, or on the Commission's 75 Stat. 523.
own initiative, after notice and hearing, be suspended or
revoked for willful failure to comply with any provision
of this Act, or with any lawful order, rule, or regulation
of the Commission promulgated thereunder.

(e) A common carrier by water may compensate a per- Compensation
son carrying on the business of forwarding to the extent of licensees.
of the value rendered such carrier in connection with any
shipment dispatched on behalf of others when, and only
when, such person is licensed hereunder and has performed with respect to such shipment the solicitation
and securing of the cargo for the ship or the booking of,
or otherwise arranging for space for, such cargo, and at
least two of the following services:

(1) The coordination of the movement of the cargo
to shipside;

(2) The preparation and processing of the ocean
bill of lading;

(3) The preparation and processing of dock re- P.L. 87-254.
ceipts or delivery orders;

(4) The preparation and processing of consular
documents or export declarations;

(5) The payment of the ocean freight charges on
such shipments:

Provided, however, That where a common carrier by Separate com-
water has paid, or has incurred an obligation to pay, pensation.
either to an ocean freight broker or freight forwarder,
separate compensation for the solicitation or securing of
cargo for the ship or the booking of, or otherwise arrang-

ing for space for, such cargo, then such carrier shall not be obligated to pay additional compensation for any other forwarding services rendered on the same cargo. Before any such compensation is paid to or received by any person carrying on the business of forwarding, such person shall, if he is qualified under the provisions of this paragraph to receive such compensation, certify in writing to the common carrier by water by which the shipment was dispatched that he is licensed by the Federal Maritime Commision as an independent ocean freight forwarder and that he performed the above specified services with respect to such shipment. Such carrier shall be entitled to rely on such certification unless it knows that the certification is incorrect.

"Title of Act". 46 U.S.C. 842.

SEC. 45. That this Act may be cited as "Shipping Act, 1916."

INTERCOASTAL SHIPPING ACT, 1933

[As amended through the 89th Congress first session]

(47 Stat. 1425, chapter 199, approved March 3, 1933)

[NOTE.—See excerpts from the Transportation Act of 1940, as amended, infra.]

AN ACT

Amending the Shipping Act, 1916, as amended, for the purpose of further regulating common carriers by water in interstate commerce of the United States engaged in transportation by way of the Panama Canal.

Be it enacted by the Senate and House of Representatives of the United States of America in Congress assembled, That when used in this Act—

The term "common carrier by water in intercoastal commerce" for the purposes of this Act shall include every common and contract carrier by water engaged in the transportation for hire of passengers or property between one State of the United States and any other State of the United States by way of the Panama Canal.

SEC. 2. That every common carrier by water in intercoastal commerce shall file with the Federal Maritime Board and keep open to public inspection schedules showing all the rates, fares, and charges for or in connection with transportation between intercoastal points on its own route; and, if a through route has been established, all the rates, fares, and charges for or in connection with transportation between intercoastal points on its own route and points on the route of any other carrier by water. The schedules filed, and kept open to public inspection as aforesaid by any such carrier shall plainly show the places between which passengers and/or freight will be carried, and shall contain the classification of freight and of passenger accommodations in force, and shall also state separately each terminal or other charge, privilege, or facility, granted or allowed, and any rules or regulations which in anywise change, affect, or determine any part of the aggregate of such aforesaid rates, fares, or charges, or the value of the service rendered to the passenger consignor, or consignee, and shall include the terms and conditions of any passenger ticket, bill of lading, contract of affreightment, or other document evidencing the transportation agreement. The terms and conditions as filed with the Federal Maritime Board shall

46 U.S.C. 843.
47 Stat. 1425.

"Common carrier by water in intercoastal commerce."

Rate schedules.
46 U.S.C. 844.
47 Stat. 1425.
72 Stat. 977.
P.L. 85–810.

be framed under glass and posted in a conspicuous place on board each vessel where they may be seen by passengers and others at all times. Such carriers in establishing and fixing rates, fares, or charges may make equal rates, fares, or charges for similar service between all ports of origin and all ports of destination, and it shall be unlawful for any such carrier, either directly or indirectly, through the medium of any agreement, conference, association, understanding, or otherwise, to prevent or attempt to prevent any such carrier from extending service to any publicly owned terminal located on any improvement project authorized by the Congress at the same rates which it charges at its nearest regular port of call. Such schedules shall be plainly printed, and copies shall be kept posted in a public and conspicuous place at every wharf, dock, and office of such carrier where passengers or freight are received for transportation, in such manner that they shall be readily accessible to the public and can be conveniently inspected. In the event that any such schedule includes the terms and conditions of any passenger ticket, bill of lading contract of affreightment or other document evidencing the transportation agreement, as herein provided, copies of such terms and conditions shall be made available to any shipper, consignee, or passenger upon request. Such terms and conditions, if filed as permitted by this section and framed under glass and posted in a conspicuous place on board each vessel where they may be seen by passengers and others at all times, may be incorporated by reference in a short form of same actually issued for the transportation, or in a dock receipt or other document issued in connection therewith, by notice printed on the back of each document that all parties to the contract are bound by the terms and conditions as filed with the Federal Maritime Board and posted on board each vessel, and when so incorporated by reference every carrier and any other person having any interest or duty in respect of such transportation shall be deemed to have such notice thereof as if all such terms and conditions had been set forth in the short form document.

Changes.

No change shall be made in the rates, fares, or charges, or classifications, rules, or regulations, which have been filed and posted as required by this section, except by the publication, filing, and posting as aforesaid of a new schedule or schedules which shall become effective not earlier than thirty days after date of posting and filing thereof with the board, and such schedule or schedules shall plainly show the changes proposed to be made in the schedule or schedules then in force and the time when the rates, fares, charges, classifications, rules, or regulations as changed are to become effective: *Provided*, That the board may, in its discretion and for good cause, allow changes upon less than the period of thirty days herein specified: *And provided further*, That schedules or changes which provide for extension of actual service to additional ports at rates of said carrier already in effect for similar service at the nearest port of call to said additional ports shall become effective immediately upon notice to the board.

From and after ninety days following enactment hereof no person shall engage in transportation as a common carrier by water in intercoastal commerce unless and until its schedules as provided by this section have been duly and properly filed and posted; nor shall any common carrier by water in intercoastal commerce charge or demand or collect or receive a greater or less or different compensation for the transportation of passengers or property or for any service in connection therewith than the rates, fare, and/or charges which are specified in its schedules filed with the board and duly posted and in effect at the time; nor shall any such carrier refund or remit in any manner or by any device any portion of the rates, fares, or charges so specified, nor extend or deny to any person any privilege or facility, except in accordance with such schedules.

No transportation unless and until filing and posting.

Rates larger than schedule.

The board shall by regulations prescribe the form and manner in which the schedules required by this section shall be published, filed, and posted; and the board is authorized to reject any schedule filed with it which is not in consonance with this section and with such regulations. Any schedule so rejected by the board shall be void and its use shall be unlawful.

Regulations, publications, filing, posting.

Penalty.
P.L. 89–71.
79 Stat. 213.

Whoever violates any provision of this section shall be liable to a penalty of not more than $1,000 for each day such violation continues, to be recovered by the United States in a civil action.

Sec. 3. Whenever there shall be filed with the board any schedule stating a new individual or joint rate, fare, or charge, or any new individual or joint classification, or any new individual or joint regulation or practice affecting any rate, fare, or charge, the board shall have, and it is hereby given, authority, either upon complaint or upon its own initiative without complaint, and if it so orders without answer or other formal pleading by the interested carrier or carriers, but upon reasonable notice, to enter upon a hearing concerning the lawfulness of such rate, fare, charge, classification, regulation, or practice: *Provided, however*, That there shall be no suspension of a tariff schedule or service which extends to additional ports, actual service at rates of said carrier for similar service already in effect at the nearest port of call to said additional port.

46 U.S.C. 845.
47 Stat. 1426.
53 Stat. 1182.
Hearings.

Changes in rates.

Hearing on new schedules.

[§ 3 to end]

Suspension of rates pending hearing.

Duration of suspension.

Pending such hearing and the decision thereon the board, upon filing with such schedule and delivering to the carrier or carriers affected thereby a statement in writing of its reasons for such suspension, may from time to time suspend the operation of such schedule and defer the use of such rate, fare, charge, classification, regulation, or practice, but not for a longer period than four months beyond the time when it would otherwise go into effect; and after full hearing whether completed before or after the rate, fare, charge, classification, regulation, or practice goes into effect, the board may make such order with reference thereto as would be proper in a proceeding initiated after it had become effective. If the proceeding has not been concluded and an order made within the period of suspension, the proposed change of rate, fare, charge, classification, regulation, or practice shall go into effect at the end of such period. At any hearing under this paragraph the burden of proof to show that the rate, fare, charge, classification, regulation, or practice is just and reasonable shall be upon the carrier or carriers. The board shall give preference to the hearing and decision of such questions and decide the same as speedily as possible.

46 U.S.C. 845a.
52 Stat. 964.

SEC. 4. Whenever the Commission finds that any rate, fare, charge, classification, tariff, regulation, or practice demanded, charged, collected, or observed by any carrier subject to the provisions of this Act is unjust or unreasonable, it may determine, prescribe, and order enforced a just and reasonable maximum or minimum, or maximum and minimum rate, fare, or charge, or a just and reasonable classification, tariff, regulation or practice: *Provided,* That the minimum-rate provision of this section shall not apply to common carriers on the Great Lakes.

Commission may enforce just and reasonable minimum and maximum rates.

Great Lakes excluded.

46 U.S.C. 845b.
52 Stat. 964.

SEC. 5. The provisions of this Act are extended and shall apply to every common carrier by water in interstate commerce, as defined in section 1 of the Shipping Act, 1916.

46 U.S.C. 846.
47 Stat. 1427.

SEC. 6. That nothing in this Act shall prevent the carriage, storage, or handling of property free or at reduced rates, for the United States, State, or municipal Governments, or for charitable purposes.

46 U.S.C. 847.
47 Stat. 1427.

SEC. 7. The provisions of the Shipping Act, 1916, as amended, shall in all respects, except as amended by this Act, continue to be applicable to every carrier subject to the provisions of this Act.

46 U.S.C. 848.
47 Stat. 1427.

SEC. 8. That this Act may be cited as the Intercoastal Shipping Act, 1933.

TRANSPORTATION ACT OF 1940

(excerpts)

[54 STAT. 898, AT 950]

[PUBLIC—NO. 785—76TH CONGRESS]

[CHAPTER 722—3D SESSION]

[S. 2009]

AN ACT

To amend the Act to regulate commerce, approved February 4, 1887, as amended, so as to provide for unified regulation of carriers by railroad, motor vehicle, and water, and for other purposes.

*   *   *   *   *

TITLE II—REGULATION OF WATER CARRIERS IN INTERSTATE AND FOREIGN COMMERCE

SEC. 201. * * * PART III OF INTERSTATE COMMERCE ACT.

"PART III

*   *   *   *   *

"REPEALS

"SEC. 320. (a) The Shipping Act, 1916, as amended, and the Intercoastal Shipping Act, 1933, as amended, are hereby repealed insofar as they are inconsistent with any provision of this part and insofar as they provide for the regulation of, or the making of agreements relating to, transportation of persons or property by water in commerce which is within the jurisdiction of the Commission under the provisions of this part; and any other provisions of law are hereby repealed insofar as they are inconsistent with any provision of this part. *[margin: 39 Stat. 728. 47 Stat. 1475. 46 U.S.C. 801–842, 843–848; Supp. V, 804–848. 49 U.S.C. 920.]*

"(b) Nothing in subsection (a) shall be construed to repeal— *[margin: 49 Stat. 1987. 46 U.S.C., Supp. V, 1115.]*

"(1) section 205 of the Merchant Marine Act, 1936, as amended, or any provision of law providing penalties for violations of such section 205;

231

[§ 320
(b)(2)]

47 Stat. 1425,
1427.
46 U.S.C. 844,
847 ; Supp. V,
844, 847.

39 Stat. 728.
46 U.S.C. 801.

41 Stat. 999.
46 U.S.C. 883,
884 ; Supp. V,
883, 884.

39 Stat. 728.
46 U.S.C. 814.

Laws not
affected.

Repeal.
45 Stat. 980.
49 U.S.C.,
Supp. V,
153(e).
Continuance of
certificates of
public con-
venience, etc.

Transfer of
personnel.

49 U.S.C. 921.

"(2) the third sentence of section 2 of the Intercoastal Shipping Act, 1933, as amended, as extended by section 5 of such Act, or any provision of law providing penalties for violations of such section 2;

"(3) the provisions of the Shipping Act, 1916, as amended, insofar as such Act provides for the regulation of persons included within the term 'other person subject to this Act', as defined in such Act;

"(4) sections 27 and 28 of the Merchant Marine Act, 1920, as amended.

"(c) Nothing in subsection (a) shall be construed to affect the provisions of section 15 of the Shipping Act, 1916, so as to prevent any water carrier subject to the provisions of this part from entering into any agreement under the provisions of such section 15 with respect to transportation not subject to the provisions of this part in which such carrier may be engaged.

"(d) Nothing in this part shall be construed to affect any law of navigation, the admiralty jurisdiction of the courts of the United States, liabilities of vessels and their owners for loss or damage, or laws respecting seamen, or any other maritime law, regulation, or custom not in conflict with the provisions of this part.

"(e) Subsection (e) of section 3 of the Inland Waterways Corporation Act of June 7, 1924, as amended (U.S.C., title 49, sec. 153(e)), is hereby repealed as of October 1, 1940: *Provided, however,* That (1) any certificate of public convenience and necessity granted to any carrier pursuant to the provisions of such subsection (e) shall continue in effect as though issued under the provisions of section 309 of the Interstate Commerce Act, as amended; and (2) through routes and joint rates, and rules, regulations, and practices relating thereto, put into effect pursuant to the provisions of such subsection (e) shall, after the repeal of such subsection (e), be held and considered to have been put into effect pursuant to the provisions of the Interstate Commerce Act, as amended.

"TRANSFER OF EMPLOYEES, RECORDS, PROPERTY, AND APPROPRIATIONS

"SEC. 321. (a) Such officers and employees of the United States Maritime Commission as the President shall determine to have been employed in the administration of the provisions of law repealed by section 320, and whose retention by the United States Maritime Commission is not necessary, in the opinion of the President, for the performance of other duties, are transferred to the Interstate Commerce Commission upon such date or dates as the President shall specify by Executive order. Such transfer of such personnel shall be without reduction in classification or compensation, except that this requirement shall not operate after the end of the fiscal year during which such transfer is made to prevent the adjustment of classification or compensation to conform to the duties to which such transferred personnel may be assigned.

"(b) All files, reports, records, tariff schedules, property (including office furniture and equipment), contracts, agreements, documents, or papers kept or used by, made to, or filed with the United States Maritime Commission under or in the administration of any provision or law repealed by this part, are hereby transferred to the jurisdiction and control of the Interstate Commerce Commission, and may be used for such purposes as the Interstate Commerce Commission may deem necessary in the administration of this part; except that in the case of files, reports, records, tariff schedules, contracts, agreements, documents, or papers, the retention of which is necessary for purposes of the administration by the United States Maritime Commission of matters within its jurisdiction, the furnishing to the Interstate Commerce Commission of copies thereof shall constitute sufficient compliance with the provisions of this subsection. *Records, property, etc.*

"(c) All appropriations and unexpended balances of appropriations available for expenditure by the United States Maritime Commission in the administration of any provision of law repealed by this part shall be available for expenditure by the Interstate Commerce Commission for any objects of expenditure authorized by this part, in the discretion of the Interstate Commerce Commission, without regard to the requirement of apportionment under the Anti-Deficiency Act of February 27, 1906. *Funds available.*

"EXISTING ORDERS, RULES, TARIFFS, AND SO FORTH; PENDING MATTERS

"Sec. 322. (a) Notwithstanding the provisions of section 320, or any other provision of this part, all orders, rules, regulations, permits, tariffs (including rates, fares, charges, classifications, rules, and regulations relating thereto), contracts, or agreements, to the extent that they were issued, authorized, approved, entered into, or filed under any provision of law repealed by this part, and are still in effect, shall continue in force and effect according to the terms thereof as though this part had not been enacted, except that the Commission may modify, set aside, or rescind any such order, rule, regulation, permit, *34 Stat. 48. 31 U.S.C. 665.* *49 U.S.C. 922.* *Continuance of certain orders, etc.* *Exception.*

[§ 322(a)
to end]

tariff, contract, or agreement to the extent that it finds the same to be in violation of any provision of this part or inconsistent with the national transportation policy declared in this Act.

Proceedings, hearings, or investigations.

"(b) Any proceeding, hearing, or investigation commenced or pending before the United States Maritime Commission at the time this section takes effect, to the extent that it relates to the administration of any provision of law repealed by this part, shall be continued or otherwise acted upon by the Commission as though such proceeding, hearing, or investigation had been instituted under the provisions of this part.

Judicial proceedings.

"(c) Any pending judicial proceeding arising under any provision of law repealed by the provisions of this part shall be continued, heard, and determined in the same manner and with the same effect as if this part had not been enacted; except that in the case of any such proceeding to which the United States Maritime Commission is a party, the court, upon motion or supplemental petition, may direct that the Commission be substituted for the United States Maritime Commission as a party to the proceeding or made an additional party thereto."

*   *   *   *   *

TIME EFFECTIVE

Time effective.

SEC. 202. Part III of the Interstate Commerce Act shall take effect on the date of the enactment of this Act, except that sections 304(c), 305 to 308, inclusive, 309 (a) and (f), 313 to 318, inclusive, 320, 321, and 322 shall take effect on the 1st day of January 1941: *Provided, however*, That the Interstate Commerce Commission shall, if found by it necessary or desirable in the public interest, by general or special order postpone the taking effect of any of the provisions above enumerated to such time, but not beyond the 1st day of April 1942, as the Commission shall prescribe.

Postponement permitted.

Approved September 18, 1940.

AMERICAN CARRIAGE OF GOODS BY SEA ACT, 1936

Effective July 15, 1936

Be it enacted by the Senate and House of Representatives of the United States of America in Congress assembled:

That every bill of lading or similar document of title which is evidence of a contract for the carriage of goods by sea to or from ports of the United States, in foreign trade, shall have effect subject to the provisions of this act.

TITLE I

Section 1. When used in this Act—

(a) The term "carrier" includes the owner or the charterer who enters into a contract of carriage with a shipper.

(b) The term "contract of carriage" applies only to contracts of carriage covered by a bill of lading or any similar document of title, insofar as such document relates to the carriage of goods by sea, including any bill of lading or any similar document as aforesaid issued under or pursuant to a charterparty from the moment at which such bill of lading or similar document of title regulates the relations between a carrier and a holder of the same.

(c) The term "goods" includes goods, wares, merchandise, and articles of every kind whatsoever, except live animals and cargo which by the contract of carriage is stated as being carried on deck and is so carried.

(d) The term "ship" means any vessel used for the carriage of goods by sea.

(e) The term "carriage of goods" covers the period from the time when the goods are loaded on to the time when they are discharged from the ship.

RISKS

Sec. 2. Subject to the provisions of section 6, under every contract of carriage of goods by sea, the carrier in relation to the loading, handling, stowage, carriage, custody, care, and discharge of such goods, shall be subject to the responsibilities and liabilities and entitled to the rights and immunities hereinafter set forth.

RESPONSIBILITIES AND LIABILITIES

Sec. 3. (1) The carrier shall be bound, before and at the beginning of the voyage, to exercise due diligence to—

(a) Make the ship seaworthy;

(b) Properly man, equip, and supply the ship;

(c) Make the holds, refrigerating and cooling chambers, and all other parts of the ship in which goods are carried, fit and safe for their reception, carriage, and preservation.

(2) The carrier shall properly and carefully load, handle, stow, carry, keep, care for, and discharge the goods carried.

(3) After receiving the goods into his charge the carrier, or the master or agent of the carrier, shall, on demand of the shipper, issue to the shipper a bill of lading showing among other things—

(a) The leading marks necessary for identification of the goods as the same are furnished in writing by the shipper before the loading of such goods starts, provided such marks are stamped or otherwise shown clearly upon the goods if uncovered, or on the cases or coverings in which such goods are contained, in such a manner as should ordinarily remain legible until the end of the voyage.

(b) Either the number of packages or pieces, or the quantity or weight, as the case may be, as furnished in writing by the shipper.

(c) The apparent order and condition of the goods:

Provided, That no carrier, master, or agent of the carrier, shall be bound to state or show in the bill of lading any marks, number, quantity, or weight which he has reasonable ground for suspecting not accurately to represent the goods actually received, or which he has had no reasonable means of checking.

(4) Such a bill of lading shall be *prima facie* evidence of the receipt by the carrier of the goods as therein described in accordance with paragraph (3) (a), (b), and (c) of this section:

Provided, That nothing in this Act shall be construed as repealing or limiting the application of any part of the Act, as amended, entitled "An Act relating to bills of lading in interstate and foreign commerce," approved August 29, 1916 (U.S. C., title 49, secs. 81–124, commonly known as the "Pomerene Bills of Lading Act)."

(5) The shipper shall be deemed to have guaranteed to the carrier the accuracy at the time of shipment of the marks, number, quantity, and weight, as furnished by him; and the shipper shall indemnify the carrier against all loss, damages, and expenses arising or resulting from inaccuracies in such particulars. The right of the carrier to such indemnity shall in no way limit his responsibility and liability under the contract of carriage to any person other than the shipper.

(6) Unless notice of loss or damage and the general nature of such loss or damage be given in writing to the carrier or his agent at the port of discharge before or at the time of the removal of the goods into the custody of the person entitled to delivery thereof under the contract of carriage, such removal shall be *prima facie* evidence of the delivery by the carrier of the goods as described in the bill of lading.

If the loss or damage is not apparent, the notice must be given within three days of the delivery.

Said notice of loss or damage may be endorsed upon the receipt for the goods given by the person taking delivery thereof.

The notice in writing need not be given if the state of the goods has at the time of their receipt been the subject of joint survey or inspection.

In any event, the carrier and the ship shall be discharged from all liability in respect of loss or damage unless suit is brought within one year after delivery of the goods or the date when the goods should have been delivered:

Provided, That if a notice of loss or damage, either apparent or concealed, is not given as provided for in this section, that fact shall not affect or prejudice the right of the shipper to bring suit within one year after the delivery of the goods or the date when the goods should have been delivered.

In case of any actual or apprehended loss or damage the carrier and the receiver shall give all reasonable facilities to each other for inspecting and tallying the goods.

(7) After the goods are loaded the bill of lading to be issued by the carrier, master, or agent of the carrier to the shipper shall, if the shipper so demands, be a "shipped" bill of lading:

Provided, That if the shipper shall have previously taken up any document of title to such goods, he shall surrender the same as against the issue of the "shipped" bill of lading, but at the option of the carrier such document of title may be noted at the port of shipment by the carrier, master, or agent with the name or names of the ship or ships upon which the goods have been shipped and the date or dates of shipment, and when so noted the same shall for the purpose of this section be deemed to constitute a "shipped" bill of lading.

(8) Any clause, covenant, or agreement in a contract of carriage relieving the carrier or the ship from liability for loss or damage to or in connection with the goods, arising from negligence, fault, or failure in the duties and obligations provided in this section or lessening such liability otherwise than as provided in this Act, shall be null and void and of no effect.

A benefit of insurance in favor of the carrier, or similar clause, shall be deemed to be a clause relieving the carrier from liability.

RIGHTS AND IMMUNITIES

Sec. 4. (1) Neither the carrier nor the ship shall be liable for loss or damage arising or resulting from unseaworthiness unless caused by want of due diligence on the part of the carrier to make the ship seaworthy, and to secure that the ship is properly manned, equipped, and supplied, and to make the holds, refrigerating and cool chambers, and all other parts of the ship in which goods are carried fit and safe for their reception, carriage, and preservation in accordance with the provisions of paragraph (1) of section 3. Whenever loss or damage has resulted from unseaworthiness, the burden of proving the exercise of due diligence shall be on the carrier or other persons claiming exemption under this section.

(2) Neither the carrier nor the ship shall be responsible for loss or damage arising or resulting from—

(a) Act, neglect, or default of the master, mariner, pilot or the servants of the carrier in the navigation or in the management of the ship;

(b) Fire, unless caused by the actual fault or privity of the carrier;

(c) Perils, dangers, and accidents of the sea or other navigable waters;

(d) Act of God;

(e) Act of war;

(f) Act of public enemies;

(g) Arrest or restraint of princes, rulers, or people, or seizure under legal process;

(h) Quarantine restrictions;

(i) Act or omission of the shipper or owner of the goods, his agent, or representative;

(j) Strikes or lockouts or stoppage or restraint of labor from whatever cause, whether partial or general; *Provided,* That nothing herein contained shall be construed to relieve a carrier from responsibility for the carrier's own acts;

(k) Riots and civil commotions;

(l) Saving or attempting to save life or property at sea;

(m) Wastage in bulk or weight or any other loss or damage arising from inherent defect, quality, or vice of the goods;

(n) Insufficiency or packing;

(o) Insufficiency of inadequacy of marks;

(p) Latent defects not discoverable by due diligence; and

(q) Any other cause arising without the actual fault and privity of the carrier and without the fault or neglect of the agents or servants of the carrier, but the burden of proof shall be on the person claiming the benefit of this exception to show that neither the actual fault or privity of the carrier nor the fault or neglect of the agents or servants of the carrier contributed to the loss or damage.

(3) The shipper shall not be responsible for loss or damage sustained by the carrier or the ship arising or resulting from any cause without the act, fault, or neglect of the shipper, his agents, or his servants.

(4) Any deviation in saving or attempting to save life or property at sea, or any reasonable deviation shall not be deemed to be an infringement or breach of this Act or the contract of carriage, and the carrier shall not be liable for any loss or damage resulting therefrom:

Provided, however, That if the deviation is for the purpose of loading or unloading cargo or passengers it shall, prima facie, be regarded as unreasonable.

(5) Neither the carrier nor the ship shall in any event be or become liable for any loss or damage to or in connection with the transportation of goods in an amount exceeding $500 per package lawful money of the United States, or in case of goods not shipped in packages, per customary freight unit, or the equivalent of that sum in other currency, unless the nature and value of such goods have been declared by the shipper before shipment and inserted in the bill of lading.

This declaration, if embodied in the bill of lading, shall be *prima facie* evidence, but shall not be conclusive on the carrier.

By agreement between the carrier, master, or agent of the carrier and the shipper, another maximum amount than that mentioned in this paragraph may be fixed:

Provided, That such maximum shall not be less than the figure above named. In no event shall the carrier be liable for more than the amount of damage actually sustained.

Neither the carrier nor the ship shall be responsible in any event for loss or damage to or in connection with the transportation of the goods if the nature or value thereof has been knowingly and fraudulently misstated by the shipper in the bill of lading.

(6) Goods of an inflammable, explosive, or dangerous nature to the shipment whereof the carrier, master, or agent of the carrier, has not consented with knowledge of their nature and character, may at any time before discharge be landed

at any place or destroyed or rendered innocuous by the carrier without compensation, and the shipper of such goods shall be liable for all damages and expenses directly or indirectly arising out of or resulting from such shipment.

If any such goods shipped with such knowledge and consent shall become a danger to the ship or cargo, they may in like manner be landed at any place, or destroyed or rendered innocuous by the carrier without liability on the part of the carrier except to general average, if any.

SURRENDER OF RIGHTS AND IMMUNITIES AND INCREASE OF RESPONSIBILITIES AND LIABILITIES

Sec. 5. A carrier shall be at liberty to surrender in whole or in part all or any of his rights and immunities or to increase any of his responsibilities and liabilities under this Act, provided such surrender or increase shall be embodied in the bill of lading issued to the shipper.

The provisions of this Act shall not be applicable to charter-parties, but if bills of lading are issued in the case of a ship under a charter-party they shall comply with the terms of this Act.

Nothing in this Act shall be held to prevent the insertion in a bill of lading of any lawful provision regarding general average.

SPECIAL CONDITIONS

Sec. 6. Notwithstanding the provisions of the preceding sections, a carrier, master or agent of the carrier, and a shipper shall, in regard to any particular goods be at liberty to enter into any agreement in any terms as to the responsibility and liability of the carrier for such goods, and as to the rights and immunities of the carrier in respect of such goods, or his obligation as to seaworthiness (so far as the stipulation regarding seaworthiness is not contrary to public policy), or the care or diligence of his servants or agents in regard to the loading, handling, stowage, carriage, custody, care and discharge of the goods carried by sea:

Provided, That in this case no bill of lading has been or shall be issued and that the terms agreed shall be embodied in a receipt which shall be a nonnegotiable document and shall be marked as such.

Any agreement so entered into shall have full legal effect:

Provided, That this section shall not apply to ordinary commercial shipments made in the ordinary course of trade but only to other shipments where the character or condition of the property to be carried or the circumstances, terms, and conditions under which the carriage is to be performed are such as reasonably to justify a special agreement.

Sec. 7. Nothing contained in this Act shall prevent a carrier or a shipper from entering into any agreement, stipulation, condition, reservation, or exemption as to the responsibility and liability of the carrier or the ship for the loss or damage to or in connection with the custody and care and handling of goods prior to the loading on and subsequent to the discharge from the ship on which the goods are carried by sea.

Sec. 8. The provisions of this Act shall not affect the rights and obligations of the carrier under the provisions of the Shipping Act, 1916, or under the provisions of sections 4281 to 4289, inclusive, of the Revised Statutes of the United States, or of any amendments thereto; or under the provisions of any other enactment for the time being in force relating to the limitation of the liability of the owners of seagoing vessels.

TITLE II

Sec. 9. Nothing contained in this Act shall be construed as permitting a common carrier by water to discriminate between competing shippers similarly placed in time and circumstances, either (a) with respect to their right to demand and receive bills of lading subject to the provisions of this Act; or (b) when issuing such bills of lading, either in the surrender of any of the carrier's rights and immunities or in the increase of any of the carrier's responsibilities and liabilities pursuant to section 5, title I, of this Act; or (c) in any other way prohibited by the Shipping Act, 1916, as amended.

Sec. 10. Section 25 of the Interstate Commerce Act is hereby amended by adding the following proviso at the end of paragraph 4 thereof: "*Provided, however,* That insofar as any bill of lading authorized hereunder relates to the carriage of goods by sea, such bill of lading shall be subject to the provisions of the Carriage of Goods by Sea Act."

Sec. 11. Where under the customs of any trade the weight of any bulk cargo inserted in the bill of lading is a weight ascertained or accepted by a third party other than the carrier or the shipper, and the fact that the weight is so ascertained or accepted is stated in the bill of lading, then, notwithstanding anything in this Act, the bill of lading shall not be deemed to be *prima facie* evidence against the carrier of the receipt of goods of the weight so inserted in the bill of lading, and the accuracy thereof at the time of shipment shall not be deemed to have been guaranteed by the shipper.

Sec. 12. Nothing in this Act shall be construed as superseding any part of the (Harter) Act or of any other law which would be applicable in the absence of this Act, insofar as they relate to the duties, responsibilities, and liabilities of the ship or carrier prior to the time when the goods are loaded on or after the time they are discharged from the ship.

Sec. 13. This Act shall apply to all contracts for carriage of goods by sea to or from ports of the United States in foreign trade.

As used in this Act the term "United States" includes its districts, territories, and possessions:

Provided, however, That the Philippine Legislature may by law exclude its application to transportation to or from ports of the Philippine Islands.

The term "foreign trade" means the transportation of goods between the ports of the United States and ports of foreign countries.

Nothing in this Act shall be held to apply to contracts for carriage of goods by sea between any port of the United States or its possessions, and any other port of the United States or its possessions:

Provided, however, That any bill of lading or similar document of title which is evidence of a contract for the carriage of goods by sea between such ports, containing an express statement that it shall be subject to the provisions of this Act, shall be subjected hereto as fully as if subject hereto by the express provisions of this Act:

Provided further, That every bill of lading or similar document of title which is evidence of a contract for the carriage of goods by sea from ports of the United States, in foreign trade, shall contain a statement that it shall have effect subject to the provisions of this Act.

Sec. 14. Upon the certification of the Secretary of Commerce that the foreign commerce of the United States in its competition with that of foreign nations is prejudiced by the provisions, or any of them, of Title I of this Act, or by the laws of any foreign country or countries relating to the carriage of goods by sea, the President of the United States may, from time to time, by proclamation, suspend any or all provisions of Title I of this Act for such periods of time or indefinitely as may be designated in the proclamation. The President may at any time rescind such suspension of Title I hereof, and any provisions thereof which may have been suspended shall thereby be reinstated and again apply to contracts thereafter made for the carriage of goods by sea. Any proclamation of suspension or rescission of any such suspension shall take effect on a date named therein, which date shall be not less than ten days from the issue of the proclamation.

Any contract for the carriage of goods by sea, subject to the provisions of this Act, effective during any period when Title I hereof, or any part thereof, is suspended, shall be subject to all provisions of law now or hereafter applicable to that part of Title I which may have thus been suspended.

Sec. 15. This Act shall take effect ninety days after the date of its approval, but nothing in this Act shall apply during a period not to exceed 1 year following its approval to any contract for the carriage of goods by sea, made before the date on which this Act is approved, nor to any bill of lading or similar document of title issued, whether before or after such date of approval in pursuance of any such contract as aforesaid.

Sec. 16. This Act may be cited as the "Carriage of Goods by Sea Act."

THE HARTER ACT

Act of Congress, Approved February 13, 1893

An Act relating to navigation of vessels, bills of lading, and to certain obligations, duties, and rights in connection with the carriage of property.

Be it enacted by the Senate and House of Representatives of the United States of America in Congress assembled,

Section 1. That it shall not be lawful for the manager, agent, master or owner of any vessel transporting merchandise or property from or between ports of the United States and foreign ports to insert in any bill of lading or shipping document any clause, covenant, or agreement whereby it, he, or they shall be relieved from liability for loss or damage arising from negligence, fault, or failure in proper loading, stowage, custody, care, or proper delivery of any and all lawful merchandise or property committed to its or their charge. Any and all words or clauses of such import inserted in bills of lading or shipping receipts shall be null and void and of no effect.

Sec. 2. That it shall not be lawful for any vessel transporting merchandise or property from or between ports of the United States of America and foreign ports, her owner, master, agent or manager to insert in any bill of lading or shipping document any covenant or agreement whereby the obligations of the owner or owners of said vessel to exercise due diligence, properly equip, man, provision, and outfit said vessel, and to make said vessel seaworthy and capable of performing her intended voyage, or whereby the obligations of the master, officers, or servants to carefully handle and stow her cargo and to care for and properly deliver same, shall in any wise be lessened, weakened, or avoided.

Sec. 3. That if the owner of any vessel transporting merchandise or property to or from any port in the United States of America shall exercise due diligence to make the said vessel in all respects seaworthy and properly manned, equipped, and supplied, neither the vessel, nor owners, agents, or charterers shall become or be held responsible for damage or loss resulting from faults or errors in navigation or in the management of said vessel, nor shall the vessel, her owner or owners, charterers, agent, or master be held liable for losses arising from dangers of the sea or other navigable waters, acts of God, or public enemies, or the inherent defect, quality, or vice of the thing carried, or from insufficiency of package, or seizure under legal process, or for loss resulting from any act or omission of the shipper or owner of the goods, his agent or representative, or from saving or attempting to save life or property at sea, or from any deviation in rendering such service.

Sec. 4. That it shall be the duty of the owner or owners, master, or agent of any vessel transporting merchandise or property from or between ports of the United States and foreign ports to issue to shippers of any lawful merchandise a bill of lading, or shipping document, stating among other things, the marks necessary for identification, number of packages, or quantity, stating whether it be carrier's or shipper's weight and apparent order or condition of such merchandise or property delivered to and received by the owner, master, or agent of the vessel for transportation, and such document shall be prima facie evidence of the receipt of the merchandise therein described.

Sec. 5. That for a violation of any of the provisions of this Act the agent, owner, or master of the vessel guilty of such violation, and who refuses to issue on demand the bill of lading herein provided for, shall be liable to a fine not exceeding two thousand dollars. The amount of the fine and costs for such violation shall be a lien upon the vessel, whose agent, owner, or master is guilty of such violation, and such vessel may be libeled therefor in any district court of the United States, within whose jurisdiction the vessel may be found. One-half of such penalty shall go to the party injured by such violation and the remainder to the Government of the United States.

Sec. 6. That this Act shall not be held to modify or repeal sections forty-two hundred and eight-one, forty-two hundred and eighty-two, and forty-two hundred and eighty-three of the Revised Statutes of the United States, or any other statute defining the liability of vessels, their owners, or representatives.

Sec. 7. Sections one and four of this Act shall not apply to the transportation of live animals.

Sec. 8. That this Act shall take effect from and after the first day of July, eighteen hundred and ninety-three.

FEDERAL BILLS OF LADING ACT, 1916,[1]

AS AMENDED TO AUGUST, 1952.

KNOWN AS THE POMERENE ACT.

U. S. Code, Title 49, §§ 81-124.

39 Stat. L. 538 [S. 19;] amended March 4, 1927,
44 Stat. L. 1540 [S. 3286.]

AN ACT RELATING TO BILLS OF LADING IN INTERSTATE AND FOREIGN COMMERCE.

Be it enacted by the Senate and House of Representatives of the United States of America in Congress assembled,

Bills of Lading—Issued in Interstate and Foreign Commerce Governed Hereby.

U. S. Code, Title 49, Section 81.

SECTION 1.[2] That bills of lading issued by any common carrier for the transportation of goods in any Territory of the United States, or the District of Columbia, or from a place in a State to a place in a foreign country, or from a place in one State to a place in another State, or from a place in one State to a place in the same State through another State or foreign country, shall be governed by this act.

(Compare Uniform Act, section 1).

---

[1] The federal act is derived from the Uniform Bill of Lading Act adopted by the Conference of Commissioners on Uniform State Laws and the American Bar Association in 1909. Nine sections are altered; and 15 of the Uniform Act sections are omitted—chiefly those stating criminal sanctions.

In 1951 the Conference of Commissioners on Uniform State Laws and the American Bar Association adopted a text called the Uniform Commercial Code, with a federal as well as a state version. That Code would, if enacted, replace the series of Uniform Acts relating to Bills of Lading, Negotiable Instruments, Warehouse Receipts, etc. with a single combined text. At this writing—October 1952—the new Code had not been introduced as a Bill in Congress, nor enacted by any state. The reader is cautioned to inform himself about the progress of the new Code in the legislatures.

[2] These section numbers should preferably be used in discussion and citation, being more permanent than current re-numberings of codes and compilations.

243

Straight Bills Defined.
U. S. Code, Title 49, Section 82.

SEC. 2. That a bill in which it is stated that the goods are consigned or destined to a specific person is a straight bill.

(Compare Uniform Act, section 4).

Order Bills Defined—Negotiability.
U. S. Code, Title 49, Section 83.

SEC. 3. That a bill in which it is stated that the goods are consigned or destined to the order of any person named in such bill is an order bill. Any provision in such a bill or in any notice, contract, rule, regulation, or tariff that it is non-negotiable shall be null and void and shall not affect its negotiability within the meaning of this act unless upon its face and in writing agreed to by the shipper.

(Compare Uniform Act, section 5).

Issue of Order Bills in Parts for Continental Use Forbidden— Proviso—For Insular and Foreign Use Permitted.
U. S. Code, Title 49, Section 84.

SEC. 4. That order bills issued in a State for the transportation of goods to any place in the United States on the Continent of North America, except Alaska and Panama, shall not be issued in parts or sets. If so issued, the carrier issuing them shall be liable for failure to deliver the goods described therein to anyone who purchases a part for value in good faith, even though the purchase be after the delivery of the goods by the carrier to a holder of one of the other parts: *Provided, however,* That nothing contained in this section shall be interpreted or construed to forbid the issuing of order bills in parts or sets for such transportation of goods to Alaska, Panama, Porto Rico, the Philippines, Hawaii, or foreign countries, or to impose the liabilities set forth in this section for so doing.

(Compare Uniform Act, section 6).

Duplicates—Character to Be Noted—Liability for Failure— Insular and Foreign Trades Excepted.
U. S. Code, Title 49, Section 85.

SEC. 5. That when more than one order bill is issued in a State for the same goods to be transported to any place in the United States on the Continent of North America, except Alaska and Panama, the word "duplicate," or some other word or words indicating that the document is not an original bill, shall be placed plainly upon the face of every such bill except the one first issued. A carrier shall be liable for the damage caused by his failure so to do to anyone who has purchased the bill for value in good faith as an original, even though the purchase be after the delivery of the goods by the carrier to the holder of the original bill: *Provided, however,* That nothing contained in this section shall in such case for such transportation of goods to Alaska, Panama, Porto Rico, the Philippines, Hawaii, or foreign countries be interpreted or construed so as to require the placing of the word "duplicate" thereon, or to impose the liabilities set forth in this section for failure so to do.

(Compare Uniform Act, section 7).

Straight Bills Not Negotiable—Memoranda.
U. S. Code, Title 49, Section 85.

SEC. 6. That a straight bill shall have placed plainly upon its face by the carrier issuing it "non-negotiable" or "not negotiable."

This section shall not apply, however, to memoranda or acknowledgments of an informal character.

(Compare Uniform Act, section 8).

Negotiability of Order Bills—Insertion of Name of Notify Party.
U. S. Code, Title 49, Section 87.

SEC. 7. That the insertion in an order bill of the name of a person to be notified of the arrival of the goods shall not limit the negotiability of the bill or constitute notice to a purchaser thereof of any rights or equities of such person in the goods.

(Compare Uniform Act, section 9).

Carrier to Deliver Goods on Demand—Conditions—Lawful Excuse.
U. S. Code, Title 49, Section 88.

SEC. 8. That a carrier, in the absence of some lawful excuse, is bound to deliver goods upon a demand made either by the consignee named in the bill for the goods or, if the bill is an order bill, by the holder thereof, if such a demand is accompanied by—

(a) An offer in good faith to satisfy the carrier's lawful lien upon the goods;

(b) Possession of the bill of lading and an offer in good faith to surrender, properly indorsed, the bill which was issued for the goods, if the bill is an order bill; and

(c) A readiness and willingness to sign, when the goods are delivered, an acknowledgement that they have been delivered, if such signature is requested by the carrier.

In case the carrier refuses or fails to deliver the goods, in compliance with a demand by the consignee or holder so accompanied, the burden shall be upon the carrier to establish the existence of a lawful excuse for such refusal or failure.

(Compare Uniform Act, section 11).

To Whom Carrier Shall Deliver Goods.
U. S. Code, Title 49, Section 89.

SEC. 9. That a carrier is justified, subject to the provisions of the three following sections, in delivering goods to one who is—

(a) A person lawfully entitled to the possession of the goods, or

(b) The consignee named in a straight bill for the goods, or

(c) A person in possession of an order bill for the goods, by the terms of which the goods are deliverable to his order; or which has been indorsed to him, or in blank by the consignee, or by the mediate or immediate indorsee of the consignee.

(Compare Uniform Act, section 12).

Liability for Unlawful Delivery—Exceptions.
U. S. Code, Title 49, Section 90.

SEC. 10. That where a carrier delivers goods to one who is not lawfully entitled to the possession of them, the carrier shall be liable to anyone having a right of property or possession in the goods if he delivered the goods otherwise than as authorized by subdivisions (b) and (c) of the preceding section; and, though he delivered the goods as authorized by either of said subdivisions, he shall be so liable if prior to such delivery he—

(a) Had been requested, by or on behalf of a person having a right of property or possession in the goods, not to make such delivery, or

(b) Had information at the time of the delivery that it was to a person not lawfully entitled to the possession of the goods.

Such request or information, to be effective within the meaning of this section, must be given to an officer or agent of the carrier, the actual or apparent scope of whose duties includes action upon such a request or information, and must be given in time to enable the officer or agent to whom it is given, acting with reasonable diligence, to stop delivery of the goods.

(Compare Uniform Act, section 13).

Failure to Cancel Bill on Delivery of Goods—Liability to Subsequent Holder.

U. S. Code, Title 49, Section 91.

SEC. 11. That except as provided in section twenty-six, and except when compelled by legal process, if a carrier ·delivers goods for which an order bill had been issued, the negotiation of which would transfer the right to the possession of the goods, and fails to take up and cancel the bill, such carrier shall be liable for failure to deliver the goods to anyone who for value and in good faith purchases such bill, whether such purchaser acquired title to the bill before or after the delivery of the goods by the carrier and notwithstanding delivery was made to the person entitled thereto.

(Compare Uniform Act, section 14).

Delivery of Part of the Goods.

U. S. Code, Title 49, Section 92.

SEC. 12. That except as provided in section twenty-six, and except when compelled by legal process, if a carrier delivers part of the goods for which an order bill had been issued and fails either—

(a) To take up and cancel the bill, or

(b) To place plainly upon it a statement that a portion of the goods has been delivered with a description which may be in general terms either of the goods or packages that have been so delivered or of the goods or packages which still remain in the carrier's possession, he shall be liable for failure to deliver all the goods specified in the bill to anyone who for value and in good faith purchases it, whether such purchaser acquired title to it before or after the delivery of any portion of the goods by the carrier, and notwithstanding such delivery was made to the person entitled thereto.

(Compare Uniform Act, section 15).

Alteration of a Bill.

U. S. Code, Title 49, Section 93.

SEC. 13. That any alteration, addition, or erasure in a bill after its issue without authority from the carrier issuing the same, either in writing or noted on the bill, shall be void, whatever be the nature and purpose of the change, and the bill shall be enforceable according to its original tenor.

(Compare Uniform Act, section 16).

Lost, Stolen, or Destroyed Bill—Delivery of Goods Under Order of Court.

U. S. Code, Title 49, Section 94.

SEC. 14. That where an order bill has been lost, stolen, or destroyed, a court of competent jurisdiction may order the delivery of the goods upon satisfactory proof of such loss, theft, or destruction and upon the giving of a bond, with sufficient surety, to be approved by the court, to protect the carrier or any person injured by such delivery from any liability or loss incurred by reason of the original bill remaining outstanding. The court may also in its discretion order the payment of the carrier's reasonable costs and counsel fees. *Provided*, a voluntary indemnifying bond without order of court shall be binding on the parties thereto.

The delivery of the goods under an order of the court, as provided in this section, shall not relieve the carrier from liability to a person to whom the order bill has been or shall be negotiated for value without notice of the proceedings or of the delivery of the goods.

(Compare Uniform Act, section 17).

Liability Under " Duplicates."

U. S. Code, Title 49, Section 96.

SEC. 15. That a bill, upon the face of which the word " duplicate " or some other word or words indicating that the document is not an original bill is placed,* plainly shall impose upon the carrier issuing the same the liability of one who represents and warrants that such bill is an accurate copy of an original bill properly issued, but no other liability.

(Compare Uniform Act, section 18).

* So in original. The comma should follow the word *plainly*.

Carrier's Title to Goods.

U. S. Code, Title 49, Section 96.

SEC. 16. That no title to goods or right to their possession asserted by a carrier for his own benefit shall excuse him from liability for refusing to deliver the goods according to the terms of a bill issued for them, unless such title or right is derived directly or indirectly from a transfer made by the consignor or consignee after the shipment, or from the carrier's lien.

(Compare Uniform Act, section 19).

Conflicting Claimants.

U. S. Code, Title 49, Section 97.

SEC. 17. That if more than one person claim the title or possession of goods, the carrier may require all known claimants to interplead, either as a defense to an action brought against him for nondelivery of the goods or as an original suit, whichever is appropriate.

(Compare Uniform Act, section 20).

Carrier May Decline to Deliver Goods Until Adverse Claim Disposed Of.

U. S. Code, Title 49, Section 98.

SEC. 18. That if some one other than the consignee or the person in possession of the bill has a claim to the title or possession of the goods, and the carrier has information of such claim, the carrier shall be excused from liability for refusing to deliver the goods, either to the consignee or person in possession of the bill or to the adverse claimant, until the carrier has had a reasonable time to ascertain the validity of the adverse claim or to bring legal proceedings to compel all claimants to interplead.

(Compare Uniform Act, section 21).

Defenses Against Holders of Bills—Claim of Third Person.

U. S. Code, Title 49, Section 99.

SEC. 19. That except as provided in the two preceding sections and in section nine, no right or title of a third person, unless enforced by legal process, shall be a defense to an action brought by the consignee of a straight bill or by the holder of an order bill against the carrier for failure to deliver the goods on demand.

(Compare Uniform Act, section 22).

When Goods Are Loaded by a Carrier.

U. S. Code, Title 49, Section 100.

SEC. 20. That when goods are loaded by a carrier such carrier shall count the packages of goods, if package freight, and ascertain the kind and quantity if bulk freight, and such carrier shall not, in such cases, insert in the bill of lading or in any notice, receipt, contract, rule, regulation, or tariff, " Shipper's weight, load, and count," or other words of like purport indicating that the goods were loaded by the shipper and description of them made by him or in case of bulk freight and freight not concealed by packages the description made by him. If so inserted, contrary to the provisions of this section, said words shall be treated as null and void and as if not inserted therein.

(Compare Uniform Act, section, 23, sentence 2).

When Goods are Loaded by a Shipper—Conditions—Shipper's Scales.

U. S. Code, Title 49, Section 101.

SEC. 21. That when package freight or bulk freight is loaded by a shipper and the goods are described in a bill of lading merely by a statement of marks or labels upon them or upon packages containing them, or by a statement that the goods are said to be goods of a certain kind or quantity, or in a certain condition, or it is stated in the bill of lading that packages are said to contain goods of a certain kind or quantity or in a certain condition, or that the contents or condition of the contents of packages are unknown, or words of like purport are contained in the bill of lading, such statements, if true, shall not make liable the carrier issuing the bill of lading, although the goods are not of the kind or quantity or in the condition which the marks or labels upon them indicate, or of the kind or quantity or in the condition they were said to be by the consignor. The carrier may also by inserting in the bill of lading the words " Shipper's weight, load, and count," or other words of like purport indicate that the goods were loaded by the shipper and the description of them made by him; and if such statement be true, the carrier shall not be liable for damages caused by the improper loading or by the non-receipt or by the misdescription of the goods described in the bill of lading: *Provided, however,* Where the shipper of bulk freight installs and maintains adequate facilities for weighing such freight, and the same are available to the carrier, then the carrier, upon written request of such shipper and when given a reasonable opportunity so to do, shall ascertain the kind and quantity of bulk freight within a reasonable time after such written request, and the carrier shall not in such cases insert in the bill of lading the words " Shipper's weight," or other words of like purport, and if so inserted contrary to the provisions of this section, said words shall be treated as null and void and as if not inserted therein.

(Compare Uniform Act, section 23 sentence 3 and New York Act, section 209 (3).

Liability of Carrier Issuing Bill for Goods Not Received, Misdescribed or Received Later than Date Stated.
U. S. Code, Title 49, Section 102.

SEC. 22. That if a bill of lading has been issued by a carrier or on his behalf by an agent or employee the scope of whose actual or apparent authority includes the receiving of goods and issuing bills of lading therefor for transportation in commerce among the several States and with foreign nations, the carrier shall be liable to (a) the owner of goods covered by a straight bill subject to existing right of stoppage in transitu or (b) the holder of an order bill, who has given value in good faith, relying upon the description therein of the goods, or upon the shipment being made upon the date therein shown, for damages caused by the nonreceipt by the carrier of all or part of the goods upon or prior to the date therein shown, or their failure to correspond with the description thereof in the bill at the time of its issue. (As amended March 4, 1927.)

(Compare Uniform Act, section 23 and New York Act, section 209 (1)).

Garnishment.
U. S. Code, Title 49, Section 103.

SEC. 23. That if goods are delivered to a carrier by the owner or by a person whose act in conveying the title to them to a purchaser for value in good faith would bind the owner, and an order bill is issued for them, they can not thereafter, while in the possession of the carrier, be attached by garnishment or otherwise or be levied upon under an execution unless the bill be first surrendered to the carrier or its negotiation enjoined. The carrier shall in no such case be compelled to deliver the actual possession of the goods until the bill is surrendered to him or impounded by the court.

(Compare Uniform Act, section 24).

Injunction.
U. S. Code, Title 49, Section 104.

SEC. 24. That a creditor whose debtor is the owner of an order bill shall be entitled to such aid from courts of appropriate jurisdiction by injunction and otherwise in attaching such bill or in satisfying the claim by means thereof as is allowed at law or in equity in regard to property which cannot readily be attached or levied upon by ordinary legal process.

(Compare Uniform Act, section 25).

Carrier's Lien.
U. S. Code, Title 49, Section 105.

SEC. 25. That if an order bill is issued the carrier shall have a lien on the goods therein mentioned for all charges on those goods for freight, storage, demurrage and terminal charges, and expenses necessary for the preservation of the goods or incident to their transportation subsequent to the date of the bill and all other charges incurred in transportation and delivery, unless the bill expressly enumerates other charges for which a lien is claimed. In such case there shall also be a lien for the charges enumerated so far as they are allowed by law and the contract between the consignor and the carrier.

(Compare Uniform Act, section 26).

Bill for Goods Sold Under Carriers Lien.
U. S. Code, Title 49, Section 106.

SEC. 26. That after goods have been lawfully sold to satisfy a carrier's lien, or because they have not been claimed, or because they are perishable or hazardous, the carrier shall not thereafter be liable for failure to deliver the goods themselves to the consignee or owner of the goods, or to a holder of the bill given for the goods when they were shipped, even if such bill be an order bill.

(Compare Uniform Act, section 27).

Order Bill Endorsed in Blank—Negotiation by Delivery.
U. S. Code, Title 49, Section 107.

SEC. 27. That an order bill may be negotiated by delivery where, by terms of the bill, the carrier undertakes to deliver the goods to the order of a specified person, and such person or a subsequent indorsee of the bill has indorsed it in blank.

(Compare Uniform Act, section 28).

Negotiation by Endorsement.
U. S. Code, Title 49, Section 108.

SEC. 28. That an order bill may be negotiated by the indorsement of the person to whose order the goods are deliverable by the tenor of the bill. Such indorsement may be in blank or to a specified person. If indorsed to a specified person it may be negotiated again by the indorsement of such person in blank or to another specified person. Subsequent negotiation may be made in like manner.

(Compare Uniform Act, section 29).

Existing Equities Upon Transfer of Straight Bill.
U. S. Code, Title 49, Section 109.

SEC. 29. That a bill may be transferred by the holder by delivery, accompanied with an agreement, express or implied, to transfer the title to the bill or to the goods represented thereby. A straight bill cannot be negotiated free from existing equities, and the indorsement of such a bill gives the transferee no additional right.

(Compare Uniform Act, section 30).

Order Bill Negotiable by Possessor.
U. S. Code, Title 49, Section 110.

SEC. 30. That an order bill may be negotiated by any person in possession of the same, however such possession may have been acquired, if by the terms of the bill the carrier undertakes to deliver the goods to the order of such person, or if at the time of negotiation the bill is in such form that it may be negotiated by delivery.

(Compare Uniform Act, section 31).

Title Acquired by Transferee of Order Bill.
U. S. Code, Title 49, Section 111.

SEC. 31. That a person to whom an order bill has been duly negotiated acquires thereby—

(a) Such title to the goods as the person negotiating the bill to him had or had ability to convey to a purchaser in good faith for value, and also such title to the goods as the consignee and consignor had or had power to convey to a purchaser in good faith for value; and

(b) The direct obligation of the carrier to hold possession of the goods for him according to the terms of the bill as fully as if the carrier had contracted directly with him.

(Compare Uniform Act, section 32).

Transferee's Title—Notification to Carrier—Lawful Notification.
U. S. Code, Title 49, Section 112.

SEC. 32. That a person to whom a bill had been transferred, but not negotiated, acquires thereby as against the transferor the title to the goods, subject to the terms of any agreement with the transferor. If the bill is a straight bill such person also acquires the right to notify the carrier of the transfer to him of such bill and thereby to become the direct obligee of whatever obligations the carrier owed to the transferor of the bill immediately before the notification.

Prior to the notification of the carrier by the transferor or transferee of a straight bill the title of the transferee to the goods and the right to acquire the obligation of the carrier may be defeated by garnishment or by attachment or execution upon the goods by a creditor of the transferor, or by a notification to the carrier by the transferor or a subsequent purchaser from the transferor of a subsequent sale of the goods by the transferor.

A carrier has not received notification within the meaning of this section unless an officer or agent of the carrier, the actual or apparent scope of whose duties includes action upon such a notification, has been notified; and no notification shall be effective until the officer or agent to whom it is given has had time, with the exercise of reasonable diligence, to communicate with the agent or agents having actual possession or control of the goods.

(Compare Uniform Act, section 33).

Compelling Indorsement.
U. S. Code, Title 49, Section 113.

SEC. 33. That where an order bill is transferred for value by delivery, and the indorsement of the transferor is essential for negotiation, the transferee acquires a right against the transferor to compel him to indorse the bill, unless a contrary intention appears. The negotiation shall take effect as of the time when the indorsement is actually made. This obligation may be specifically enforced.

(Compare Uniform Act, section 34).

Warranties.
U. S. Code, Title 49, Section 114.

SEC. 34. That a person who negotiates or transfers for value a bill by indorsement or delivery, unless a contrary intention appears, warrants—

(a) That the bill is genuine;

(b) That he has a legal right to transfer it;

(c) That he has knowledge of no fact which would impair the validity or worth of the bill;

(d) That he has a right to transfer the title to the goods, and that the goods are merchantable or fit for a particular purpose whenever such warranties would have been implied if the contract of the parties had been to transfer without a bill the goods represented thereby.

(Compare Uniform Act, section 35).

Liability of Indorser.
U. S. Code, Title 49, Section 115.

SEC. 35. That the indorsement of a bill shall not make the indorser liable for any failure on the part of the carrier or previous indorsers of the bill to fulfill their respective obligations.

(Compare Uniform Act, section 36).

Pledgee's Warranty.
U. S. Code, Title 49, Section 116.

SEC. 36. That a mortgagee or pledgee or other holder of a bill for security who in good faith demands or receives payment of the debt for which such bill is security, whether from a party to a draft drawn for such debt or from any other person, shall not be deemed by so doing to represent or warrant the genuineness of such bill or the quantity or quality of the goods therein described.

(Compare Uniform Act, section 37).

Validity of Negotiations for Value Given in Good Faith Without Notice of Fraud, Etc.
U. S. Code, Title 49, Section 117.

SEC. 37. That the validity of the negotiation of a bill is not impaired by the fact that such negotiation was a breach of duty on the part of the person making the negotiation, or by the fact that the owner of the bill was deprived of the possession of the same by fraud, accident, mistake, duress, loss, theft, or conversion, if the person to whom the bill was negotiated, or a person to whom the bill was subsequenty negotiated, gave value therefor in good faith, without notice of the breach of duty, or fraud, accident, mistake, duress, loss, theft, or conversion.

(Compare Uniform Act, section 38).

Goods Sold, Mortgaged or Pledged—Subsequent Negotiation of Order Bill.
U. S. Code, Title 49, Section 118.

SEC. 38. That where a person, having sold, mortgaged, or pledged goods which are in a carrier's possession and for which an order bill has been issued, or having sold, mortgaged, or pledged the order bill representing such goods, continues in possession of the order bill, the subsequent negotiation thereof by that persons under any sale, pledge, or other disposition thereof to any person receiving the same in good faith, for value and without notice of the previous sale, shall have the same effect as if the first purchaser of the goods or bill had expressly authorized the subsequent negotiation.

(Compare Uniform Act, section 39).

Seller's Lien—Stoppage in Transit.
U. S. Code, Title 49, Section 119.

SEC. 39. That where an order bill has been issued for goods no seller's lien or right of stoppage in transitu shall defeat the rights of any purchaser for value in good faith to whom such bill has been negotiated, whether such negotiation be prior or subsequent to the notification to the carrier who issued such bill of the seller's claim to a lien or right of stoppage in transitu. Nor shall the carrier be obliged to deliver or justified in delivering the goods to an unpaid seller unless such bill is first surrendered for cancellation.

(Compare Uniform Act, section 42).

Rights of Lien Holder.
U. S. Code, Title 49, Section 120.

SEC. 40. That, except as provided in section thirty-nine, nothing in this act shall limit the rights and remedies of a mortgagee or lien holder whose mortgage or lien on goods would be valid, apart from this act, as against one who for value and in good faith purchased from the owner, immediately prior to the time of their delivery to the carrier, the goods which are subject to the mortgage or lien and obtained possession of them.

(Compare Uniform Act, section 43).

Forging or Counterfeiting Bills—Penalty.
U. S. Code, Title 49, Section 121.

SEC. 41. That any person who, knowingly or with intent to defraud, falsely makes, alters, forges, counterfeits, prints, or photographs any bills of lading purporting to represent goods received for shipment among the several States or with foreign nations, or with like intent utters or publishes as true and genuine any such falsely altered, forged, counterfeited, falsely printed or photographed bill of lading, knowing it to be falsely altered, forged, counterfeited, falsely printed or photographed, or aids in making, altering, forging, counterfeiting, printing or photographing, or uttering or publishing the same, or issues or aids in issuing or procurring the issue of, or negotiates or transfers for value a bill which contains a false statement as to the receipt of the goods, or as to any other matter, or who, with intent to defraud, violates, or fails to comply with, or aids in any violation of, or failure to comply with any provision of this act, shall be guilty of a misdemeanor, and, upon conviction, shall be punished for each offense by imprisonment not exceeding five years, or by a fine not exceeding $5,000, or both.

(Compare Uniform Act, sections 45-50).

Definitions.

U. S. Code, Title 49, Section 122.

SEC. 42. First.* That in this act, unless the context or subject matter otherwise requires—

" Action " includes counterclaim, set-off, and suit in equity.

" Bill " means bill of lading governed by this act.

" Consignee " means the person named in the bill as the person to whom delivery of the goods is to be made.

" Consignor " means the person named in the bill as the person from whom the goods have been received for shipment.

" Goods " means merchandise or chattels in course of transportation or which have been or are about to be transported.

" Holder " of a bill means a person who has both actual possession of such bill and a right of property therein.

" Order " means an order by indorsement on the bill.

" Person " includes a corporation or partnership, or two or more persons having a joint or common interest.

To "purchase" includes to take as mortgagee and to take as pledgee.

" State " includes any Territory, District, insular possession, or isthmian possession.

(Compare Uniform Act, section 53 (1).

Not Retroactive.

U. S. Code, Title 49, Section 123.

SEC. 43. That the provisions of this act do not apply to bills made and delivered prior to the taking effect thereof.

(Compare Uniform Act, section 54).

Sections Severable.

U. S. Code, Title 49, Section 124.

SEC. 44. That the provisions and each part thereof and the sections and each part thereof of this act are independent and severable,

---

* So in original; the federal act does not include paragraph " second " of the Uniform and New York Acts.

and the declaring of any provision or part thereof, or provisions or part thereof, or section or part thereof, or sections or part thereof, unconstitutional shall not impair or render unconstitutional any other provision or part thereof or section or part thereof.

Effective Date, Jan. 1, 1917.

Sec. 45. That this act shall take effect and be in force on and after the first day of January next after its passage.

Approved by the President, August 29, 1916.

Amendment approved by the President, March 4, 1927.

INTERNATIONAL CONVENTION FOR THE UNIFICATION OF CERTAIN RULES RELATING TO (OCEAN) BILLS OF LADING.

(The Hague Rules)

(Translation by the U. S. State Department)

ARTICLE I.—DEFINITIONS.

In this convention the following words are employed with the meanings set out below:

(a) "Carrier" includes the owner of the vessel or the charterer who enters into a contract of carriage with a shipper.

(b) "Contract of carriage" applies only to contracts of carriage covered by a bill of lading or any similar document of title, in so far as such document relates to the carriage of goods by sea; it also applies to any bill of lading or any similar document as aforesaid issued under or pursuant to a charter party from the moment at which such instrument regulates the relations between a carrier and a holder of the same.

(c) "Goods" includes goods, wares, merchandise, and articles of every kind whatsoever except live animals and cargo which by the contract of carriage is stated as being carried on deck and is so carried.

(d) "Ship" means any vessel used for the carriage of goods by sea.

(e) "Carriage of goods" covers the period from the time when the goods are loaded on to the time they are discharged from the ship.

ARTICLE II.—RISKS.

Subject to the provisions of Article VI, under every contract of carriage of goods by sea the carrier, in relation to the loading, handling, stowage, carriage, custody, care, and discharge of such goods shall be subject to the responsibilities and liabilities, and entitled to the rights and immunities hereinafter set forth.

260

ARTICLE III.—RESPONSIBILITIES AND LIABILITIES.

1. The carrier shall be bound before and at the beginning of the voyage to exercise due diligence to—

(a) Make the ship seaworthy;

(b) Properly man, equip, and supply the ship;

(c) Make the holds, refrigerating and cool chambers, and all other parts of the ship in which goods are carried, fit and safe for their reception, carriage, and preservation.

2. Subject to the provisions of Article IV the carrier shall properly and carefully load, handle, stow, carry, keep, care for, and discharge the goods carried.

3. After receiving the goods into his charge the carrier or the master or agent of the carrier shall, on demand of the shipper, issue to the shipper a bill of lading showing among other things:

(a) The leading marks necessary for identification of the goods as the same are furnished in writing by the shipper before the loading of such goods starts, provided such marks are stamped or otherwise shown clearly upon the goods if uncovered, or on the cases or coverings in which such goods are contained, in such a manner as should ordinarily remain legible until the end of the voyage;

(b) Either the number of packages or pieces, or the quantity, or weight, as the case may be, as furnished in writing by the shipper;

(c) The apparent order and condition of the goods;

Provided that no carrier, master, or agent of the carrier shall be bound to state or show in the bill of lading any marks, number, quantity, or weight which he has reasonable grounds for suspecting not accurately to represent the goods actually received or which he has had no reasonable means of checking.

4. Such a bill of lading shall be *prima facie* evidence of the receipt by the carrier of the goods as therein described in accordance with paragraph 3 (a), (b), and (c).

5. The shipper shall be deemed to have guaranteed to the carrier the accuracy at the time of shipment of the marks, number, quantity, and weight, as furnished by him, and the shipper shall indemnify the carrier against all loss, damages, and expenses arising or resulting from inaccuracies in such particulars. The right of the carrier to such indemnity shall in no way limit his responsibility and liability under the contract of carriage to any person other than the shipper.

6. Unless notice of loss or damage and the general nature of such loss or damage be given in writing to the carrier or his agent at the port of discharge before or at the time of the removal of the goods into the custody of the person entitled to delivery thereof under the contract of carriage, such removal shall be *prima facie* evidence of the delivery by the carrier of the goods as described in the bill of lading.

If the loss or damage is not apparent, the notice must be given within three days of the delivery of the goods.

The notice in writing need not be given if the state of the goods has at the time of their receipt been the subject of joint survey or inspection.

In any event the carrier and the ship shall be discharged from all liability in respect of loss or damage unless suit is brought within one year after delivery of the goods or the date when the goods should have been delivered.

In the case of any actual or apprehended loss or damage the carrier and the receiver shall give all reasonable facilities to each other for inspecting and tallying the goods.

7. After the goods are loaded the bill of lading to be issued by the carrier, master, or agent of the carrier to the shipper shall, if the shipper so demands, be a " shipped " bill of lading, provided that if the shipper shall have previously taken up any document of title to such goods, he shall surrender the same as against the issue of the " shipped " bill of lading. At the option of the carrier such document of title may be noted at the port of shipment by the carrier, master, or agent with the name or names of the ship or ships upon which the goods have been shipped and the date or dates of shipment, and when so noted, if it shows the particulars mentiond in para-

graph 3 of Article III, it shall for the purpose of this article be deemed to constitute a " shipped " bill of lading.

8. Any clause, covenant, or agreement in a contract of carriage relieving the carrier or the ship from liability for loss or damage to or in connection with goods arising from negligence, fault, or failure in the duties and obligations provided in this article, or lessening such liability otherwise than as provided in this convention, shall be null and void and of no effect. A benefit of insurance in favor of the carrier or similar clause shall be deemed to be a clause relieving the carrier from liability.

Article IV.—Rights and Immunities.

1. Neither the carrier nor the ship shall be liable for loss or damage arising or resulting from unseaworthiness unless caused by want of due diligence on the part of the carrier to make the ship seaworthy and to secure that the ship is properly manned, equipped, and supplied and to make the holds, refrigerating and cool chambers, and all other parts of the ship in which goods are carried fit and safe for their reception, carriage, and preservation in accordance with the provisions of paragraph 1 of Article III. Whenever loss or damage has resulted from unseaworthiness, the burden of proving the exercise of due diligence shall be on the carrier or other person claiming exemption under this article.

2. Neither the carrier nor the ship shall be responsible for loss or damage arising or resulting from:

(a) Act, neglect, or default of the master, mariner, pilot, or the servants of the carrier in the navigation or in the management of the ship.

(b) Fire, unless caused by the actual fault or privity of the carrier.

(c) Perils, dangers, and accidents of the sea or other navigable waters.

(d) Act of God.

(e) Act of war.

(f) Act of public enemies.

(g) Arrest or restraint of princes, rulers, or people or seizure under legal process.

(h) Quarantine restrictions.

(i) Act or omission of the shipper or owner of the goods, his agent, or representative.

(j) Strikes or lockouts or stoppage or restraint of labor from whatever cause, whether partial or general.

(k) Riots and civil commotions.

(l) Saving or attempting to save life or property at sea.

(m) Wastage in bulk or weight or any other loss or damage arising from inherent defect, quality, or vice of the goods.

(n) Insufficiency of packing.

(o) Insufficiency or inadequacy or marks.

(p) Latent defects not discoverable by due diligence.

(Q) Any other cause arising without the actual fault or privity of the carrier, or without the fault or neglect of the agents or servants of the carrier, but the burden of proof shall be on the person claiming the benefit of this exception to show that neither the actual fault or privity of the carrier nor the fault or neglect of the agents or servants of the carrier contributed to the loss or damage.

3. The shipper shall not be responsible for loss or damage sustained by the carrier or the ship arising or resulting from any cause without the act, fault, or neglect of the shipper, his agents, or his servants.

4. Any deviation in saving or attempting to save life or property at sea or any reasonable deviation shall not be deemed to be an infringement or breach of this convention or of the contract of carriage, and the carrier shall not be liable for any loss or damage resulting therefrom.

5. Neither the carrier nor the ship shall in any event be or become liable for any loss or damage to or in connection with goods in an amount exceeding 100 pounds sterling per package or unit or

the equivalent of that sum in other currency unless the nature and value of such goods have been declared by the shipper before shipment and inserted in the bill of lading.

This declaration if embodied in the bill of lading shall be *prima facie* evidence but shall not be binding or conclusive on the carrier.

By agreement between the carrier, master, or agent of the carrier and the shipper another maximum amount than that mentioned in this paragraph may be fixed, provided that such maximum shall not be less than the figure above named.

Neither the carrier nor the ship shall be responsible in any event for loss or damage to, or in connection with, goods if the nature or value thereof has been knowingly misstated by the shipper in the bill of lading.

6. Goods of an inflammable, explosive, or dangerous nature to the shipment whereof the carrier, master, or agent of the carrier has not consented with knowledge of their nature and character may at any time before discharge be landed at any place or destroyed or rendered innocuous by the carrier without compensation, and the shipper of such goods shall be liable for all damages and expenses directly or indirectly arising out of or resulting from such shipment. If any such goods shipped with such knowledge and consent shall become a danger to the ship or cargo, they may in like manner be landed at any place or destroyed or rendered innocuous by the carrier without liability on the part of the carrier except to general average, if any.

ARTICLE V.—SURRENDER OF RIGHTS AND IMMUNITIES AND INCREASE
OF RESPONSIBILITIES AND LIABILITIES.

A carrier shall be at liberty to surrender in whole or in part all or any of his rights and immunities, or to increase any of his responsibilities and liabilities under this convention provided such surrender or increase shall be embodied in the bill of lading issued to the shipper.

The provisions of this convention shall not be applicable to charter parties, but if bills of lading are issued in the case of a ship under a charter-party they shall comply with the terms of this convention. Nothing in these rules shall be held to prevent the insertion in a bill of lading of any lawful provision regarding general average.

Article VI.—Special Conditions.

Notwithstanding the provisions of the preceding articles, a carrier, master, or agent of the carrier and a shipper shall in regard to any particular goods be at liberty to enter into any agreement in any terms as to the responsibility and liability of the carrier for such goods, and as to the rights and immunities of the carrier in respect of such goods, or concerning his obligation as to seaworthiness so far as this stipulation is not contrary to public policy, or concerning the care or diligence of his servants or agents in regard to the loading, handling, stowage, carriage, custody, care, and discharge of the goods carried by sea, provided that in this case no bill of lading has been or shall be issued and that the terms agreed shall be embodied in a receipt which shall be a nonnegotiable document and shall be marked as such.

Any agreement so entered into shall have full legal effect:

Provided that this article shall not apply to ordinary commercial shipments made in the ordinary course of trade, but only to other shipments where the character or condition of the property to be carried or the circumstances, terms, and conditions under which the carriage is to be performed are such as reasonably to justify a special agreement.

Article VII.—Limitations on the Application of the Rules.

Nothing herein contained shall prevent a carrier or a shipper from entering into any agreement, stipulation, condition, reservation, or exemption as to the responsibility and liabiilty of the carrier or the ship for the loss or damage to, or in connection with, the custody and care and handling of goods prior to the loading on, and subsequent to the discharge from, the ship on which the goods are carried by sea.

Article VIII.—Limitation of Liability.

The provisions of this convention shall not affect the rights and obligations of the carrier under any statute for the time being in force relating to the limitation of the liability of owners of seagoing vessels.*

* See Part III.

ARTICLE IX.

The monetary units mentioned in this convention are to be taken to be gold value.*

Those contracting states in which the pound sterling is not a monetary unit reserve to themselves the right of translating the sums indicated in this convention in terms of pound sterling into terms of their own monetary system in round figures.†

The national laws may reserve to the debtor the right of discharging his debt in national currency according to the rate of exchange prevailing on the day of the arrival of the ship at the port of discharge of the goods concerned.

ARTICLE X.

The provisions of this convention shall apply to all bills of lading issued in any of the contracting States.‡

ARTICLE XI.

After an interval of not more than two years from the day on which the convention is signed, the Belgian Government shall place itself in communication with the governments of the high contracting parties which have declared themselves prepared to ratify the convention, with a view to deciding whether it shall be put into force. The ratifications shall be deposited at Brussels at a date to be fixed by agreement among the said governments. The first deposit of ratifications shall be recorded in a proces-verbal signed by the representatives of the powers which take part therein and by the Belgian Minister for Foreign Affairs.

The subsequent deposits of ratifications shall be made by means of a written notification, addressed to the Belgian Government and accompanied by the instrument of ratification.

---

* The effect of this provision is nullified for the United States by the declaration of the First Understanding, p. 77, infra.

† Congress has declared the equivalent to be $500 lawful money of the United States. See Section 4 (5) of the Act of April 15, 1936, p. 8, supra.

‡ The declaration of the Second Understanding is that in the event of a conflict, the text of the Act of April 15, 1936, shall prevail over the text of the Convention. The Act applies only to bills of lading in foreign commerce; hence the Convention also applies only to foreign commerce of the United States and not to domestic commerce, where its use is optional.

A duly certified copy of the proces-verbal relating to the first deposit of ratifications, of the notifications referred to in the previous paragraph, and also of the instruments of ratification accompanying them, shall be immediately sent by the Belgian Government through the diplomatic channel to the powers who have signed this convention or who have acceded to it. In the cases contemplated in the preceding paragraph the said Government shall inform them at the same time of the date on which it received the notification.

ARTICLE XII.

Nonsignatory States may accede to the present convention whether or not they have been represented at the International Conference at Brussels.

A State which desires to accede shall notify its intention in writing to the Belgian Government, forwarding to it the document of accession, which shall be deposited in the archives of the said Government.

The Belgian Government shall immediately forward to all the States which have signed or acceded to the convention a duly certified copy of the notification and of the act of accession, mentioning the date on which it received the notification.

ARTICLE XIII.

The high contracting parties may at the time of signature, ratification, or accession declare that their acceptance of the present convention does not include any or all the self-governing dominions, or of the colonies, overseas possessions, protectorates, or territories under their sovereignty or authority, and they may subsequently accede separately on behalf of any self-governing dominion, colony, overseas possession, protectorate, or territory excluded in their declaration. They may also denounce the convention separately in accordance with its provisions in respect of any self-governing dominion, or any colony, overseas possession, protectorate, or territory under their sovereignty or authority.

ARTICLE XIV.

The present convention shall take effect, in the case of the States which have taken part in the first deposit of ratifications, one year after the date of the proces-verbal recording such deposit. As respects the States which ratify subsequently or which accede, and also in cases in which the convention is subsequently put into effect in accordance with Article XIII, it shall take effect six months after the notifications specified in paragraph 2 of Article XI, and paragraph 2 of Article XII, have been received by the Belgian Government.

Article XV.

In the event of one of the contracting States wishing to denounce the present convention, the denunciation shall be notified in writing to the Belgian Government, which shall immediately communicate a duly certified copy of the notification to all the other States informing them of the date on which it was received.

The denunciation shall only operate in respect of the State which made the notification, and on the expiry of one year after the notification has reached the Belgian Government.

Article XVI.

Any one of the contracting States shall have the right to call for a fresh conference with a view to considering possible amendments.

A State which would exercise this right should notify its intention to the other States through the Belgian Government which would make arrangements for convening the conference.

Done at Brussels, in a single copy August 25, 1924.

Signatures ad referendum: Chile,* Estonia,* France, Germany, Great Britain. Hungary, Italy, Japan,* with reservations, Poland and Danzig, Rumania, Spain, United States of America, Yugoslavia.*

* Not yet (1951) implemented by ratification.

90TH CONGRESS
2D SESSION

S. 3235

A BILL

To authorize and foster joint rates for international transportation of property, to facilitate the transportation of such property, and for other purposes.

By Mr. MAGNUSON

MARCH 27, 1968

Read twice and referred to the Committee on Commerce

Be it enacted by the Senate and House of Representatives of the United States of America in Congress assembled, That this Act may be cited as the "Trade Simplification Act of 1968".

DECLARATION OF POLICY

Sec. 2. The Congress hereby declares that it is the policy of the United States to facilitate the movement of freight in international commerce and for this purpose to foster the use of joint rates by carriers by land, water, and air in the international transportation of property between places in the United States and places in foreign countries. All Federal departments and agencies concerned are directed to cooperate to the fullest extent in carrying out this policy.

DEFINITIONS

Sec. 3. As used in this Act—

(1) "Agency" means the Civil Aeronautics Board, the Federal Maritime Commission, or the Interstate Commerce Commission.

(2) "Carrier" means a common carrier subject to the jurisdiction of an agency, or a transporter of property by land, water, or air for hire between points both of which are outside the United States.

(3) "Common carrier subject to the jurisdiction of an agency" means:

(a) An air carrier, foreign air carrier, or air freight forwarder holding a certificate, permit, or operating authorization from the Civil Aeronautics Board;

(b) A common carrier by water (including a non-vessel operating common carrier by water) subject to the jurisdiction of the Federal Maritime Commission; or

(c) A common carrier subject to parts I, II, III, or IV of the Interstate Commerce Act.

(4) "International transportation" means the transportation of property by land, water or air carrier or by any combination thereof between places in the United States, on the one hand, and places in a foreign country, on the other.

270

(5) "Joint rate" means a rate jointly offered for a through service, and expressed as a single, comprehensive rate, by (a) two or more carriers, at least one of which shall be a common carrier subject to the jurisdiction of an agency, or (b) one common carrier subject to the jurisdiction of more than one agency, or (c) one common carrier subject to the jurisdiction of an agency and also performing transportation wholly outside the United States: Provided however, That an ocean rate and a charge for pickup or delivery service in the port area of origin or delivery cannot be combined to form a joint rate.

(6) "United States" includes the several States, the District of Columbia, the Commonwealth of Puerto Rico, and the territories and possessions of the United States.

ESTABLISHMENT OF JOINT RATES

Sec. 4. (a) A common carrier subject to the jurisdiction of an agency may agree to establish joint rates for international transportation which shall become effective upon compliance with section 5 of this Act. Subject to section 8 of this Act, the division of revenues, the apportionment of liability, and the pooling or interchange of equipment, or other operating matters may be fixed by the carriers participating in the joint rate arrangement. No joint rate arrangement shall prohibit any party thereto from entering into similar arrangements with other carriers.

(b) Notwithstanding any other provision of law, nothing shall prevent a freight forwarder subject to the provisions of part IV of the Interstate Commerce Act from entering into joint rates for international transportation with other common carriers.

TARIFFS

Sec. 5. Joint rates established under this Act shall be set forth in a tariff, filed, posted, and published concurrently by every participating common carrier subject to the jurisdiction of an agency with the agency having jurisdiction over that carrier. No tariffs or joint rates filed or established under this Act shall be of any lawful force and effect unless such rates or tariffs, as the case may be, are in effect with all agencies involved, and the use of any tariff or rate not so in effect shall be unlawful. The tariff, copies of which shall be available for public inspection, shall set forth all rates and charges under joint rates authorized by section 4, and all classifications, rules, regulations, practices, and services in connection therewith. Each agency may require a common carrier subject to its jurisdiction to set forth in a tariff or file with it for informational purposes the division of revenue to be collected by such carrier. The names of the several carriers which are parties to any joint tariff established under this Act shall be specified therein, and each of the parties thereto, other than the one filing the same, shall file with each agency having jurisdiction over any one of such carriers, such evidence of concurrence therein or acceptance thereof as may be required or approved under such rules and regulations as may be established under section 7 of this Act, and where such evidence of concurrence is filed, it shall not be necessary for the carriers filing the same to also file copies of the tariffs in which they are named as parties. Copies of such tariff shall be made available by the carriers to any person and a reasonable charge may be made therefor. Except as permitted under the rules and regulations promulgated under section 7 of this Act, no new joint rate shall be established nor shall any change be made in any tariff setting forth a joint rate on less than thirty days notice.

ADHERENCE TO TARIFF

Sec. 6. International transportation under joint rates shall be performed strictly in accordance with the tariff, and no common carrier subject to the jurisdiction of an agency shall demand, or collect any greater, less, or different compensation for international transportation than that specified in the tariff in which it participates. A carrier violating this section shall be subject to a civil penalty, to be imposed by the agency having jurisdiction over it, not to exceed $5,000 for each such violation, which may, in the discretion [S. 3235 — 2] of such agency, be remitted or mitigated by it. Every shipment violating this section shall constitute a separate offense.

FORMS OF TARIFFS

Sec. 7. The Interstate Commerce Commission, the Civil Aeronautics Board, and the Federal Maritime Commission, shall, after consultation with the Secretary of Transportation, jointly promulgate a single set of rules and regulations as to the form and manner of filing, posting, and publishing of tariffs setting forth joint rates established under this Act, and the conditions, if any, under which new rates may be established and changes in tariffs may be made on less than thirty days notice. An agency having jurisdiction of a carrier participating in a joint rate may reject a tariff which does not comply with the rules and regulations. The rules and regulations shall encourage to the maximum extent possible the use of simplified forms of tariffs, simplified classifications, and coordinated commodity descriptions.

JURISDICTION AND AUTHORITY OF AGENCIES

Sec. 8. Each agency may exercise for the purpose of this Act the jurisdiction and authority which it possesses under existing law, including the jurisdiction and authority each agency has to suspend, investigate, approve, or disapprove rates and practices. The jurisdiction and authority each agency has to approve with or without conditions or disapprove agreements among carriers subject to its jurisdiction is hereby extended to an agreement between such a carrier or carriers and a carrier or carriers of a different mode relating to joint rates and practices, or the interchange or pooling of equipment and facilities. Divisions of joint rates and practices in connection with joint rates, are regarded for purposes of this Act as rates and practices of carriers and, as such, are subject to all applicable statutory provisions governing the lawfulness of rates and practices. An order of an agency directed to or arising out of a carrier's participation in joint rates is subject to judicial review and enforcement as provided under existing law with regard to other orders of that agency. The agencies may hold joint hearings pursuant to rule or order on any matter within this Act.

THROUGH BILL OF LADING

Sec. 9. A carrier participating in a joint rate may issue a through bill of lading assuming responsibility from place of origin to place of destination. The through bill of lading may be in the form desired by participating carriers, if otherwise lawful, and may include or be designed to be accompanied by waybills or transportation documents prescribed or recommended by international agreement, by law or regulation of foreign governments, or by international organizations.

DAMAGES; OVERCHARGES AND UNDERCHARGES; VENUE

Sec. 10. For the purpose of determining (1) the rights and obligations of the shipper and the carrier in the event of loss or damage to goods or undercharges or overcharges, and (2) jurisdiction over actions brought in connection therewith, shipments under tariffs established pursuant to this Act shall be treated as if they were shipments moving under separate tariffs established pursuant to the Interstate Commerce Act, the Shipping Act, 1916, or the Federal Aviation Act of 1958.

INTERNATIONAL COOPERATION; REPORTS

Sec. 11. The Secretary of Transportation, in consultation with each agency and the Secretary of State, shall encourage and foster the adoption of procedures and documents facilitating prompt and efficient international transportation of goods within and without the United States. From time to time, the Secretary of Transportation shall report to the Congress on use of joint rates under this Act, on obstacles to employment of such rates, and on facilitation of international movements.

EFFECT ON EXISTING LAW

Sec. 12. This Act shall be deemed to be supplementary to the jurisdiction and authority which each agency possesses under existing law and nothing in this Act shall be construed to repeal, modify, or change any provision of the Interstate Commerce Act, the Federal Aviation Act of 1958, or the Shipping Act, 1916, or any other provision of law except to the extent that the provisions of such Acts or other laws or rules and regulations issued thereunder are clearly inconsistent with this Act: Provided, however, That in construing such Acts or other provisions of law each agency shall exercise its jurisdiction and authority under existing law so as to implement the policy set forth in section 2 of this Act.

AMENDMENTS

Sec. 13. (a) Section 1003 of the Federal Aviation Act of 1958, as amended (49 U.S.C. 1483), is amended by adding the following subsection at the end thereof:

"(f) This section does not apply to joint rates for international transportation of property; however, joint rates for international transportation of property established and filed under this section before the effective date of this amendment are not affected."

(b) Section 412 (b) of the Federal Aviation Act of 1958, as amended (49 U.S.C. 1382 (b)), is amended by inserting, immediately before the word "between", the following: "(other than an agreement relating to joint rates for international transportation of property)".

EFFECTIVE DATE

Sec. 14. This Act shall be effective ninety days after the date of enactment. Not later than the effective date, the agencies shall publish rules for the filing of tariffs.

ISO RECOMMENDATION NO. R668
DIMENSIONS AND RATINGS OF FREIGHT CONTAINERS

1. DEFINITIONS

1.1   A *freight container* is an article of transport equipment

    (a)   of a permanent character and accordingly strong enough to be suitable for repeated use;

    (b)   specially designed to facilitate the carriage of goods, by one or more modes of transport, without intermediate reloading;

    (c)   fitted with devices permitting its ready handling, particularly its transfer from one mode of transport to another;

    (d)   so designed as to be easy to fill and empty;

    (e)   having an internal volume of 1 m^3 (35.3 ft^3) or more.

The term *freight container* includes neither vehicles nor conventional packing.

1.2   *Rating*, for the purposes of this document, means the maximum gross weight and is the maximum permissible combined weight of the freight container and of its contents.

2. CLASSIFICATION AND DESIGNATION OF FREIGHT CONTAINERS

2.1   Two series of freight containers are approved :

    —   those of series 1, having a uniform cross-section of   2435 mm × 2435 mm (8 ft × 8 ft), are shown in Table 1;

    —   those of series 2, having a uniform height of 2100 mm (6 ft 10 1/2 in), are shown in Table 2.

The actual dimensions of both series 1 and series 2 freight containers, and their tolerances, are given in Table 3.

Table 1

NOMINAL DIMENSIONS OF SERIES 1 FREIGHT CONTAINERS

Freight Container Designation	SERIES 1					
	Height		Width		Nominal Length	
	mm	ft.	mm	ft.	mm	ft.
1A	2435	8	2435	8	12000	40*
1B	2435	8	2435	8	9000	30
1C	2435	8	2435	8	6000	20
1D	2435	8	2435	8	3000	10
1E	2435	8	2435	8	2000	6 2/3
1F	2435	8	2435	8	1500	5

*In certain countries there are legal limitations to this length.

Table 2

NOMINAL DIMENSIONS OF SERIES 2 FREIGHT CONTAINERS

Freight Container Designation	SERIES 2					
	Height		Width		Length	
	mm	ft.	mm	ft.	mm	ft.
2A	2100	6'11"	2300	7'7"	2920	9'7"
2B	2100	6'11"	2100	6'11"	2400	7'11"
2C	2100	6'11"	2300	7'7"	1450	4'9"

3. OVERALL DIMENSIONS AND RATINGS

3.1   The overall external dimensions, tolerances and ratings are given in Table 3.

Table 3

ACTUAL DIMENSIONS, PERMISSIBLE TOLERANCES AND RATINGS

The dimensions and tolerances apply when measured at the temperature of 20°C. (68°F.); measurements taken at other temperatures shall be adjusted accordingly.

Freight container designation	Height					Width					Length					Rating (maximum gross weight) tonnes*
	mm	Tolerances mm	ft	in	Tolerances in	mm	Tolerances mm	ft	in	Tolerances in	mm	Tolerances mm	ft	in	Tolerances in	
1A	2435	+3 / -2	8	10.5	0 / -0.1875	2435	+3 / -2	8		0 / -0.1875	12190	+2 / -8	40		0 / -0.375	30
1B	2435	+3 / -2	8		0 / -0.1875	2435	+3 / -2	8		0 / -0.1875	9125	0 / -10	29	11.25	0 / -0.375	25
1C	2435	+3 / -2	8		0 / -0.1875	2435	+3 / -2	8		0 / -0.1875	6055	+3 / -3	19	10.5	0 / -0.25	20
1D	2435	+3 / -2	8		0 / -0.1875	2435	+3 / -2	8		0 / -0.1875	2990	+1 / -4	9	9.75	0 / -0.1875	10
1E	2435	+3 / -2	8		0 / -0.1875	2435	+3 / -2	8		0 / -0.1875	1965	+3 / -2	6	5.5	0 / -0.1875	7
1F	2435	+3 / -2	8		0 / -0.1875	2435	+3 / -2	8		0 / -0.1875	1460	0 / -3	4	9.5	0 / -0.125	5
2A	2100	0 / -5	6	10.5	+0.1875 / 0	2300	0 / -5	7	6.5	+0.1875 / 0	2920	0 / -5	9	7	0 / -0.1875	7
2B	2100	0 / -5	6	10.5	+0.1875 / 0	2100	0 / -5	6	10.5	+0.1875 / 0	2400	0 / -5	7	10.5	0 / -0.1875	7
2C	2100	0 / -5	6	10.5	+0.1875 / 0	2300	0 / -5	7	6.5	+0.1875 / 0	1450	0 / -5	4	9	+0.0625 / -0.125	7

ISO RECOMMENDATION NO. R830
TERMINOLOGY RELATING TO FREIGHT CONTAINERS

1. Definitions

1.1 Freight container

By freight container is meant an article of transport equipment
   (a) of a permanent character and accordingly strong enough to be suitable for repeated use;
   (b) specially designed to facilitate the carriage of goods, by one or more modes of transport, without intermediate reloading;
   (c) fitted with devices permitting its ready handling, particularly its transfer from one mode of transport to another;
   (d) so designed as to be easy to fill and empty;
   (e) having an internal volume of 35.3 ft^3 (1 m^3) or more.
The term freight container does not include vehicles or conventional packing.

1.2 General purpose freight container

Freight container of rectangular shape, weatherproof, for transporting and storing a number of unit loads, packages or bulk material; that confines and protects the contents from loss or damage; that can be separated from the means of transport, handled as a unit load and transhipped without rehandling the contents (see Figs 1 and 2).

2. Characteristics of Freight Containers

2.1 Non-collapsible freight container

Freight container of rigid construction, the components of which are permanently assembled.

2.2 Collapsible freight container

Freight container of rigid construction, the major components of which can easily be folded or disassembled and then reassembled.

3. Freight Container Weights

3.1 Maximum gross weights

Maximum allowable total weight of freight container and its payload.

3.2 Tare weight

Weight of empty freight container.

3.3 Maximum payload

Maximum allowable weight of payload (maximum gross weight less tare weight).

3.4 Actual gross weight

Total weight of the freight container and its payload.

3.5 Actual payload

Difference between the actual gross weight and the tare weight of the freight container.

4. Freight Container Static and Dynamic Loads

4. 1   Floor load

Static and dynamic loads imposed on the floor by the payload and the wheels of handling equipment when used.

4. 2   End load

Static and dynamic loads imposed by the payload on the freight container walls and doors which are perpendicular to the longitudinal axis of the freight container.

4. 3   Side load

Static and dynamic loads imposed by the payload on the freight container walls and doors which are parallel to the longitudinal axis of the freight container.

4. 4   Roof load

External static and dynamic loads imposed on the roof of a freight container.

4. 5   Superimposed load

External static and dynamic loads imposed vertically downwards on the structure of the freight container.

5. Freight Container Dimensions and Volume

5. 1   Dimensions

Height, width and length of a freight container, measured parallel to each of its axis and expressed in this order.

5. 2   Overall external dimensions

Maximum external overall dimensions of a freight container, including any permanent attachment.

5. 3   Displacement

Volume of a freight container as determined by the multiplication of its overall external dimensions.

5. 4   Internal unobstructed dimensions

Dimensions determined on the greatest unobstructed rectangular parallelepiped that can be inscribed in the freight container, discounting corner fittings.

5. 5   Unobstructed capacity

Volume determined by the multiplication of the internal unobstructed dimensions.

5. 6   Capacity

Total internal volume.

6. Freight Container Components

6.1 Corner structures

Vertical frame component located at the corners of the freight container, integral with the corner fittings and connecting the roof and floor structures (see Fig. 3).

6.2 Corner fittings

Fittings located at the corners of the freight container which normally provide means for handling, stacking and securing the freight container (see Figs. 3 and 4).

6.3 Lifting or securing eye

System attached to the freight container consisting essentially of rings or loops intended to facilitate its lifting or its securing (see Fig. 2).

6.4 End frame

Each of the structures of the freight container perpendicular to its longitudinal axis consisting of the corner structures and the end members of the base and of the roof (see Fig. 5).

6.5 End wall

Assembly surrounded by the end frame which encloses either end of the freight container (see Fig. 5).

6.6 Side frame

Each of the structures parallel to the longitudinal axis of the freight container, consisting of the corner structures and of the bottom side rails and roof rails (see Fig. 6).

6.7 Side wall

Assembly surrounded by the side frame either side of the freight container (see Fig. 6).

6.8 Roof rails

Longitudinal structural member situated at the top edge on either side of the freight container (see Fig. 6).

6.9 Bottom side rails

Structural members situated on the longitudinal sides of the base (see Fig. 6).

6.10 End door

Door located in an end wall (see Fig. 1).

6.11 Side door

Door located in a side wall (see Fig. 7).

6.12 Roof

Assembly forming the top closure of the freight container limited by the end frames and the roof rails (see Fig. 6).

6.13 Base

    Assembly of which the principal components are
      (a) the two bottom longitudinal members,
      (b) the two bottom end members,
      (c) the floor, and
      (d) possibly, the cross members.
    (See Fig. 8).

6.14 Cross members

    Transverse components attached to the bottom side rails and supporting the floor (see Fig. 8).

6.15 Floor

    Component supporting the payload (see Fig. 8).

6.16 Skids

    Beams on which certain freight containers are mounted to facilitate handling (see Fig. 2).

6.17 Fork pockets

    Openings arranged for the entry of the forks of handling devices (see Fig. 2).

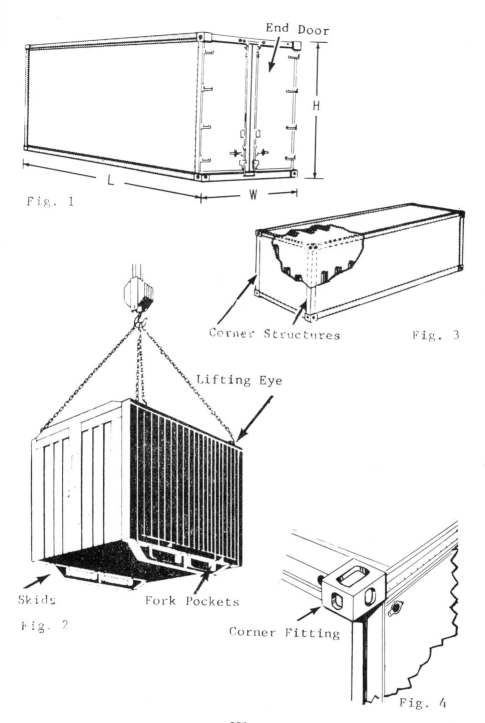

Fig. 1

End Door

H

L

W

Corner Structures   Fig. 3

Lifting Eye

Skids   Fork Pockets

Fig. 2

Corner Fitting

Fig. 4

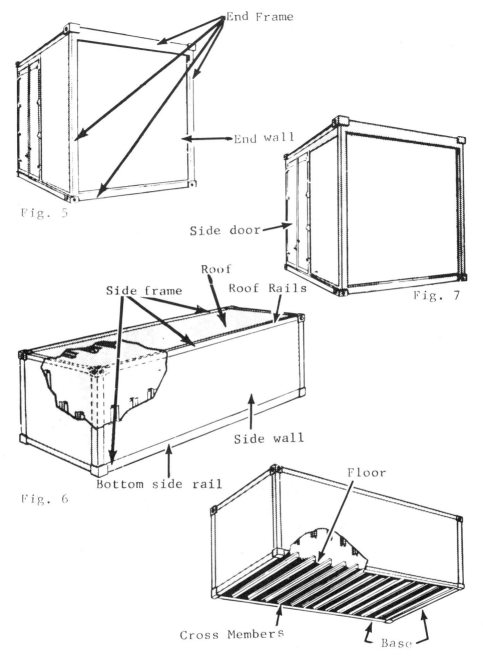

End Frame

End wall

Fig. 5

Side door

Fig. 7

Roof

Side frame

Roof Rails

Side wall

Bottom side rail

Fig. 6

Floor

Cross Members

Base

Fig. 8

International Road Transport Union

HISTORY OF IRU

One of the most remarkable phenomena of the post-war period has been the rapid development of trade at a rate which, in spite of all forecasts, considerably exceeds that of industrial production. In Europe, the rate of growth of goods transport in particular is worth noting for the kilometric tonnage is doubling about every 15-20 years. In addition with the dismantling of Customs barriers, international transport is developing even more rapidly than national transport, so that in the course of the last fifteen years the volume of international transport had been doubling every five years.

In these circumstances, it is obvious that it was necessary to set up an international organization bringing together the organizations directly concerned in the steady expansion of road transport. On the initiative of the professional associations of the transport industries (both for hire and reward and on own account) of Belgium, Denmark, France, Great Britain, Norway, the Netherlands, Sweden and Switzerland and under the impulsion of a number of personalities such as Dr F. Ernest Spat (Netherlands), Mr. Per E. Eriksson (Sweden) and Mr. Claude Leblanc (France), **the International Road Transport Union (IRU) came into being on 23rd March 1948 in Geneva,** under the auspices of the international governmental and non-governmental organizations.

On 16 February 1949, the Economic and Social Council of the United Nations (UNO) decided to grant IRU a category B **Consultative Status.**

On 2nd March 1959, the Council of Europe took the same step with IRU in view of the growing importance of its work in the field of transport.

STRUCTURE OF IRU

IRU, as the only international organization representing road transport, though it had at first dealt with purely European problems, was soon called on to let overseas countries have the benefit of its experience. At present time, it has 39 active members (National Road Transport Federations) in 23 countries and 25 associate members (associations, groups and undertakings concerned with international road transport) including 5 national associations outside Europe : Australia, Brazil, Canada, India and Japan.

IRU's basic objective, as laid down in its Constitution, is to promote the development of international road transport in the interests of road carriers and the economy as a whole. With this end in view, IRU has adopted an organization particularly suited to its task.

The Council of Direction, which takes the place of the General Assembly, is the supreme organ of the Union. Its basic task is to lay down the general policy of IRU to decide on any action falling within its scope, to reach decisions on applications for membership, to elect the members of the Presidential Executive and to amend the Constitution. It consists of **three Sections :** Section I (passenger transport for hire or reward), Section II (goods transport for hire or reward), Section III (transport on own account).

The Presidential Executive of IRU is the executive organ of the Council and supervises its general administration. It consists of the President, whose period of office does not, in principle, exceed two years, and 8 Vice-Presidents. **The Secretariat's offices** are in Geneva. It has the task of carrying out the decisions taken by the Presidential Executive and the Council and all other administrative work and of representing the Union. It ensures the liaison between the various organs of the IRU and its members and maintains relations with the other international organizations.

In addition IRU has set up four consultative **bodies :**
— the International Technical Commission,
— the Commission on Labour Questions,
— the Commission on Road Safety,
as well as
— an Audit Commission.

It has also formed two Liaison Committees for the European Economic Community :
— the Liaison Committee for road carriers for hire or reward,
— the Liaison Committee for transport on own account.

Groups of experts make their contribution in specialized fields : combined transport, customs problems, legal problems, costing, road infrastructure, etc.

Besides the ordinary meetings of the Commissions, the Sections, the Presidential Executive, and the Council, IRU organizes a **Congress** every two years, at which representatives from the principal international governmental and non-governmental organizations take part : Paris (1948), Geneva (1950), Madrid (1952), Rome (1954), Stockholm (1956), Vienna (1958), Dubrovnik (1960), Munich (1962), Cannes (1964) and London (1966).

ACTIVITIES OF IRU

Leon Say, the economist, stated, a century ago, that transport was « at the basis of all human activity » while Aldous Huxley wrote that « carrying little bits of matter from one point of the globe to another was the whole work of man ».

These statements, though apparently daring, are, in fact, absolutely correct. Road transport in particular has been proved to be the most adaptable servant of the economy and one of its best stimulants. Internationally, it is of increasing importance, while technical developments have been given it potentialities hitherto unimagined. It is found at the sources of trade through its diversity and its capacity for adaptation. Its essentially dynamic role makes it a real and powerful factor for progress.

Consequently **the tasks of IRU** are many and varied. They follow essentially from its general aims which can be considered from two aspects :

a) IRU is the representative, in all spheres as an international non-governmental organization, of road transport for hire or reward and of road transport on own account. It is its task to maintain the interests of road transport on the basis of a joint policy agreed by the member associations.

IRU coordinates viewpoints and brings together the sometimes divergent conceptions of transport for hire or reward and transport on own account. It aims at improving the conditions of existence of its members and their employees, at rationalising their work and at encouraging the adoption of the most up-to-date techniques (road stations, combined transport, etc.).

b) As it holds a Consultative Status at the United Nations and the Council of Europe, IRU is under an obligation to provide objective collaboration with these organizations and also with governmental bodies like the European Conference of Ministers of Transport (ECMT) or the European Economic Community (EEC), the meetings of which it attends.

It has also the duty of bringing the views of its members to the notice of government without, however, losing sight of the general interest.

Within these general objectives of IRU, its most urgent tasks are as follows :

Crossing frontiers

IRU is trying to promote the steady development of international road transport and to smooth away the obstacles which still hinder it. The reduction of Customs formalities for motorcoaches and for lorries is one of its main concerns.

The creation of the TIR carnet, under the auspices of the Economic Commission for Europe of the United Nations, has very very much simplified frontier controls and has led to an extraordinary upsurge of the international transport of goods by road.

IRU has also brought an International Consignment Note (CMR) into common and steadily expanding use in 10 European countries.

Further since 1st January 1965, IRU has organized a system of aid for passenger transport in difficulties abroad. To this end it has set up an International Mutual Assistance (AMI) network which provides on easy terms : emergency repairs for breakdowns, the provision of replacements of vehicles and drivers, lodging and if necessary hospital treatment for passengers and drivers.

IRU is doing everything in its power to safeguard the principle of freedom of transit.

Licences

IRU is taking every opportunity to secure an increase in the licences for international transport issued by the Public Authorities, which are still far from sufficient to meet the requirements of an ever-growing demand.

Fiscal charges

IRU, for increasing the commercial exchanges, is concerned to secure the steady lowering of fiscal charges which burden road transport and to do away with double taxation.

Labour relations

IRU is paying special attention to the harmonization of the relations between its members and their employees.

Through its Commission on Labour Questions, IRU takes place in the work of the International Labour Office (ILO), the Economic Commission for Europe of the United Nations and the European Economic Community.

PUBLICATIONS

IRU has made a series of studies of the problems of particular interest to road transport. Of these, the most important are :

— General transport policy and the European Economic Community (1958),

— Tariff policy in the transport of goods by road (1958),

— Transport on own account and European policy of transport (1962),

— General transport policy and its consequences (1962),
— Considerations on infrastructure policy (1965).
— Study and research on road technic and the economy of infrastructure (1966).

RELATIONS

IRU maintains permanent contacts with the Economic Commission for Europe of the United Nations and with the Council of Europe. It also maintains close relationships with other governmental organizations such as :
— the European Conference of Ministers of Transport (ECMT),
— the Organization for Economic Cooperation and Development (OECD),

— the European Coal and Steel Community (ECSC),
and also with the international non-governmental organizations such as :
— the International Chamber of Commerce (ICC).

— the International Road Federation (IRF),
— the International Union of Railways (UIC),
— the International Organization for Standardization (ISO),
— the International Federation of International Furniture Removers (FIDI),
— the International Institute of Refrigeration (IIR),
— the International Container Bureau (BIC),
— the International Federation of Forwarding Agents Associations (FIATA),

— Transfrigoroute Europe,
— International Cargo Handling Coordination Association (ICHCA), etc.

* * *

IRU has been granted B Consultative Status to the Economic and Social Council of the **United Nations** and as a result is called in for consultation by :

— the Economic Commission for Asia and the Far East (ECAFE),
— the Economic Commission for Latin America (ECLA),
— the Economic Commission for Africa (ECA)
and above all to the Economic Commission for Europe (ECE).

* * *

FUTURE PROSPECTS

IRU is the only international organization for road transport and as such should be a world organization. This is, indeed, its aim and it intends to broaden the scope of its activities and the radius of its action outside Europe, by bringing its aid to all the countries anxious to give road transport the place they deserve in their economies.

In conclusion, IRU is contributing to the organization of international transport both in the interests of its membership — which is daily becoming more aware of its primordial role and its force — and also in the interests of the economy as a whole, that is to say, the community in general.

CONVENTION ON THE CONTRACT
FOR THE INTERNATIONAL CARRIAGE OF GOODS BY ROAD
(CMR)

PREAMBLE

THE CONTRACTING PARTIES,

HAVING RECOGNIZED the desirability of standardizing the conditions governing the contract for the international carriage of goods by road, particularly with respect to the documents used for such carriage and to the carrier's liability,

HAVE AGREED AS FOLLOWS :

CHAPTER I

Scope of application

Article 1

1. This Convention shall apply to every contract for the carriage of goods by road in vehicles for reward, when the place of taking over of the goods and the place designated for delivery, as specified in the contract, are situated in two different countries, of which at least one is a contracting country, irrespective of the place of residence and the nationality of the parties.

2. For the purposes of this Convention, « vehicles » means motor vehicles, articulated vehicles, trailers and semi-trailers as defined in article 4 of the Convention on Road Traffic dated 19th September 1949.

3. This Convention shall apply also where carriage coming within its scope is carried out by States or by governmental institutions or organizations.

4. this Convention shall not apply :
a) to carriage performed under the terms of any international postal convention ;

b) to funeral consignments

c) to furniture removal.

5. The Contracting Parties agree not to vary any of the provisions of this Convention by special agreements between two or more of them, except to make it inapplicable to their frontier traffic or to authorize the use in transport operations entirely confined to their territory of consignment notes representing a title to the goods.

Article 2

1. Where the vehicle containing the goods is carried over part of the journey by sea, rail, inland waterways or air, and, except where the provisions of article 14 are applicable, the goods are not unloaded from the vehicle, this Convention shall nevertheless apply to the whole of the carriage. Provided that to the extent that it is proved that any loss, damage or delay in delivery of the goods which occurs during the carriage by the other means of transport was not caused by an act or omission of the carrier by road, but by some event which could only have occurred in the course of and by reason of the carriage by that other means of transport, the liability of the carrier by road shall be determined not by this Convention but in the manner in which the liability of the carrier by the other means of transport would have been determined if a contract for the carriage of the goods alone had been made by the sender with the carrier by the other means of transport in accordance with the conditions prescribed by law for the carriage of goods by that means of transport. If, however there are no such prescribed conditions, the liability of the carrier by road shall be determined by this Convention.

2. If the carrier by road is also himself the carrier by the other means of transport, his liability shall also be determined in accordance with the provisions of paragraph 1 of this article, but as if, in his capacities as carrier by road and as carrier by the other means of transport, he were two separate persons.

CHAPTER II

Persons for whom the carrier is responsible

Article 3

For the purposes of this Convention the carrier shall be responsible for the acts and omissions of his agents and servants and of any other persons of whose services he makes use for the performance of the carriage, when such agents, servants or other persons are acting within the scope of their employment, as if such acts or omissions were his own.

CHAPTER III

Conclusion and performance of the contract of carriage

Article 4

The contract of carriage shall be confirmed by the making out of consignment note. The absence, irregularity or loss of the consignment note shall not affect the existence or the validity of the contract of carriage which shall remain subject to the provisions of this Convention.

Article 5

1. The consignment note shall be made out in three original copies signed by the sender and by the carrier. These signatures may be printed or replaced by the stamps of the sender and the carrier if the law of the country in which the consignment note has been made out so permits. The first copy shall be handed to the sender, the second shall accompany the goods and the third shall be retained by the carrier.

2. When the goods which are to be carried have to be loaded in different vehicles, or are of different kinds or are divided into different lots, the sender or the carrier shall have the right to require a separate consignment note to be made out for each vehicle used, or for each kind or lot of goods.

Article 6

1. The consignment note shall contain the following particulars :
a) the date of the consignment note and the place at which it is made out ;
b) the name and address of the sender ;
c) the name and adress of the carrier ;
d) the place and the date of taking over of the goods and the place designated for delivery ;
e) the name and address of the consignee :
f) the description in common use of the nature of the goods and the method of packing, and, in the case of dangerous goods, their generally recognized description ;
g) the number of packages and their special marks and numbers ;
h) the gross weight of the goods or their quantity otherwise expressed ;
i) charges relating to the carriage (carriage charges, supplementary charges, customs duties and other charges incurred from the making of the contract to the time of delivery) ;
j) the requisite instructions for Customs and other formalities ;
k) a statement that the carriage is subject notwithstanding any clause to the contrary, to the provisions of this Convention.

2. Where applicable, the consignment note shall also contain the following particulars :
a) a statement that trans-shipment is not allowed ;
b) the carges which the sender undertakes to pay ;
c) the amount of « cash on delivery » charges ;

d) a declaration of the value of the goods and the amount representing special interest in delivery ;
e) the sender's instructions to the carrier regarding insurance of the goods ;

f) the agreed time-limit within which the carriage is to be carried out ;
g) a list of the documents handed to the carrier.

3. The parties may enter in the consignment note any other particulars which they may deem useful.

Article 7

1. The sender shall be responsible for all expenses, loss and damage sustained by the carrier by reason of the inaccuracy of :

a) the particulars specified in article 6, paragraph 1, b, d, e, f, g, h, and j ;
b) the particulars specified in article 6, paragraph 2 ;
c) any other particulars or instructions given by him to enable the consignment note to be made out or for the purpose of their being entered therein.

2. If, at the request of the sender, the carrier enters in the consignment note the particulars referred to in paragraph 1 of this article, he shall be deemed, unless the contrary is proved, to have done so on behalf of the sender.

3. If the consignment note does not contain the statement specified in article 6, paragraph 1 k), the carrier shall be liable for all expenses, loss and damage sustained through such omission by the person entitled to dispose of the goods.

Article 8

1. On taking over the goods, the carrier shall check :
a) the accuracy of the statements in the consignment note as to the number of packages and their marks and numbers, and
b) the apparent condition of the goods and their packaging.

2. Where the carrier has no reasonable means of checking the accuracy of the statements referred to in paragraph 1 a) of this article he shall enter his reservations in the consignment note together with the grounds on which they are based. He shall likewise specify the grounds for any reservations which he makes with regard to the apparent condition of the goods and their packaging. Such reservations shall not bind the sender unless he has expressly agreed to be bound by them in the consignment note.

3. The sender shall be entitled to require the carrier to check the gross weight of the goods or their quantity otherwise expressed. He may also require the contents of the packages to be checked. The carrier shall be entitled to claim the cost of such checking. The result of the checks shall be entered in the consignment note.

Article 9

1. The consignment note shall be **prima facie** evidence of the making of the contract of carriage, the conditions of the contract and the receipt of the goods by the carrier.

2. If the consignment note contains no specific reservations by the carrier, it shall be presumed, unless the contrary is proved, that the goods and their packaging appeared to be in good condition when the carrier took them over and that the number of packages, their marks and numbers corresponded with the statements in the consignment note.

Article 10

The sender shall be liable to the carrier for damage to persons, equipment or other goods, and for any expenses due to defective packing of the goods, unless the defect was apparent or known to the carrier at the time when he took over the goods and he made no reservations concerning it.

Article 11

1. For the purposes of the Customs or other formalities which have to be completed before delivery of the goods, the sender shall attach the necessary documents to the consignment note or place them at the disposal of the carrier and shall furnish him with all the information which he requires.

2. The carrier shall not be under any duty to enquire into either the accuracy or the adequacy of such documents and information. The sender shall be liable to the carrier for any damage caused by the absence, inadequacy or irregularity of such documents and information, except in the case of some wrongful act or neglect on the carrier.

3. The liability of the carrier for the consequences arising from the loss or incorrect use of the documents specified in and accompanying the consignment note or deposited with the carrier shall be that of an agent, provided that the compensation payable by the carrier shall not exceed that payable in the event of loss of the goods.

Article 12

1. The sender has the right to dispose of the goods, in particular by asking the carrier to stop the goods in transit, to change the place at which delivery is to take place or to deliver the goods to a consignee other than the consignee indicated in the consignment note.

2. This right shall cease to exist when the second copy of the consignment note is handed to the consignee or when the consignee exercises his right under article 13, paragraph 1 ; from that time onwards the carrier shall obey the orders of the consignee.

3. The consignee shall, however, have the right of disposal from the time when the consignment note is drawn up, if the sender makes an entry to that effect in the consignment note.

4. If in exercising his right of disposal the consignee has ordered the delivery of the goods to another person, that other person shall not be entitled to name other consignees.

5. The exercise of the right of disposal shall be subject to the following conditions :
a) that the sender or, in the case referred to in paragraph 3 of this article, the consignee who wishes to exercise the right produces the first copy of the consignment note on which the new instructions to the carrier have been entered and indemnifies the carrier against all expenses, loss and damage involved in carrying out such instructions ;

b) that the carrying out of such instructions is possible at the time when the instructions reach the person who is to carry them out and does not either interfere with the normal working of the carrier's undertaking or prejudice the senders or consignees of other consignments ;
c) that the instructions do not result in a division of the consignment.

6. When, by reason of the provisions of paragraph 5 (b) of this article, the carrier cannot carry out the instructions which he receives, he shall immediately notify the person who gave him such instructions.

7. A carrier who has not carried out the instructions given under the conditions provided for in this article, or who has carried them out without requiring the first copy of the consignment note to be produced, shall be liable to the person entitled to make a claim for any loss or damage caused thereby

Article 13

1. After arrival of the goods at the place designated for delivery, the consignee shall be entitled to require the carrier to deliver to him, against a receipt, the second copy of the consignment note and the goods. If the loss of the goods is established or if the goods have not arrived after the expiry of the period provided for in article 19, the consignee shall be entitled to enforce in his own name against the carrier any rights arising from the contract of carriage.

2. The consignee who avails himself of the rights granted to him under paragraph 1 of this article shall pay the charges shown to be due on the consignment note, but in the event of dispute on this matter the carrier shall not be required to deliver the goods unless security has been furnished by the consignee.

Article 14

1. If for any reason it is or becomes impossible to carry out the contract in accordance with the terms laid down in the consignment note before the goods reach the place designated for delivery, the carrier shall ask for instructions from the person entitled to dispose of the goods in accordance with the provisions of article 12.

2. Nevertheless, if circumstances are such as to allow the carriage to be carried out under conditions differing from those laid down in the consignment note and if the carrier has been unable to obtain instructions in reasonable time from the person entitled to dispose of the goods in accordance with the provisions of article 12, he shall take such steps as seem to him to be in the best interests of the person entitled to dispose of the goods.

Article 15

1. Where circumstances prevent delivery of the goods after their arrival at the place designated for delivery, the carrier shall ask the sender for his instructions. If the consignee refuses the goods the sender shall be entitled to dispose of them without being obliged to produce the first copy of the consignment note.

2. Even if he has refused the goods, the consignee may nevertheless require delivery so long as the carrier has not received instructions to the contrary from the sender.

3. When circumstances preventing delivery of the goods arise after the consignee, in exercise of his rights under article 12, paragraph 3, has given an order for the goods to be delivered to another person, paragraphs 1 and 2 of this article shall apply as is the consignee were the sender and that other person were the consignee.

Article 16

1. The carrier shall be entitled to recover the cost of his request for instructions and any expenses entailed in carrying out such instructions, unless such expenses were caused by the wrongful act or neglect of the carrier.

2. In the cases referred to in article 14, paragraph 1, and in article 15, the carrier may immediately unload the goods for account of the person entitled to dispose of them and there-upon the carriage shall be deemed to be at an end. The carrier shall then hold the goods on behalf of the person so entitled. He may, however, entrust them to a third party, and in that case be shall not be under liability except for the exercise of reasonable care in the choice of such third party. The charges due under the consignment note and all other expenses shall remain chargeable against the goods.

3. The carrier may sell the goods, without awaiting instructions from the person entitled to dispose of them, if the goods are perishable or their condition warrants such a course, or when the storage expenses would be out of proportion to the value of the goods. He may also proceed to the sale of the goods in other cases if after the expiry of a reasonable period he has not received from the person entitled to dispose of the goods instructions to the contrary which he may reasonably be required to carry out.

4. If the goods have been sold pursuant to this article, the proceeds of sale, after deduction of the expenses chargeable against the goods, shall be placed at the disposal of the person entitled to dispose of the goods. If these charges exceed the proceeds of sale, the carrier shall be entitled to the difference.

5. The procedure in the case of sale shall be determined by the law or custom of the place where the goods are situated.

CHAPTER IV

Liability of the carrier

Article 17

1. The carrier shall be liable for the total or partial loss of the goods and for damage there to occurring between the time when he takes over the goods and the time of delivery, as well as for any delay in delivery.

2. The carrier shall however be relieved of liability if the loss, damage or delay was caused by the wrongful act or neglect of the claimant, by the instructions of the claimant given otherwise than as the result of a wrongful act or neglect on the part of the carrier, by inherent vice of the goods or through circumstances which the carrier could not avoid and the consequences of which he was unable to prevent.

3. The carrier shall not be relieved of liability by reason of the defective condition of the vehicle used by him in order to perform the carriage, or by reason of the wrongful act or neglect of the person from whom he may have hired the vehicle or of the agents or servants of the latter.

4. Subject to article 18, paragraphs 2 to 5, the carrier shall be relieved of liability when the loss or damage arises from the special risks inherent in one or more of the following circumstances :

a) use of open unsheeted vehicles, when their use has been expressly agreed and specified in the consignment note ;

b) the lack of, or defective condition of packing in the case of goods which, by their nature, are liable to wastage or to be damaged when not packed or when not properly packed ;

c) handling, loading, stowage or unloading of the goods by the sender, the consignee or persons acting on behalf of the sender or the consignee ;

d) the nature of certain kinds of goods which particulary exposes them to total or partial loss or to damage, especially through breakage, rust, decay, desiccation, leakage, normal wastage, or the action of moth or vermin ;

e) insufficiency or inadequacy of marks or numbers on the packages ;

f) the carriage of livestock.

5. Where under this, article the carrier is not under any liability in respect of some of the factors causing the loss, damage or delay, he shall only be liable to the extent that those factors for which he is liable under this article have contributed to the loss, damage or delay.

Article 18

1. The burden of proving that loss, damage or delay was due to one of the causes specified in article 17, paragraph 2, shall rest upon the carrier.

2. When the carrier establishes that in the circumstances of the case, the loss or damage could be attributed to one or more of the special risks referred to in article 17, paragraph 4, it shall be presumed that it was so caused. The claimant shall however be entitled to prove that the loss or damage was not, in fact, attributable either wholly or partly to one of these risks.

3. This presumption shall not apply in the circumstances set out in artcile 17, paragraph 4 (a), if there has been an abnormal shortage, or a loss of any package.

4. If the carriage is performed in vehicles specially equipped to protect the goods from the effects of heat, cold, variations in temperature or the humidity of the air, the carrier shall not be entitled to claim the benefit of article 17, paragraph 4 (d), unless he proves that all steps incumbent on him in the circumstances with respect to the choice, maintenance and use of such equipment were taken and that he complied with any special instructions issued to him.

5. The carrier shall not be entitled to claim the benefit of article 17, paragraph 4 (f), unless he proves that all steps normally incumbent on him in the circumstances were taken and that he complied with any special instructions issued to him.

Article 19

Delay in delivery shall be said to occur when the goods have not been delivered within the agreed time-limit or when, failing an agreed time-limit, the actual duration of the carriage having regard to the circumstances of the case, and in particular, in the case of partial loads, the time required for making up a complete load in the normal way, exceeds the time it would be reasonable to allow a diligent carrier.

Article 20

1. The fact that goods have not been delivered within thirty days following the expiry of the agreed time-limit, or if there is no agreed time-limit, within sixty days from the time when the carrier took over the goods, shall be conclusive evidence of the loss of the goods, and the person entitled to make a claim may thereupon treat them as lost.

2. The person so entitled may, on receipt of compensation for the missing goods, request in writing that he shall be notified immediately should the goods be recovered in the course of the year following the payment of compensation. He shall be given a written acknowledgment of such request.

3. Within the thirty days following receipt of such notification, the person entitled as aforesaid may require the goods to be delivered to him against payment of the charges shown to be due on the consignment note and also against refund of the compensation he received less any charges included therein but without prejudice to any claims to compensation for delay in delivery under article 23 and, where applicable, article 26.

4. In the absence of the request mentioned in paragraph 2 or of any instructions given within the period of thirty days specified in paragraph 3, or if the goods are not recovered until more than one year after the payment of compensation, the carrier shall be entitled to deal with them in accordance with the law of the place where the goods are situated.

Article 21

Should the goods have been delivered to the consignee without collection of the« cash on delivery » charge which should have been collected by the carrier under the terms of the contract of carriage, the carrier shall be liable to the sender for compensation not exceeding the amount of such charge without prejudice to his right of action against the consignee.

Article 22

1. When the sender hands goods of a dangerous nature to the carrier, he shall inform the carrier of the exact nature of the danger

and indicate, if necessary, the precautions to be taken. If this information has not been entered in the consignment note, the burden of proving, by some other means, that the carrier knew the exact nature of the danger constituted by the carriage of the said goods shall rest upon the sender or the consignee.

2. Goods of a dangerous nature which, in the circumstances referred to in paragraph 1 of this article, the carrier did not know were dangerous, may, at any time or place, be unloaded, destroyed or rendered harmless by the carrier without compensation ; further, the sender shall be liable for all expenses, loss or damage arising out of their handing over for carriage or of their carriage.

Article 23

1. When, under the provisions of this Convention, a carrier is liable for compensation in respect of total or partial loss of goods, such compensation shall be calculated by reference to the value of the goods at the place and time at which they were accepted for carriage.

2. The value of the goods shall be fixed according to the commodity exchange price or, if there is no such price, according to the current market price or, if there is no commodity exchange, price or current market price, by reference to the normal value of goods of the same kind and quality.

3. Compensation shall not, however, exceed 25 francs per kilogram of gross weight short. « Franc » means the gold franc weighing 10/31 of a gramme and being of millesimal fineness 900.

4. In addition, the carriage charges, Customs duties and other charges incurred in respect of the carriage of the goods shall be refunded in full in case of total loss and in proportion to the loss sustained in case of partial loss, but no further damages shall be payable.

5. In the case of delay, if the claimant proves that damage has resulted therefrom the carrier shall pay compensation for such damage not exceeding the carriage charges.

6. Higher compensation may only be claimed where the value of the goods or a special interest in delivery has been declared in accordance with articles 24 and 26.

Article 24

The sender may, against payment of a surcharge to be agreed upon, declare in the consignment note a value for the goods exceeding the limit laid down in article 23, paragraph 3, and in that case the amount of the declared value shall be substituted for that limit.

Article 25

1. In case of damage, the carrier shall be liable for the amount by which the goods have diminished in value, calculated by reference to the value of the goods fixed in accordance with article 23, paragraphs 1, 2 and 4.

2. The compensation may not, however, exceed :
a) if the whole consignment has been damaged, the amount payable in the case of total loss ;
b) if part only of the consignment has been damaged, the amount payable in the case of loss of the part affected.

Article 26

1. The sender may, against payment of a surcharge to be agreed upon, fix the amount of a special interest in delivery in the case of loss or damage or of the agreed time-limit being exceeded, by entering such amount in the consignment note.

2. If a declaration of a special interest in delivery has been made, compensation for the additional loss or damage proved may be claimed, up to the total amount of the interest declared, independently of the compensation provided for in articles 23, 24 and 25.

Article 27

1. The claimant shall be entitled to claim interest on compensation payable. Such interest, calculated at five per centum per annum, shall accrue from the date on which the claim was sent in writing to the carrier or, if no such claim has been made, from the date on which legal proceedings were instituted.

2. When the amounts on which the calculation of the compensation is based are not expressed in the currency of the country in which payment is claimed, conversion shall be at the rate of exchange applicable on the day and at the place of payment of compensation.

Article 28

1. In cases where, under the law applicable, loss, damage or delay arising out of carriage under this Convention gives rise to an extracontractual claim, the carrier may avail himself of the provisions of this Convention which exclude his liability or which fix or limit the compensation due.

2. In cases where the extra-contractual liability for loss damage or delay of one of the persons for whom the carrier is responsible under the terms of article 3 is in issue, such person may also avail himself of the provisions of this Convention which exclude the

liability of the carrier or which fix or limit the compensation due.

Article 29

1. The carrier shall not be entitled to avail himself of the provisions of this chapter which exclude or limit his liability or which shift the burden of proof if the damage was caused by his wilful misconduct or by such default on his part as, in accordance with the law of the court or tribunal seized of the case, is considered as equivalent to wilful misconduct.

2. The same provision shall apply if the wilful misconduct or default is commited by the agents or servants of the carrier or by any other persons of whose services he makes use for the performance of the carriage when such agents, servants or other persons are acting within the scope of their employment. Furthermore, in such a case such agents, servants or other persons shall not be entitled to avail themselves, with regard to their personal liability, of the provisions of this chapter referred to in paragraph 1.

CHAPTER V

Claims and actions

Article 30

1. If the consignee takes delivery of the goods without duly checking their condition with the carrier or without sending him reservations giving a general indication of the loss or damage, not later than the time of delivery in the case of apparent loss or damage, and within seven days of delivery, Sundays and Public Holidays excepted, in the case of loss or damage which is not apparent, the fact of his taking delivery shall be **prima facie** evidence that he has received the goods in the condition described in the consignment note. In the case of loss or damage which is not apparent the reservations referred to shall be made in writing.

2. When the condition of the goods has been duly checked by the consignee and the carrier, evidence contradicting the result of this checking shall only be admissible in the case of loss or damage which is not apparent and provided that the consignee has duly sent reservations in writing to the carrier within seven days, Sundays and Public holidays excepted from the date of checking.

3. No compensation shall be payable for delay in delivery unless a reservation has been sent in writing to the carrier within twenty-one days from the time that the goods were placed at the disposal of the consignee.

4. In calculating the time-limits provided for in this article the date of delivery, or the date of checking, or the date when the goods were placed at the disposal of the consignee, as the case may be, shall not be included.

5. The carrier and the consignee shall give each other every reasonable facility for making the requisite investigations and checks.

Article 31

1. In legal proceedings arising out of carriage under this Convention, the plaintiff may bring an action in any court or tribunal of a contracting country designated by agreement between the parties and, in addition, in the courts or tribunals of a country within whose territory :
a) the defendant is ordinarily resident, or has his principal place of business, or the branch or agency through which the contract of carriage was made, or
b) the place where the goods were taken over by the carrier or the place designated for delivery is situated,
and in no other courts or tribunals.

2. Where in respect of a claim referred to in paragraph 1 of this article an action is pending before a court or tribunal competent under that paragraph, or where in respect of such a claim a judgment has been entered by such a court or tribunal no new action shall be started between the same parties on the same grounds unless the judgment of the court or tribunal before which the first action was brought is not enforceable in the country in which the fresh proceedings are brought.

3. When a judgment entered by a court or tribunal of a contracting country in any such action as is referred to in paragraph 1 of this article has become enforceable in that country, it shall also become enforceable in each of the other contracting States, as soon as the formalities required in the country concerned have been complied with. These formalities shall not permit the merits of the case to be re-opened.

4. The provisions of paragraph 3 of this article shall apply to judgments after ·trial, judgments by default and settlements confirmed by an order of the court, but shall not apply to interim judgments or to awards of damages, in addition to costs against a plaintiff who wholly or partly fails in his action.

5. Security for costs shall not be required in proceedings arising out of carriage under this Convention from nationals of contracting countries resident or having their place of business in one of those countries.

Article 32

1. The period of limitation for an action arising out of carriage under this Convention

shall be one year. Nevertheless, in the case of wilful misconduct, or such default as in accordance with the law of the court or tribunal seised of the case, is considered as equivalent to wilful misconduct, the period of limitation shall be three yaers. The period of limitation shall begin to run :

a) in the case of partial loss, damage or delay in delivery, from the date of delivery ;

b) in the case of total loss, from the thirtieth day after the expiry of the agreed time-limit or where there is no agreed time-limit from the sixtieth day from the date on which the goods were taken over by the carrier ;

c) in all other cases, on the expiry of a period of three months after the making of the contract of carriage.

The day on which the period of limitation begins to run shall not be included in the period.

2. A written claim shall suspend the period of limitation until such date as the carrier rejects the claim by notification in writing and returns the documents attached thereto. If a part of the claim is admitted the period of limitation shall start to run again only in respect of that part of the claim still in dispute. The burden of proof of the receipt of the claim, or of the reply and of the return of the documents, shall rest with the party relying upon these facts. The running of the period of limitation shall not be suspended by further claims having the same object.

3. Subject to the provisions of paragraph 2 above, the extension of the period of limitation shall be governed by the law of the court or tribunal seised of the case. That law shall also govern the fresh accrual of rights of action.

4. A right of action which has become barred by lapse of time may not be exercised by way of counter-claim or set-off.

Article 33

The contract of carriage may contain a clause conferring competence on an arbitration tribunal if the clause conferring competence on the tribunal provides that the tribunal shall apply this Convention.

CHAPTER VI

Provisions relating to carriage performed by successive carriers

Article 34

If carriage governed by a single contract is performed by successive road carriers, each of them shall be responsible for the performance of the whole operation, the second carrier and each succeeding carrier becoming a party to the contract of carriage, under the terms of the consignment note, by reason of his acceptance of the goods and the consignment note.

Article 35

1. A carrier accepting the goods from a previous carrier shall give the latter a dated and signed receipt. He shall enter his name and address on the second copy of the consignment note. Where applicable, he shall enter on the second copy of the consignment note on the receipt reservations of the kind provided for in article 8, paragraph 2.

2. The provisions of article 9 shall apply to the relations between successive carriers.

Article 36

Except in the case of a counter-claim or a set-off raised in an action concerning a claim based on the same contract of carriage, legal proceedings in respect of liability for loss, damage or delay may only be brought against the first carrier, the last carrier or the carrier who was performing that portion of the carriage during which the event causing the loss, damage or delay occured ; an action may be brought at the same time against several of these carriers.

Article 37

A carrier who has paid compensation in compliance with the provisions of this Convention, shall be entitled to recover such compensation, together with interest thereon and all costs and expenses incurred by reason of the claim, from the other carriers who have taken part in the carriage, subject to the following provisions :

a) the carrier responsible for the loss or damage shall be solely liable for the compensation whether paid by himself or by another carrier ;

b) when the loss or damage has been caused by the action of two or more carriers, each of them shall pay an amount proportionate to his share of liability ; should it be impossible to apportion the liability, each carrier shall be liable in proportion to the share of the payment for the carriage which is due to him ;

c) if it cannot be ascertained to which carriers liability is attributable for the loss or damage, the amount of the compensation shall be apportioned between all the carriers as laid down in b) above.

Article 38

If one of the carriers is insolvent, the share of the compensation due from him and unpaid by him shall be divided among the other carriers in proportion to the share of the payment for the carriage due to them.

Article 39

1. No carrier against whom a claim is made under articles 37 and 38 shall be entitled to dispute the validity of the payment made by the carrier making the claim if the amount of the compensation was determined by judicial authority after the first mentioned carrier had been given due notice of the proceedings and afforded an opportunity of entering an appearance.

2. A carrier wishing to take proceedings to enforce his right of recovery may make his claim before the competent court or tribunal of the country in which one of the carriers concerned is ordinarily resident, or has his principal place of business or the branch or agency through which the contract of carriage was made. All the carriers concerned may be made defendants in the same action.

3. The provisions of article 31, paragraphs 3 and 4, shall apply to judgments entered in the proceedings referred to in articles 37 and 38.

4. The provisions of article 32 shall apply to claims between carriers. The period of limitation shall, however, begin to run either on the date of the final judicial decision fixing the amount of compensation payable under the provisions of this Convention, or, if there is no such judicial decision, from the actual date of payment.

Article 40

Carriers shall be free to agree among themselves on provisions other than those laid down in articles 37 and 38.

CHAPTER VII

Nullity of stipulations contrary to the Convention

Article 41

1. Subject to the provisions of article 40, any stipulation which would directly or indirectly derogate from the provisions of this Convention shall be null and void. The nullity of such a stipulation shall not involve the nullity of the other provisions of the contract.

2. In particular, a benefit of insurance in favour of the carrier or any other similar clause, or any clause shifting the burden of proof shall be null and void.

CHAPTER VIII

Final provisions

Article 42

1. This Convention is open for signature or accession by countries members of the Economic Commission for Europe and countries admitted to the Commission in a consultative capacity under paragraph 8 of the Commission's terms of reference.

2. Such countries as may participate in certain activities of the Economic Commission for Europe in accordance with paragraph 11 of the Commission's terms of reference may become Contracting Parties to this Convention by acceding thereto after its entry into force.

3. The Convenion shall be open for signature until 31 August 1956 inclusive. Thereafter, it shall be open for accession.

4. This Convention shall be ratified.

5. Ratification or accession shall be effected by the deposit of an instrument with the Secretary-General of the United-Nations.

Article 43

1. This Convention shall come into force on the ninetieth day after five of the countries referred to in article 42, paragraph 1, have deposited their instruments of ratification or accession.

2. For any country ratifying or acceding to it after five countries have deposited their instruments of ratification or accession, this Convention shall enter into force on the ninetieth day after the said country has deposited its instrument of ratifiaction or accession.

Article 44

1. Any Contracting Party may denounce this Convention by so notifying the Secretary-General of the United-Nations.

2. Denunciation shall take effect twelve months after the date of receipt by the Secretary-General of the notification of denunciation.

Article 45

If, after the entry into force of this Convention, the number of Contracting Parties is reduced, as a result of denunciations, to less than five, the Convention shall cease to be in force from the date on which the last of such denunciations takes effect.

Article 46

1. Any country may, at the time of depositing its instrument of ratification or accession or at any time thereafter, declare, by notification adressed to the Secretary-General of the United Nations that this Convention shall extend to all or any of the territories for the international relations of which it is responsible. The Convention shall extend to the territory or territories named in the notification as from the ninetieth day after its receipt by the Secretary-General or, if on that day the Convention has not yet entered into force, at the time of its entry into force.

2. Any country which has made a declaration under the preceding paragraph extending this Convention to any territory for whose international relations it is responsible may denounce the Convention separately in respect of that territory in accordance with the provisions of article 44.

Article 47

Any dispute between two or more Contracting Parties relating to the interpretation or application of this Convention, which the parties are unable to settle by negociation or other means may, at the request of any one of the Contracting Parties concerned, be referred for settlement to the International Court of Justice.

Article 48

1. Each Contracting Party may, at the time of signing, ratifying, or acceding to, this Convention, declare that it does not consider itself as bound by article 47 of the Convention. Other Contracting Parties shall not be bound by article 47 in respect of any Contracting Party which has entered such a reservation.

2. Any Contracting Party having entered a reservation as provided for in paragraph 1 may at any time withdraw such reservation by notifying the Secretary-General of the United Nations.

3. No other reservation to this Convention shall be permitted.

Article 49

1. After this Convention has been in force for three years, any Contracting Party may, by notification to the Secretary-General of the United Nations, request that a conference be convened for the purpose of reviewing the Convention. The Secretary-General shall notify all Contracting Parties of the request and a review conference shall be convened by the Secretary-General if, within a period of four months following the date of notification by the Secretary-General, not less than one-fourth of the Contracting Parties notify him of their concurrence with the request.

2. Il a conference is convened in accordance with the preceding paragraph, the Secretary-General shall notify all the Contracting Parties and invite them to submit within a period of three months such proposals as they may wish the Conference to consider. The Secretary-General shall circulate to all Contracting Parties the provisional agenda for the conference together with the texts of such proposals at least three months before the date on which the conference is to meet.

3. The Secretary-General shall invite to any conference convened in accordance with this article all countries referred to in article 42, paragraph 1, and countries which have become Contracting Parties under article 42, paragraph 2.

Article 50

In addition to the notifications provided for in article 49, the Secretary-General of the United Nations shall notify the countries referred to in article 42, paragraph 1, and the countries which have become Contracting Parties under article 42, paragraph 2, of :

a) ratifications and accessions under article 42 ;
b) the dates of entry into force of this Convention in accordance with article 43 ;
c) denunciations under article 44 ;
d) the termination of this Convention in accordance with article 45 ;
e) notifications received in accordance with article 46 ;
f) declarations and notifications received in accordance with article 48, paragraphs 1 and 2.

The IRU/TIR carnet, for simplifying Customs formalities for the international transport of goods by road, is issued by the Secretariat-General of IRU to its national associations. More and more commercial vehicles, carrying the TIR registration plate, have, from 1950 up to the present, been crossing the frontiers of 20 European countries : Austria, Belgium, Bulgaria, Czechoslovakia, Denmark, Finland, France, Federal German Republic, Hungary, Italy, Luxemburg, Netherlands, Norway, Poland, Rumania, Sweden, Spain, Switzerland, United Kingdom, and Yugoslavia.

TIR CONVENTION

Customs Convention on the international transport of goods under cover of TIR carnets, on 15 January 1959.

CHAPTER I

DEFINITIONS

Article 1

For the purpose of this Convention :

a) the term « import or export duties and taxes » shall mean not only Customs duties but also all duties and taxes whatsoever chargeable by reason of importation or exportation ;

b) the term « road vehicle » shall mean not only any road motor vehicle but also any trailer or semi-trailer designed to be drawn by such a vehicle ;

c) the term « container » shall mean an article of transport equipment (lift-van, movable tank or other similar structure) ;

    I. of a permanent character and accordingly strong enough to be suitable for repeated use ;

    II. specially designed to facilitate the carriage of goods, by one or more modes of transport, without intermediate reloading ;

    III. fitted with devices permitting its ready handling particularly its transfer from one mode of transport to another ;

    IV. so designed as to be easy to fill and empty ; and

    V. having an internal volume of one cubic metre or more ;

the term « container » includes neither vehicles nor conventional packing ;

d) the term « Customs office of departure » shall mean any inland or frontier Customs office of a Contracting Party where the system provided by this Convention begins to apply to an international transport by road vehicle of a load or part-load of goods ;

e) the term « Customs office of destination » shall mean any inland or frontier Customs office of a Contracting Party where the system provided by this Convention ceases to apply to an international transport by road vehicle of a load or part-load of goods ;

f) the term « Customs office **en route** » shall mean any frontier Customs office of a Contracting Party which a road vehicle merely passes through in the course of an international transport under the system provided by this Convention ;

g) the term « persons » shall mean both natural and legal persons ;

h) the term « heavy or bulky goods » shall mean any object which, in the opinion of the Customs authorities of the Customs office of departure, cannot readily be dismantled for transport and of which

    I. the weight exceeds 7000 kg ; or

    II. one dimension exceeds 5 metres ; or

    III. two dimensions exceed 2 metres ; or

    IV. the height, taking account of the loading position, exceeds 2 metres.

CHAPTER II

SCOPE

Article 2

This Convention shall apply to the transport of goods without intermediate reloading across one or more frontiers between a Customs office of departure of one Contracting Party and a Customs office of destination of another Contracting Party, or of the same Contracting Party, in road vehicles, or in containers carried on such vehicles, notwithstanding that such vehicles are carried on another means of transport for part of the journey between the offices of departure and destination.

Article 3

For the provisions of this Convention to become applicable :

a) transport must be performed under the conditions set forth in Chapter III by means of road vehicles or containers previously approved ; however, in the territory of Contracting Parties who have entered no reservation in accordance with paragraph 1 of Article 45 of this Convention, it may also, save in the cases covered by paragraph 2 of that Article, be performed by means of other road vehicles under the conditions set forth in Chapter IV ;

b) transport must be guaranteed by associations approved in accordance with the provisions of Article 5 and be performed under cover of a document known as the TIR carnet.

CHAPTER III

PROVISIONS CONCERNING TRANSPORT IN SEALED ROAD VEHICLES OR SEALED CONTAINERS

Article 4

Provided the conditions laid down in this Chapter and in Chapter V are fulfilled, goods carried in sealed road vehicles or in sealed containers carried on road vehicles

a) shall not be subjected to the payment or deposit of import or export duties and taxes at Customs offices **en route ;** and

b) shall not, as a general rule, be subjected to Customs examination at such offices.

However, in order to prevent abuse, the Customs authorities may, in exceptional cases and particularly when irregularity is suspected, carry out at such offices a summary or full examination of the goods.

Article 5

1. Subject to such conditions and guarantees as it shall determine, each Contracting Party may authorize associations to issue TIR carnets either directly or through corresponding associations, and to act as guarantors.

2. An association shall not be approved in any country unless its guarantee covers the responsibilities incurred in that country in connexion with operations under cover of TIR carnets issued by foreign associations affiliated to the same international organization as that to which it is itself affiliated.

Article 6

1. The guaranteeing association shall undertake to pay the import or export duties and taxes due, any interest due thereon, any other charges, and any pecuniary penalties incurred by the holder of the TIR carnet and the persons participating in the performance of the transport under the Customs laws and regulations of the country in which an offence has been committed. It shall be liable, jointly and severally with the persons from whom the sums mentioned above are due, for payment of such sums.

2. The fact that Customs authorities authorize the examination of the goods elsewhere than at a place where the business of Customs offices of departure ·or destination is usually conducted shall not affect the liability of the guaranteeing association.

3. The liability of the guaranteeing association to the authorities of a given country shall run only from the time when the TIR carnet is accepted by the Customs authorities of that country.

4. The liability of the guaranteeing association shall cover not only such goods as are enumerated in the TIR carnet, but also goods which, though not enumerated therein, are contained in the sealed section of the road vehicle or in the sealed container. It shall not extend to other goods.

5. For the purposes of determining the duties, taxes and, where applicable, pecuniary penalties mentioned in paragraph 1 of this Article, the particulars of the goods as entered in the TIR carnet shall be valid in the absence of proof to the contrary.

6. When the Customs authorities of a country have unconditionally discharged a TIR carnet they can no longer claim from the guaranteeing association paragraph 1 of this Article unless the certificate of discharge was obtained improperly or fraudulently.

7. Where a TIR carnet has not been discharged or has been discharged conditionally the competent authorities shall not have the right to claim from the guaranteeing association payment of the amounts mentioned in paragraph 1 of this Article unless, within one year of the date upon which the TIR carnet was taken on charge, they have notified the association of the non-discharge or conditional discharge. The same provision shall apply where the certificate of discharge was obtained improperly or fraudulently, save that the period shall be two years.

8. The claim for payment referred to in paragraph 1 of the present Article shall be made to the guaranteeing association within three years of the date when the association was informed that the carnet had not been discharged or had been discharged subject to a reservation or that the certificate of discharge had been obtained improperly or fraudulently. However, in cases which, during the above-mentioned period of three years, become the subject of legal proceedings, any claim for payment shall be made within one year of the date when the decision of the court becomes enforceable.

9. The guaranteeing association shall have a period of three months, from the date when a claim for payment is made upon it, in which to pay the amounts claimed. The amounts paid shall be reimbursed to the association if, within a periode twelve months from the date on which the claim for payment was made, it is established to the satisfaction of the Customs authorities that no irregularity took place as regards the transport operation in question.

Article 7

1. The TIR carnet shall conform to the standard form contained in Annex 1 to this Convention.

2. A TIR carnet shall be made out in respect of each road vehicle or container. Such carnet shall be valid for one journey only ; it shall contain such number of detachable vouchers for Customs control and discharge as are required for the transport operation concerned.

Article 8

Transport under cover of a TIR carnet may involve several Customs offices of departure and destination ; but, save as otherwise authorized by the Contracting Party or Parties concerned,

a) the Customs offices of departure shall be situated in the same country,

b) the Customs offices of destination shall be situated in not more than two countries, and

c) the total number of Customs offices of departure and destination shall not exceed four.

Article 9

At the Customs office of departure the goods, the road vehicle and, where appropriate, the container, shall be produced with the Customs authorities together with the TIR carnet for checking and the affixing of Customs seals.

Article 10

For journeys on the territory of their country, the Customs authorities may fix a time-limit and require the road vehicle to follow a stipulated itinerary.

Article 11

At each Customs office **en route** and at Customs offices of destination the road vehicle or container shall be produced with its load to the Customs authorities, together with the TIR carnet relating to the load.

Article 12

Save where they examine the goods in accordance with the last sentence of Article 4, the Customs authorities of the Customs offices **en route** of each of the Contracting Parties shall respect the seals affixed by the Customs authorities of the other Contracting Parties. They may, however, affix additional seals of their own.

Article 13

In order to prevent abuse, the Customs authorities may, if they consider it necessary,

a) in special cases require road vehicles to be escorted on the territory of their country, at the carrier's expense ;

b) require examination of road vehicles, containers and their loads to be carried out **en route.**

Loads shall be examined only in exceptional cases.

Article 14

If the Customs authorities conduct an examination of the load of a road vehicle or of a container at a Customs office **en route** or in the course of the journey, they shall record on the TIR carnet vouchers used in their country and on the corresponding counterfoils particulars of the new seals affixed.

Article 15

On arrival at the Customs office of destination, the TIR carnet shall be discharged without delay. If, however, the goods are not immediately entered under another Customs regime, the Customs authorities may reserve the right to make discharge of the carnet conditional upon a new liability being substituted for that of the association guaranteeing the said carnet.

Article 16

When it is established to the satisfaction of the Customs authorities that goods the subject of a TIR carnet have been destroyed by force majeure, exemption from payment of the duties and taxes normally chargeable shall be granted.

Article 17

1. In order to fall within the provisions of this chapter, road vehicles must fulfil the conditions as regards construction and equipment set out in Annex 3 to this Convention and containers those set out in Annex 6.

2. Road vehicles and containers shall be approved according to the procedures laid down in Annexes 4 and 7 to this Convention ; the certificates of approval shall conform to the specimens reproduced in Annexes 5 and 8.

Article 18

1. No special document shall be required for a container used under cover of a TIR carnet, provided the characteristics and value of the container are entered in the « Goods Manifest » of the TIR carnet.

2. The provisions of paragraph 1 of this Article shall not prevent a Contracting Party requiring the fulfilment at the Customs office of destination of the formalities laid down by its national regulations or taking measures to prevent the container being used for a fresh consignment of goods intended for delivery within its territory.

CHAPTER IV

PROVISIONS CONCERNING TRANSPORT OF HEAVY OR BULKY GOODS

Article 19

1. The benefit of the provisions of this Chapter shall extend only to the transport of goods which are heavy or bulky goods as defined in sub-paragraph (h) of Article 1 of this Convention.

2. The benefit of the provisions of this Chapter shall be accorded only if, in the opinion of the Customs authorities of the Customs office of departure,

a) the heavy or bulky goods and any accessories thereto can be easily identified by reference to the description given, or can be provided with identification marks, or can be sealed, so that the goods and accessories cannot be replaced in who'e or in part by others and that nothing can be removed from them ;

b) the road vehicle contains no hidden spaces where goods can be concealed.

Article 20

Provided the conditions laid down in this Chapter and in Chapter V are fulfilled, heavy or bulky goods carried under cover of a TIR carnet shall not be subjected to the payment or deposit of import or export duties and taxes at Customs offices **en route.**

Article 21

1. The provisions of Articles 5, 6 (except paragraph 4), 9, 10, 11, 15 and 16 of this Convention shall apply to the transport of heavy or bulky goods under cover of a TIR carnet.

2. The provisions of Article 7 shall also apply, but the cover and all vouchers of the TIR carnet shall bear the endorsement « Heavy or bulky goods » in bold red letters in the language in which the carnet is printed.

Article 22

The liability of the guaranteeing association shall cover not only such goods as are enumerated in the TIR carnet, but also goods which, though not enumerated in the carnet, are on the loading platform or among the goods enumerated in the TIR carnet.

Article 23

The Customs authorities of the Customs office of departure may require packing lists, photographs, blueprints etc. of the goods carried to be appended to the TIR carnet. In this case they shall visa these documents, one copy of the said documents shall be attached to the reverse of the cover page of the TIR carnet, and all the manifests of the TIR carnet shall incorporate a reference to such documents.

Article 24

Transport of heavy or bulky goods under cover of a TIR carnet shall not involve more than one Customs office of departure or more than one Customs office of destination.

Article 25

If the Customs authorities of a Customs office **en route** so require at the time of entry, the person who produces the load to the Customs office shall insert and sign a supplementary description of the goods in the TIR carnet manifests.

Article 26

The Customs authorities may, if they see fit,

a) require examination of the vehicles and their loads at Customs offices **en route** or in the course of the journey ;

b) require road vehicles to be escorted on the territory of their country at the carrier's expense.

Article 27

The Customs authorities of the Customs office **en route** of each of the Contracting Parties shall, as far as possible, respect the identification marks and seals affixed by the Customs authorities of other Contracting Parties. They may, however, affix additional identification marks or seals of their own.

Article 28

If Customs authorities conducting an examination of the load at a Customs office **en route** or in the course of the journey are obliged to remove identification marks or break seals, they shall record on the TIR carnet vouchers used in their country and on the corresponding counterfoils particulars of the non identification marks or seals affixed.

CHAPTER V

MISCELLANEOUS PROVISIONS

Article 29

1. Each of the Contracting Parties shall have the right to exclude temporarily or permanently from the operation of this Convention any person guilty of a serious offence against the Customs laws or regulations applicable to the international transport of goods by road vehicle.

2. Such exclusion shall be notified immediately to the Customs authorities of the Contracting Party on whose territory the person concerned is established or resident, and also to the guaranteeing association in the country where the offence has been committed.

Article 30

TIR carnet forms sent to the guaranteeing associations by the corresponding foreign associations or by international organizations shall be admitted free of import duties and taxes and free of import prohibitions and restrictions.

Article 31

When a road vehicle, or a combination of coupled road vehicles, is carrying out the international transport of goods under cover of a TIR carnet, a rectangular plate bearing the letters « TIR », the specifications of which are laid down in Annex 9 to this Convention, shall be affixed to the front and to the rear of the vehicle or combination of vehicles. These plates shall be so placed as to be clearly visible ; they shall be removable and capable of being sealed. The seals shall be affixed by the Customs authorities of the first Customs Office of departure and shall be removed by the Customs authorities of the last Customs office of destination.

Article 32

If seals affixed by the Customs authorities are broken **en route** otherwise than in the circumstances of Articles 14 and 28 or if any goods are destroyed or damaged without breaking of such seals, the procedure laid down in Annex 1 to this Convention for the use of the TIR carnet shall, without prejudice to the application of the provisions of national law, be followed and a certified report shall be drawn up in the form set out in Annex 2 to this Convention.

Article 33

Each Contracting Party shall send to the other Contracting Parties facsimiles of the seals it uses.

Article 34

Each Contracting Party shall send the other Contracting Parties a list of the Customs offices of departure, Customs offices **en route** and Customs offices of destination approved by it for TIR carnet traffic, indicating, where appropriate, those offices which are only open for traffic dealt with under Chapter III. The Contracting Parties of adjacent territories shall consult each other in determining the frontier offices to be included in this list.

Article 35

As regards Customs operations mentioned in this Convention, no charge shall be made for Customs attendance, save where it is provided on days or at times or places other than those normally appointed for such operations.

Article 36

Any breach of the provisions of this Convention may render the offender liable in the country where the offence was committed to the penalties prescribed by the law of that country.

Article 37

The provisions of this Convention shall preclude neither the application of restrictions and controls imposed under national regulations on grounds of public morality, public security, hygiene or public health, or for veterinary or phytopathological considerations, nor the levy of dues chargeable by virtue of such regulations.

Article 38

Nothing in this Convention shall prevent Contracting Parties which form a Customs or economic union from enacting special provisions in respect of transport operations commencing or terminating in, or passing through, their territories, provided that such provisions do not attenuate the facilities provided by this Convention.

CHAPTER VI

FINAL PROVISIONS

Article 39

1. Countries members of the Economic Commission for Europe and countries admitted to the Commission in a consultative capacity under paragraph 8 of the Commission's terms of reference may become Contracting Parties to this Convention
a) by signing it ;
b) by ratifying it after signing it subject to ratification ; or
c) by acceding to it.

2. Such countries as may participate in certain activities of the Economic Commission for Europe in accordance with paragraph 11 of the Commission's terms of reference may become Contracting Parties to this Convention by acceding thereto after its entry into force.

3. The Convention shall be open for signature until 15 April 1959 inclusive. Thereafter, it shall be open for accession.

4. Ratification or accession shall be effected by the deposit of an instrument with the Secretary-General of the United Nations.

Article 40

1. This Convention shall come into force on the ninetieth day after five of the countries referred to in Article 39, paragraph 1, have signed it without reservation of ratification or have deposited their instruments of ratification or accession.

2. For any country ratifying or acceding to it after five countries have signed it without

reservation of ratification or have deposited their instruments of ratification or accession, this Convention shall enter into force on the ninetieth day after the said country has deposited its instrument of ratification or accession.

Article 41

1. Any Contracting Party may denounce this Convention by so notifying the Secretary-General of the United Nations.

2. Denunciation shall take effect fifteen months after the date of receipt by the Secretary-General of the notification of denunciation.

3. The validity of TIR carnets issued before the date when the denunciation takes effect shall not be affected thereby and the guarantee of the association shall hold good.

Article 42

This Convention shall cease to have effect if, for any period of twelve consecutive months after its entry into force, the number of Contracting Parties is less than five.

Article 43

1. Any country may at the time of signing this Convention without reservation of ratification or of depositing its instrument of ratification or accession or at any time thereafter, declare by notification addressed to the Secretary-General of the United Nations that this Convention shall extend to all or any of the territories for the international relations of which it is responsible. The Convention shall extend to the territory or territories named in the notification as from the ninetieth day after its receipt by the Secretary-General or, if on that day the Convention has not yet entered into force, at the time of its entry into force.

2. Any country which has made a declaration under the preceding paragraph extending this Convention to any territory for whose international relations it is responsible, may denounce the Convention separately in respect of that territory, in accordance with the provisions of Article 41.

Article 44

1. Any dispute between two or more Contracting Parties concerning the interpretation or application of this Convention shall, so far as possible, be settled by negotiation between them.

2. Any dispute which is not settled by negotiation shall be submitted to arbitration if any one of the Contracting Parties in dispute so requests and shall be referred accordingly to one or more arbitrators selected by agreement between the Parties in dispute. If within three months from the date of the request for arbitration the Parties in dispute are unable to agree on the selection of an arbitrator or arbitrators, any

of those Parties may request the Secretary-General of the United Nations to nominate a single arbitrator to whom the dispute shall be referred for decision.

3. The decision of the arbitrator or arbitrators appointed under the preceding paragraph shall be binding on the Contracting Parties in dispute.

Article 45

1. Any country may declare at the time of signing, ratifying, or acceding to this Convention, or notify the Secretary-General of the United Nations after becoming a Contracting Party to the Convention, that it does not consider itself bound by the provisions of Chapter IV of the Convention ; notifications addressed to the Secretary-General shall take effect on the ninetieth day after their receipt by the Secretary-General.

2. The other Contracting Parties shall not be required to extend the benefit of the provisions of Chapter IV of this Convention to persons established or resident in the territory of any Contracting Party which has entered a reservation as provided for in paragraph 1 of this Article.

3. Any country may, at the time of signing, ratifying or acceding to this Convention, declare that it does not consider itself bound by paragraphs 2 and 3 of Article 44 of the Convention. Other Contracting Parties shall not be bound by these paragraphs in respect of any Contracting Party which has entered such a reservation.

4. Any Contracting Party having entered a reservation as provided for in paragraph 1 or paragraph 3 of this Article may at any time withdraw such reservation by notifying the Secretary-General.

5. Apart from the reservations provided for in paragraphs 1 and 3 of this Article, no reservation to this Convention shall be permitted.

Article 46

1. After this Convention has been in force for three years, any Contracting Party may, by notification to the Secretary-General of the United Nations, request that a conference be convened for the purpose of reviewing the Convention. The Secretary-General shall notify all Contracting Parties of the request and a review conference shall be convened by the Secretary-General if, within a period of four months following the date of notification by the Secretary-General not less than one-third of the Contracting Parties notify him of their concurrence with the request.

2. If a conference is convened in accordance with the preceding paragraph, the Secretary-General shall notify all the Contrac-

ting Parties and invite them to submit, within a period of three months, such proposals as they may wish the conference to consider. The Secretary-General shall circulate to all Contracting Parties the provisional agenda for the conference, together with the text of such proposals, at least three months before the date on which the conference is to meet.

3. The Secretary-General shall invite to any conference convened in accordance with this Article all countries referred to in Article 39, paragraph 1, and countries which have become Contracting Parties under Article 39, paragraph 2.

Article 47

1. Any Contracting Party may propose one or more amendments to this Convention. The text of any proposed amendment shall be transmitted to the Secretary-General of the United Nations, who shall transmit it to all Contracting Parties and inform all other countries referred to in Article 39, paragraph 1.

2. Any proposed amendment circulated in accordance with the preceding paragraph shall be deemed to be accepted if no Contracting Party expresses an objection within a period of three months following the date of circulation of the proposed amendment by the Secretary-General.

3. The Secretary-General shall, as soon as possible, notify all Contracting Parties whether an objection to the proposed amendment has been expressed. If an objection to the proposed amendment has been expressed, the amendment shall be deemed not to have been accepted, and shall be of no effect whatever. If no such objection has been expressed the amendment shall enter into force for all Contracting Parties nine months after the expiry of the period of three months referred to in the preceding paragraph.

4. Independently of the amendment procedure laid down in paragraphs 1, 2 and 3 of this Article, the Annexes to this Convention may be modified by agreement between the competent administrations of all the Contracting Parties ; such agreement may provide that during a transitional period the old Annexes shall remain in force, wholly or in part, concurrently with the new Annexes. The Secretary-General shall fix the date of entry into force of the new texts resulting from such modifications.

Article 48

In addition to the notifications provided for in Articles 46 and 47, the Secretary-General of the United Nations shall notify the countries referred to in Article 39, paragraph 1, and the countries which have become Contracting Parties under Article 39, paragraph 2, of

a) signatures, ratifications and accessions under Article 39 ;

b) the dates of entry into force of this Convention, in accordance with Article 40 ;

c) denunciations under Article 41 ;

d) the termination of this Convention in accordance with Article 42 ;

e) notifications received in accordance with Article 43 ;

f) declarations and notifications received in accordance with Article 45, paragraphs 1, 3 and 4 ;

g) the entry into force of any amendment in accordance with Article 47.

Article 49

As soon as a country which is a Contracting Party to the Agreement providing for the provisional application of the Draft International Customs Conventions on Touring, on Commercial Road Vehicles, and on the International Transport of Goods by Road, done at Geneva on 16 June 1949, becomes a Contracting Party to this Convention, it shall take the measures required by Article IV of that Agreement to denounce it as regards the Draft International Customs Convention on the International Transport of Goods by Road.

Article 50

The Protocol of Signature of this Convention shall have the same force, effect and duration as the Convention itself, of which it shall be considered to be an integral part.

Article 51

After 15 April 1959, the original of this Convention shall be deposited with the Secretary-General of the United Nations, who shall transmit certified true copies to each of the countries mentioned in Article 39, paragraphs 1 and 2.

IN WITNESS WHEREOF, the undersigned, being duly authorized thereto, have signed this Convention.

DONE at Geneva, this fifteenth day of January one thousand nine hundred and fifty nine, in a single copy, in the English and French languages, each text being equally authentic.

Model of TIR Carnet

The TIR carnet shall be printed in French

PAGE 1 OF THE COVER

(Particulars of the international organizations to which the issuing association is affiliated)

TIR CARNET

No

1.

2. Valid up to and including ....

3. Issued by ....

   (name of issuing association)

4. Holder ....

   (name and address)

5. Country of departure ....

6. Country or countries of destination ....

   ....

7. Road vehicle registration No. ....

8. Certificate of approval of road vehicle/container (1)   No. ....

9.   Date : ....

303

10. Total gross weight of goods (as shown in the manifest) ....

11. Total value of goods (as shown in the manifest) ....

(to be given in the currency of the country of departure or in a currency prescribed by the competent authorities of that country)

12. Signature of authorized official of the issuing association and stamp of that association :

13. Signature of the secretary of the international organization :

(1) Strike out whichever does not apply.

PAGE 2 OF THE COVER

I, the undersigned, ...

acting on behalf of (1) ....

....(name and address of holder),

a) declare that the goods specified on the attached manifest have been loaded in the road vehicle/container (1) for the destination shown overleaf ;

b) undertake, under pain of the penalties prescribed by the laws and regulations in force in the countries through or in which the goods are to be carried, to produce, with this carnet, the said goods in full and with the seals intact, if seals have been affixed, at the Customs offices **en route** and of destination, and to observe the time-limits and itinerary as laid down ;

c) undertake to conform to the Customs laws and regulations of the countries through or in which the goods are to be carried.

At ...., on .... 19........

....
(signature of holder or agent)

(1) Strike out as necessary.

304

1. Voucher 1 (Part 1)

4. Country of consignment of the goods listed under Nos.

5. Country of destination of the goods listed under Nos

2. TIR CARNET No.

3. Goods manifest

Serial number	Marks and Nos. of packages	Number of packages	Type	Description of goods	Gross weight	Net weight, volume, number, etc.	Value	
6	7	8	9	10	11	12	13	14

15. This manifest covers in all .... packages, of which the first ....

.... (in full) .... Customs office, the next .... (in full)

are consigned to ....

to .... Customs office, and the remainder to .... Customs office

.... (place and country) .... (place and country)

16. I declare the above particulars to be true and complete.

17. At .... on ....

18. Signature of holder or agent ....

19. Customs officer's signature and stamp of the Customs office where goods are taken under Customs control : (Customs office of departure)

20. NOTE : At the last Customs office of departure the Customs officer's signature and stamp of the Customs office must be inserted at the foot of the manifest in all the vouchers to be used for the remainder of the transport operation.

21. Voucher 1 (Part 2)

22. of TIR Carnet No. .... valid up to and including ....

23. Issued by .... (name of issuing association)

24. To .... (name of holder)

25. Whose place of business is at .... (address of holder)

26. Customs offices of departure : 1. .... 2. .... 3. ....

27. Customs offices **en route** ....

28. Customs offices of destination : 1. .... 2. .... 3. ....

29. Registration No. of road vehicle .... (as shown in the manifest)

30. Certificate of approval of road vehicle/container (1) No. .... dated ....

31. CERTIFICATE for goods taken under Customs control by the Customs office of departure or Customs office of entry **en route**

32. This voucher has been registered at the Customs office at ....

33. Under No. ....
34. Time-limit assigned for journey ....

35. Customs office at which the load must be produced ....

36. Itinerary stipulated by the Customs ....

37. Seals affixed or identification marks ....

(1) Strike out whichever does not apply.

42. THIS VOUCHER MUST BE DETACHED AND KEPT BY THE CUSTOMS OFFICE OF DEPARTURE OR THE CUSTOMS OFFICE OF ENTRY EN ROUTE AS THE CASE MAY BE.

1. **Counterfoil 1**

2. of TIR Carnet No. ....

3. Taken under Customs control on ....

4. under No. ....

5. by the office at ....

38. Seals or identification marks recognized ....

39. Miscellaneous (for description of goods, if necessary) ....

40. Customs officer's signature and Customs office stamp :

41. NOTE : The Customs office of departure or Customs office of entry **en route** must repeat the particulars given in this certificate on the next voucher with even number.

6. Seals affixed or identification marks ....

7. Seals or identification marks recognized ....

8. Customs office at which the transport must be produced ....

9. At .... date ....

10. Customs officer's signature and Customs office stamp :

UNDERTAKING TO BE SIGNED, IF THE CUSTOMS AUTHORITIES SO REQUIRE BY THE PERSON PRESENTING THE LOAD TO THE CUSTOMS OFFICE

I, the undersigned, ...

undertake to observe, as regards the transport operation covered by this TIR carnet, the laws and regulations applicable and, in particular, to observe the time limit and itinerary laid down and to produce the goods in full with Customs seals intact, at the Customs office of

At .... on .... 19.....

(Signature)

1. **Voucher 2** (Part 1)

2. **TIR CARNET No.**

3. **Goods manifest**

4. Country of consignment of the goods listed under Nos. ....

5. Country of destination of the goods listed under Nos. ....

Serial number	Marks and Nos. of packages	Number of packages	Type	Description of goods	Gross weight	Net weight, volume, number, etc.	Value	
6	7	8	9	10	11	12	13	14

15. This manifest covers in all .... (in full) .... packages, of which the first .... (in full)

are consigned to .... Customs office, the next .... (in full)

to .... Customs office, and the remainder to .... Customs office
(place and country)   (place and country)

16. I declare the above particulars to be true and complete.

17. At .... on ....
(place and country)

18. Signature of holder or agent

19. Customs officer's signature and stamp of the Customs office where goods are taken under Customs control :
(Customs office of departure)

20. NOTE : At the last Customs office of departure the Customs officer's signature and stamp of the Customs office must be inserted at the foot of the manifest in all the vouchers to be used for the remainder of the transport operation.

21. **Voucher 2** (Part 2)

22. of TIR Carnet No. .... valid up to and including ....

23. Issued by .... (name of issuing association)

24. To .... (name of holder)

25. Whose place of business is at .... (address of holder)

26. Customs offices of departure : 1. .... 2. .... 3. ....

27. Customs offices **en route** ....

28. Customs offices of destination : 1. .... 2. .... 3. .... (as shown in the manifest)

29. Registration No. of road vehicle .... dated ....

30. Certificate of approval of road vehicle/container (1) No. ....

307

31. CERTIFICATE for goods taken under Customs control at Customs office of departure or Customs office of entry **en route**.
32. This voucher has been registered at the Customs office at ....
33. Under No. ....
34. Time-limit assigned for journey ....
35. Customs office at which the load must be produced ....
36. Itinerary stipulated by the Customs ....
37. Seals affixed or identification marks ....
38. Seals or identification marks recognized ....
39. Miscellaneous (for description of goods, if necessary) ....
40. Customs officer's signature and Customs office stamp :

41. NOTE : This certificate must be filled in by the Customs office which completed the preceding voucher with odd number.

(1) Strike out as necessary.

50. THIS VOUCHER MUST BE DETACHED BY THE CUSTOMS OFFICE OF EXIT EN ROUTE OR THE CUSTOMS OFFICE OF DESTINATION AS THE CASE MAY BE AND SENT AFTER COMPLETION TO THE OFFICE (OF THE SAME COUNTRY) WERE THE GOODS WERE TAKEN UNDER CUSTOMS CONTROL.

1. **Counterfoil 2**
2. of TIR Carnet No. ....
3. Arrival certified on ....
4. under No. ....
5. by the office at ....

42. CERTIFICATE of discharge by Customs office of exit **en route** or Customs office of destination.
43. (1) The road vehicle/container specified above has been produced in good condition. The seals and identification marks were intact and have been recognized.
44. (1) The road vehicle/container has proceeded on its way abroad/to the Customs office at ....
45. (1) It was ascertained that the road vehicle/container contained .... packages consigned to this office as specified in the above manifest.
46. Reservations or nature of offences ascertained ....
47. Discharge has been given (subject to the above reservations) of undertakings entered into under No. ....
48. At .... on ....
49. Customs officer's signature and Customs office stamp :

6. Seals or identification marks intact ....
7. Discharged without reservation ....
8. Reservations or nature of offences ascertained ....
9. At .... date ....
10. Customs officer's signature and Customs office stamp :

RULES FOR THE USE OF THE TIR CARNET

1. The TIR carnet shall be issued either in the country of departure or in the country in which the holder is established or resident.

2. The TIR carnet is printed in French ; however, additional pages may be inserted giving a translation, in the language of the country of issue, of the printed text of the carnet.

3. The manifest shall be completed in the language of the country of departure. The Customs authorities of the other countries traversed reserve the right to require its translation into their own language. In order to avoid unnecessary delay which might ensue from this requirement, carriers are advised to supply the driver of the vehicle with the requisite translations.

4. a) It is particularly recommended that the manifest should be typed or multigraphed in such a way that all the forms are clearly legible.
   b) When there is not enough space in the goods manifest to enter all the goods carried, separate sheets of the same model as the manifest may be attached to the latter, but all copies of the manifests must then contain the following particulars :
      i) a reference to the sheets ;
      ii) the number and type of packages and goods in bulk enumerated on the separate sheets ;
      iii) the total value and the total gross weight of the goods appearing on the said sheets.
   c) Where the Customs authorities require packing lists, photographs, blue-prints, etc., to be appended to the TIR carnet for the exact designation of the goods, such appendices shall bear the visa of the Customs authorities. One copy of these documents shall be attached overleaf to page 2 of the cover of the TIR carnet and all copies of the manifest shall include a list of such documents.

5. Weights, volume and other measurements shall be expressed in units of the metric system, and values in the currency of the country of departure or in a currency prescribed by the competent authorities of that country.

6. No erasures or over-writing shall be effected on the TIR carnet. Any correction shall be effected by deleting the incorrect particulars and adding, if necessary, the required particulars. Any correction, addition or other amendment shall be acknowledged by the person making it and visaed by the Customs authorities.

7. Page 2 of the cover of the TIR carnet and each copy of the manifest shall be dated and signed by the holder of the carnet or his agent. The person presenting the load to the Customs office shall, if the Customs authorities so require, sign the undertaking on the reverse of the vouchers with odd numbers.

8. Transport of heavy or bulky goods under cover of a TIR carnet may not involve more than one Customs office of departure or more than one Customs office of destination. Other transport under cover of a TIR carnet may involve several Customs offices of departure and destination, but, save as specially authorized :
   a) the Customs offices of departure must be situated in the same country ;
   b) the Customs offices of destination may not be situated in more than two countries ;
   c) the total number of Customs offices of departure and destination may not exceed four.

309

If there is only one Customs office of departure and one Customs office of destination, the carnet must contain at least 2 forms for the country of departure, 2 forms for the country of destination and 2 forms for each country traversed. For each extra place of loading or unloading 2 extra forms are required ; in addition, 2 further forms are required if the places of unloading are situated in two different countries.

9. If there are several Customs offices of departure or of destination, the entries concerning the goods taken under Customs control at, or intended for, each office shall be clearly separated from each other on the manifest.

10. The driver of the vehicle is advised to make sure that a voucher of the TIR carnet is detached by the Customs at each Customs office of departure, Customs office **en route** and Customs office of destination. Vouchers with odd numbers are to be used for taking the goods under Customs control and those with even numbers for discharging them.

11. In the event of Customs seals being broken or goods being destroyed or damaged accidentally **en route** the carrier shall ensure that a certified report is drawn up as quickly as possible by the authorities of the country in which the vehicle is located. The carrier shall approach the Customs authorities, if there are any near at hand, or, if not, any other competent authorities. Carriers shall accordingly provide themselves with copies of the certified report form laid down in Annex 2 to the TIR Convention ; these forms shall be printed in French and in the national language of each country traversed.

12. In the event of an accident necessitating transfer of the load to another vehicle or another container, this may only be done in the presence of one of the authorities mentioned in the previous paragraph ; the latter will draw up a certified report, testifying to the regularity of the proceedings. Unless the TIR carnet carries the words « Heavy or bulky goods », the vehicle or container substituted shall be approved and sealed and the seals used shall be described in the certified report. However, if no approved vehicle or container is available, transfer to a non-approved vehicle or container may be authorized, provided it affords adequate safeguards ; in the latter event the Customs authorities of succeeding countries will judge whether they, too, can allow the transport under cover of the TIR carnet to continue in that vehicle or container.

13. In the event of imminent danger necessitating immediate unloading of the whole or part of the load, the driver may take action on his own initiative without requesting or awaiting intervention by the authorities mentioned in paragraph 11. He must then furnish adequate proof that he was compelled to take such action in the interests of the vehicle or container or of the load. Having taken such preventive measures as the emergency may necessitate, he shall record them on page 4 of the cover of the TIR carnet and notify the authorities mentioned in paragraph 11 in order that the facts may be verified, the load checked, the vehicle or container sealed and a certified report drawn up.

14. In any of the various contingencies covered by paragraphs 11, 12 and 13, the authorities concerned shall mention the certified report on page 4 of the cover of the TIR carnet. The certified report shall be attached to the TIR carnet and accompany the load to the Customs office of destination.

PAGE 4 OF THE COVER

INCIDENTS OR ACCIDENTS EN ROUTE

310

International transport of goods by road vehicle under cover of a TIR carnet

CERTIFIED REPORT

The certified reports shall be completed on forms printed in one of the languages of the country in which the occurrence took place and in French.

1. **International transport of goods by road vehicle under cover of a TIR carnet**

2. **CERTIFIED REPORT**

3. drawn up in pursuance of paragraphs 11 - 14 of the Rules for the Use of the TIR carnet

4. WE, THE UNDERSIGNED (1) ....

5. CERTIFY that on .... one thousand nine hundred .... at .... hours

6. on the territory of .... at the place known as ....

7. we examined the road vehicle registered in ....

8. under No ....

9. carrying goods under cover of a TIR carnet,

10. issued on .... under No ....

11. by (2) ....

12. We established that :

13. the undermentioned seals of the Customs office of departure of .... and the Customs office of ....

14. were broken/missing ; (3)

15. the loading compartment of the road vehicle/the container (3) was no longer intact :

16. no goods were missing ; (3)

17. the following goods (in the order of entry in the TIR carnet manifest) were missing/destroyed (3) ....

18.

Marks and Nos of packages	Number and type of packages	Description of goods	Remarks (give particulars of quantities missing)

311

(1) Name and rank of officials and designation of authority to which they belong.
(2) Name and address of issuing association.
(3) Strike out whichever does not apply.

19. The carrier gave the following explanations (cause of seals breaking or of loss of goods, measures taken to save goods etc.) ....
....
....
....
....
....

20. We, the undersigned, certify that
21. the following measures were taken (affixing of new seals, transfer of load, etc.) ....
....
....

22. Number and particulars of new seals affixed ....
23. Particulars of road vehicle/container (1) to which load transferred ....
....

24. The said road vehicle/container (1)
25. is covered by certificate of approval. No .... (1)
26. is not covered by a certificate of approval. (1)
27. Signature and stamp of officials who drew up this certified report :

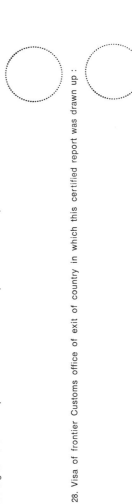

28. Visa of frontier Customs office of exit of country in which this certified report was drawn up :

(1) Strike out whichever does not apply.

312

ANNEX 3

Regulations on technical conditions applicable to road vehicles which may be accepted for international transport of goods under customs seal

Article 1

General

1. Approval for the international transport of goods by road vehicle under Customs seal may be granted only for vehicles constructed and equipped in such a manner that :

a) Customs seals can be simply and effectively affixed thereto ;

b) no goods can be removed from or introduced into the sealed part of the vehicle without obvious damage to it or without breaking the seals ;

c) they contain no concealed spaces where goods may be hidden.

2. The vehicles shall be so constructed that all spaces in the form of compartments, receptacles or other recesses which are capable of holding goods are readily accessible for Customs inspection.

3. Should any empty spaces be formed by the different layers of the sides, floor and roof of the vehicle, the inside surface shall be firmly fixed, solid and unbroken and incapable of being dismantled without leaving obvious traces.

Article 2

Structure of loading compartment

1. The sides, floor and roof of the loading compartment shall be constructed of plates, boards or panels of sufficient strength, of adequate thickness, and welded, riveted,

grooved or jointed in such a way as not to leave any gaps in the structure through which access to the contents can be obtained. The various parts shall fit each other exactly and be so arranged that it is impossible either to move or remove them without leaving visible traces or damaging the Customs seals.

2. Where assembly is effected by means of rivets, the latter may be seated on the inside or outside ; rivets joining the essential parts of the sides, floor and roof shall pass through the parts joined. Where assembly is effected otherwise than by means of rivets, those bolts or other joining devices which hold the essential parts of the sides, floor and roof shall be seated on the outside, protrude on the inside and be properly bolted, riveted or welded, while the other bolts or joining devices may be seated on the inside, provided that the nut is welded in a satisfactory manner on the outside and is not covered with non-transparent material. The assembly or metal plates or panels may also be effected by the curving or folding of their edges towards the inside of the vehicle, and these edges may be joined

— either by rivets, bolts or other joining devices passing through the edges thus curved or folded and through the devices (if any) connecting them ;

— or by metal strips curved under pressure to form clamps at the same time as the edges of the parts to be assembled and ensuring permanent compression of the joints thus made (see sketch).

HORIZONTAL SECTION

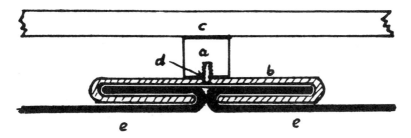

a = Wooden post

b = Metal strip bent in shape of cramp-iron

c = Outer surface of metal sheeting

c = Inner lining of grooved-and-tongued boards

d = Screws

3. Apertures for ventilation shall be allowed provided their longest side does not exceed 400 mm. If they permit direct access to the interior of the loading compartment, they shall be covered with metal gauze or perforated metal screens (maximum dimension of holes : 3 mm in both cases) and protected by welded metal lattice-work (maximum dimension of holes : 10 mm). If they do not permit direct access to the interior of the loading compartment (for example, by means of multiple-bend air ducts), they shall be provided with the same protective devices but the dimensions of the holes may be increased to 10 mm and 20 mm respectively (instead of 3 mm and 10 mm). It shall not be possible to remove these devices from outside without leaving visible traces. Metal gauze shall be of wire at least 1 mm in diameter and so made that single strands cannot be pushed together and that the size of individual holes cannot be increased without leaving visible traces.

4. Windows shall be allowed provided that they comprise a fixed glass and metal grill which cannot be removed from the outside. The holes of the grill shall not exceed 10 mm across.

5. Openings made in the floor for technical purposes, such as lubrication, maintenance and filling of the sand-box, shall be allowed only on condition that they are fitted with a cover capable of being fixed in such a way as to render the loading compartment inaccessible from the outside.

Article 3
Closing systems

1. Doors and all other closing systems of vehicles shall be fitted with a device which shall permit simple and effective Customs sealing. This device shall either be welded to the sides of doors where these are of metal, or secured by at least two bolts, riveted or welded to the nuts on the inside.

2. Hinges shall be so made and fitted that doors and other closing systems cannot be lifted off the hinge-pins, once shut ; the screws, · bolts, hinge-pins and other fasteners shall be welded to the outer parts of the hinges. These requirements shall be waived, however, where the doors and other closing systems have a locking device inaccessible from the outside which, once it is applied, prevents the doors from being lifted off the hinge-pins.

3. Doors shall be so constructed as to cover all interstices and ensure complete and effective closure.

4. The vehicle shall be provided with a satisfactory device for protecting the Customs seal, or shall be so constructed that the Customs seal is adequately protected.

Article 4
Vehicles for special use

1. The foregoing conditions shall apply to insulated vehicles, refrigerator vehicles, tank vehicles and furniture vehicles in so far as they are not incompatible with the technical requirements which such vehicles must fulfil in accordance with their use.

2. The flanges (filler caps), drain cocks and manholes of tank wagons shall be so constructed as to allow simple and effective Customs sealing.

Article 5
Sheeted vehicles

1. Where applicable, the provisions of articles 2 to 4 above shall apply to sheeted vehicles. However, the system of closing and protecting the ventilation apertures mentioned in article 2, paragraph 3, may consist externally of a perforated metal screen (maximum dimension of holes : 10 mm) and internally of metal gauze or some other very strong gauze (maximum dimension of meshes : 3 mm, the strands being such that they cannot be pushed together without leaving visible traces), the screen and the gauze being fixed to the sheet in such a way that their assembly cannot be altered without leaving obvious traces. In addition, sheeted vehicles shall conform to the following conditions.

2. The sheet shall be either of strong canvas or, provided it is not dark in colour, of plastic-covered or rubberized cloth, non-tensible and of sufficient strength. It shall be fashioned in one piece or of strips each in one piece. It shall be in good condition and made up in such a way that once the closing device has been secured, it is impossible to gain access to the load without leaving obvious traces.

3. If the sheet is made up of several strips, their edges shall be folded into one another and sewn together with two seams at least 15 mm apart. These seams shall be made as shown in sketch n° 1 attached to the present Regulations ; however, where in the case of certain parts of the sheet, such as flaps at the rear and reinforced corners, it is not possible to assemble the strips in that way, it shall be sufficient to fold the edge of the top section and make the seams as shown in sketch n° 2 attached to these Regulations. The threads used for each of the two seams shall be plainly different in colour ; one of the seams shall be visible only from the inside and the colour of the thread used for that seam shall be plainly different from the colour of the sheet itself. All seams shall be machine-sewn.

4. If the sheet is of plastic-covered cloth, and is made up of several strips, the strips may also be welded together in the manner shown in sketch n° 3 attached to these Regulations. The edges of the strips shall overlap by at least 15 mm. The strips shall be fused together over the whole width of the overlapping parts. The edge of the outer sheet shall be covered with a band of plastic at least 7 mm wide, affixed by the same welding process. The plastic band and a width of at least 3 mm on each side shall have a well-marked uniform relief stamped

on it. The strips shall be welded in such a way that they cannot be separated and then rejoined without leaving obvious traces.

5. Repairs shall be made in accordance with the method described in sketch n°. 4 attached to these Regulations ; the edges shall be folded into one another and sewn together with two visible seams at least 15 mm apart ; the colour of the thread visible from the inside shall be different from that of the thread visible from the outside and from that of the sheet itself ; all seams shall be machine-sewn. Nevertheless, sheets of plastic-covered cloth may also be repaired in accordance with the method described in paragraph 4 above.

6. Securing rings shall be so fitted that they cannot be removed from the outside. Eyelets in the sheet shall be reinforced with metal or leather. The interval between eyelets or rings shall not exceed 200 mm.

7. The sheet shall be so fixed to the sides as to render the load quite inaccessible. It shall be supported by at least three lengthwise bars or laths resting at the ends of the loading platform either on hoops or on the end walls of the platform ; where the loading platform is more than 4 metres long, at least one intermediate hoop must be provided. The hoops shall be fixed in such a way that it is impossible to alter their position from the outside.

8. The following types of fastening shall be used :

a) steel wire rope of at least 3 mm diameter ; or

b) hemp or sisal rope of at least 8 mm diameter encased in a transparent non-tensible plastic sheath ; or

c) iron bars at least 8 mm in diameter.

Steel wire ropes shall not be covered, except with a transparent non-tensible plastic sheath. Iron bars shall not be coated with non-transparent material.

9. Each wire rope or hemp or sisal rope shall be in one piece and have a metal end piece. The fastener of each metal end piece shall include a hollow rivet passing through the rope so as to allow the introduction of the string of the Customs seal. The rope shall remain visible on either side of the hollow rivet so that it is possible to ascertain whether the rope is in one piece (see sketch n°. 5 attached to these Regulations).

10. Each iron bar shall be in one piece. It shall have a hole at one end to take the closing device and, at the other end, a head forged to the bar and so constructed as to make it impossible for the bar to turn on its axis.

11. When ropes are used the sides of the vehicles shall be at least 350 mm high and the sheet shall cover the sides to a depth of at least 300 mm.

12. At the openings used for loading and unloading the vehicle, the two edges of the sheet shall have an adequate overlap. They shall likewise be fastened by a flap attached to the outside and sewn in accordance with paragraph 3 of this article. The fastenings shall be either those provided for in paragraph 8 or thongs at least 20 mm wide and 3 mm thick made of leather or non-tensible rubberised cloth. These thongs shall be attached inside the sheet and fitted with eyelets to take the wire rope or iron bar mentioned in paragraph 8.

ANNEX 4

Procedure for the approval of road vehicles complying with the technical conditions set forth in the regulations contained in annex 3

The procedure for the approval of vehicles shall be as follows :

a) Vehicles shall be approved by the competent authorities of the country in which the owner or carrier is resident or established.

b) The date and serial number of the approval decision must be specified.

c) A certificate of approval conforming to the standard form of Annex 5 shall be issued for approved vehicles. This certificate shall be printed in the language of the country of issue and in French ; and the various headings shall be numbered so that the text may be more readily understood in other languages.

d) This certificate shall be kept on the vehicle ; if necessary, photographs or diagrams taken or drawn in accordance with the directions of the Issuing Office and authenticated by that Office shall be attached to this certificate.

e) Vehicles shall be produced every two years to the competent authorities for purposes of inspection and renewal of approval where appropriate.

f) Approval shall lapse if the essential features of the vehicle are altered or on change of owner or carrier.

MODEL OF TIR CARNET

Sketch No 1

SECTION OF SHEET

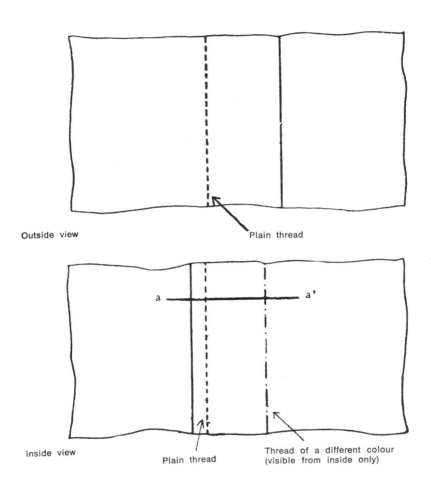

Outside view   Plain thread

Inside view

Plain thread   Thread of a different colour (visible from inside only)

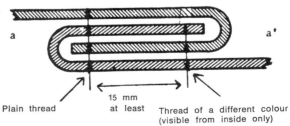

Plain thread   15 mm at least   Thread of a different colour (visible from inside only)

Section a-a'

Double flat seam for joining pieces

Sketch No 2
SECTION OF SHEET

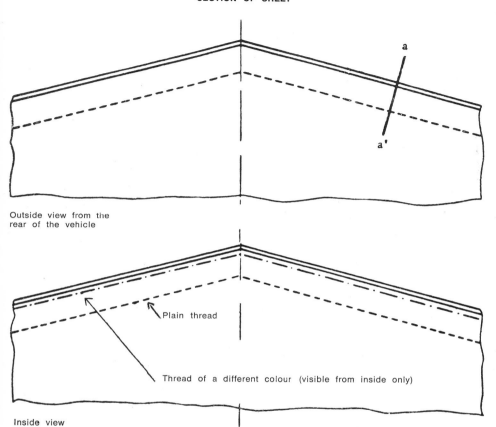

Outside view from the
rear of the vehicle

Plain thread

Thread of a different colour (visible from inside only)

Inside view

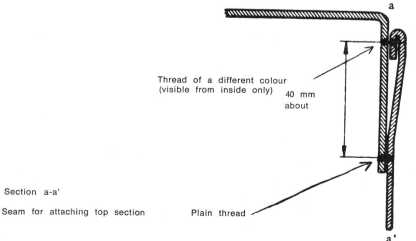

a

Thread of a different colour
(visible from inside only)   40 mm
about

Section a-a'

Seam for attaching top section   Plain thread

a'

MODEL OF TIR CARNET

Sketch No 3

SECTION OF SHEET

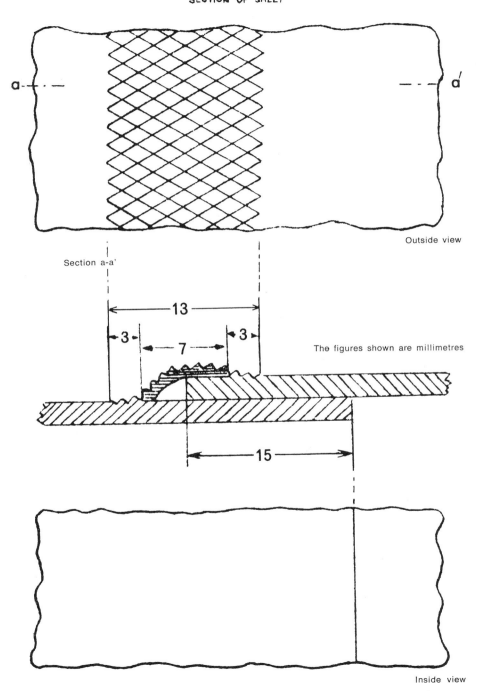

Outside view

Section a-a'

The figures shown are millimetres

Inside view

Sketch No 4

REPAIR OF THE SHEET

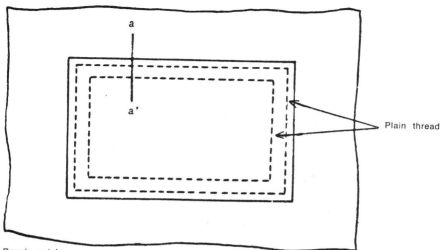

Plain thread

Repair patch,
outside view

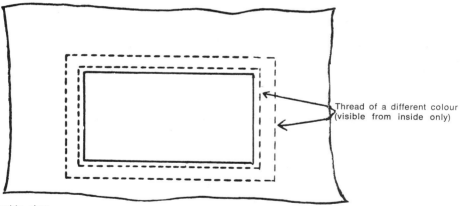

Thread of a different colour
(visible from inside only)

Inside view

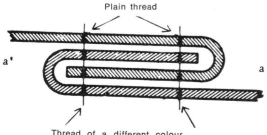

Plain thread

Section a-a'

Thread of a different colour
(visible from inside only)

Sketch No 5

SPECIMENS OF END-PIECES

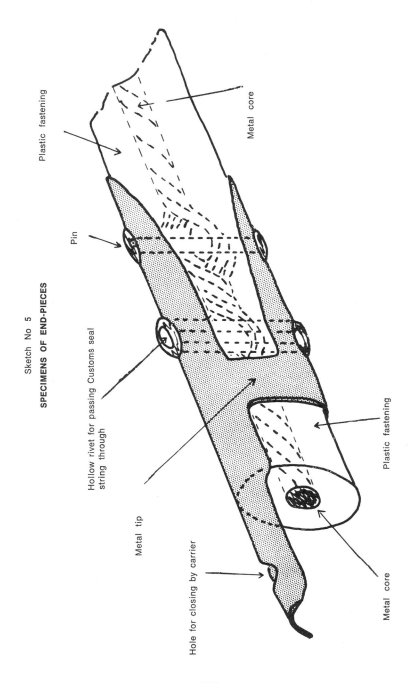

Plastic fastening

Metal core

Pin

Hollow rivet for passing Customs seal string through

Metal tip

Plastic fastening

Hole for closing by carrier

Metal core

320

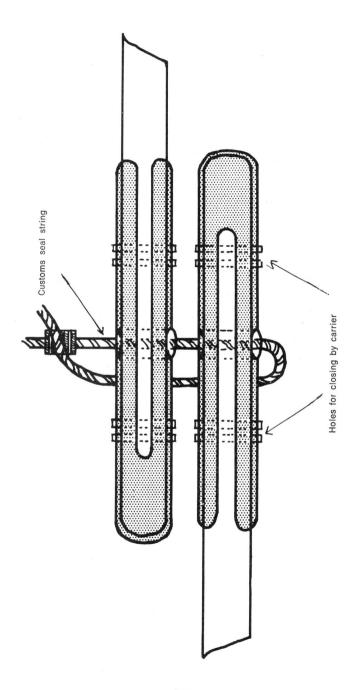

Customs seal string

Holes for closing by carrier

321

ANNEX 5

Certificate of approval of a road vehicle

Certificate No ....

1.

2. Attesting that the vehicle specified below fulfils the conditions required for admission to international transport of goods under Customs seal.

3. Valid until ....

4. This certificate must be returned to the Issuing Office when the vehicle is taken off the road, or on change of owner or carrier, on expiry of the period of validity of the certificate, or if there is any material change in any essential particulars of the vehicle.

5. Type of vehicle ....

6. Name and business address of holder (owner or carrier) ....

....

7. Name or trade-mark of the maker ....

8. Chassis number ....

9. Engine number ....

10. Registration number ....

11. Other particulars ....

12. Annexes* .... (state number)

13. Issued at .... (place) on .... (date) 19....

14. Signature and stamp of issuing office at ....

15.* N.B. Photographs or diagrams taken or drawn in accordance with the directions of the Issuing Office and authenticated by that Office shall be attached to this certificate.

322

Regulations on technical conditions applicable to containers
which may be accepted for the international transport of
goods by road vehicle under customs seal

Article 1

General

1. Approval for the international transport of goods by road vehicle under Customs seal may be granted only for containers durably marked with the name and address of the owner, with particulars of the tare and with identification marks and numbers, and constructed and equipped in such a manner that

a) Customs seals can be simply and effectively affixed thereto ;

b) no goods can be removed from or introduced into the sealed part of the container without obvious damage to it or without breaking the seals ;

c) they contain no spaces where goods may be hidden.

2. The container shall be so constructed that all spaces in the form of compartments, receptacles or other recesses which are capable of holding goods are readily accessible for Customs inspection.

3. Should any empty spaces be formed by the different layers of the sides, floor and roof of the container, the inside surface shall be firmly fixed, solid and unbroken and incapable of being dismantled without leaving obvious traces.

4. Containers to be approved in accordance with the procedure referred to in Annex 7 shall have on one of their outside walls a frame to hold the certificate of approval, which shall be covered on both sides by transparent plastic sheets hermetically sealed together. This frame shall be so designed as to protect the certificate of approval and to make it impossible to extract the certificate without breaking the seal that will be affixed in order to prevent the removal of the certificate ; it shall also adequately protect the seal.

Article 2

Structure of container

1. The sides, floor and roof of the container shall be constructed of plates, board, or panels of sufficient strength, of adequate thickness, and welded, riveted, grooved or jointed in such a way as not to leave any gaps in the structure through which access to the contents can be obtained. The various parts shall fit each other exactly and be so arranged that it is impossible either to move or remove them without leaving visible traces or damaging the Customs seals.

2. Essential joints, such as bolts, rivets, etc. shall be seated on the outside, protrude on the inside and be bolted, riveted or welded in a satisfactory manner. If the bolts holding the essential parts of the sides, floor and roof are seated on the outside, the other bolts may be seated on the inside, provided that the nut is properly welded on the outside and is not covered with non-transparent paint.

3. Apertures for ventilation shall be allowed provided their longest side does not exceed 400 mm. If they permit direct access to the interior of the container, they shall be covered with metal gauze or perforated metal screens (maximum dimension of holes : 3 mm in both cases) and protected by welded metal lattice-work (maximum dimension of holes : 10 mm). If they do not permit direct access to the interior of the container (for example, by means of multiple-bend air ducts), they shall be provided with the same protective devices but the dimensions of the holes may be increased to 10 mm and 20 mm respectively (instead of 3 mm and 10 mm). It shall not be possible to remove these devices from outside the container without leaving visible traces. Metal gauze shall be of wire at least 1 mm in diameter and so made that single strands cannot be pushed together and that the size of individual holes cannot be increased without leaving visible traces.

4. Apertures for drainage shall be allowed provided their longest side does not exceed 35 mm. They shall be covered with metal gauze or perforated metal screens (maximum dimension of holes : 3 mm in both cases) protected by welded metal lattice-work (maximum dimension of holes : 10 mm). It shall not be possible to remove these devices from outside the container without leaving visible traces.

Article 3

Closing Systems

1. Doors and all other closing systems of containers shall be fitted with a device which shall permit simple and effective Customs sealing. This device shall either be welded to the sides of doors where these are of metal, or secured by at least two bolts, riveted or welded to the nuts on the inside.

2. Hinges shall be so made and fitted that doors and other closing systems cannot be lifted off the hinge-pins, once shut ; the screws, bolts, hinge-pins and other fasteners

shall be welded to the outer parts of the hinges. These requirements shall be waived however, where the doors and other closing systems have a locking device inaccessible from the outside which, once it is applied, prevents the doors from being lifted off the hinge-pins.

3. Doors shall be so constructed as to cover all interstices and ensure complete and effective closure.

4. Containers shall be provided with a satisfactory device for protecting the Customs seal, or shall be so constructed that the Customs seal is adequately protected.

Article 4

Containers for special use

1. The foregoing conditions shall apply to insulated and refrigerator containers, tank containers, furniture containers and to containers specially built for carriage by air in so far as they are not incompatible with the technical requirements which such containers must fulfil in accordance with their use.

2. The flanges (filler caps), drain cocks and manholes of tank containers shall be so constructed as to allow simple and effective Customs sealing.

Article 5

Folding or collapsible containers

Folding or collapsible containers are subject to the same conditions as non-folding or non-collapsible containers, provided that the locking devices enabling them to be folded or collapsed allow of Customs sealing and that no part of such containers can be moved without breaking the seals.

Article 5 bis

Sheeted containers designed as loading compartments for road vehicles

Where a container is designed as the loading compartment of a road vehicle, but, instead of being closed like the other containers referred to in this Annex, is open and sheeted, it may be approved for the international transport of goods by road vehicles under Customs seal, provided that it accords with the provisions of annex 3, article 5, and also, in so far as they are applicable, with the provisions of this annex, and that the markings and certificate of approval prescribed in article 1, paragraphs 1 and 4, of this annex are visible when the container is sheeted and placed on the road vehicle.

Article 6

Transitional provisions

The provisions of Article 1, paragraph 4, and of Article 3, paragraph 4, of the present regulations, and the provisions of paragraphs 3 and 4 Article 2 which relate to the use of welded metal lattice-work for protecting apertures for ventilation, other than those with multiple-bend air ducts or apertures for drainage, shall not become mandatory before 1 January 1961, but certificates of approval issued before that date for containers which do not conform to these provisions will not be valid after 31 December 1960.

ANNEX 7

Procedure for the approval and identification of containers complying with the technical conditions set forth in the regulations contained in annex 6

The procedure for the approval of containers shall be as follows :

a) Containers may be approved by the competent authorities of the country in which the owner is resident or established or by those of the country where the container is used for the first time for transport under Customs seal.

b) The date and serial number of the approval decision must be specified.

c) A certificate of approval conforming to the standard form reproduced in Annex 8 shall be issued for approved containers. This certificate shall be printed in the language of the country of issue and in French, and the various headings shall be numbered, so that the text may be more readily understood in other languages. The certificate shall be covered on both sides by hermetically sealed transparent plastic sheets.

d) The certificate shall accompany the container ; it shall be inserted in the protective frame mentioned in Article 1 of Annex 6 and so sealed that it cannot be extracted from the protective frame without breaking the seal.

e) Containers shall be produced every two years to the competent authorities for purposes of inspection and renewal of approval where appropriate.

f) Approval shall lapse if the essential features of the container are altered or on change of ownership.

ANNEX 8

Certificate of approval of a container

Certificate No ....

1. ....

2. Attesting that the container specified below fulfils the conditions for transport under Customs seal.

3. Valid until ....

4. This Certificate must be returned to the Issuing Office when the container is taken out of service, or on change of ownership, on expiry of the period of validity of the certificate or if there is any material change in any essential particulars of the container.

5. Kind of container ....

6. Name and business address of owner ....

7. Identification marks and numbers ....

8. Tare ....

9. External dimensions in centimetres
   cm x cm x cm

10. Essential particulars of structure (nature of materials, nature of construction, parts which are reinforced, whether bolts are riveted or welded etc.) ....

11. Issued at .... (place) on .... (date) 19....

12. Signature and stamp of Issuing Office ....

325

T I R plates

1. The dimensions of the plates shall be 250 mm by 400 mm.

2. The letters TIR in capital Latin characters shall be 200 mm high and their strokes at least 20 mm wide. The letters shall be white on a blue ground.

Protocol of signature

At the time of signing the Convention of this day's date, the undersigned, duly authorized, make the following declarations :

1. The terms of this Convention set out minimum facilities. It is not the intention of the Contracting Parties to restrict the wider facilities which are granted or may be granted by certain of them in respect of the international transport of goods by road. Contracting Parties may, in particular, agree amongst themselves to consider goods which do not strictly conform to the definition in Article 1, sub-paragraph (h), of the Convention as coming under the conditions laid down in chapter IV of the Convention.

2. The provisions of this Convention shall not preclude the application of other provisions, whether national or in Conventions, governing transport.

3. As far as is practicable, the Contracting Parties shall facilitate

— operations at Customs offices relating to perishable goods, and

— the execution outside normal working days and hours of Customs formalities at Customs offices en route.

4. The Contracting Parties recognize that the satisfactory operation of this Convention requires the provision of facilities to the associations concerned for

a) the transfer of the currency necessary for the payment of import duties and import taxes and any pecuniary penalties claimed by the authorities of Contracting Parties in virtue of the provisions of this Convention ; and

b) the transfer of currency for payment for TIR carnet forms sent to the guaranteeing associations by the corresponding foreign associations or by the international organizations.

5. ad Articles 1 a), 4 and 20

The provisions contained in Articles 4 and 20 shall not preclude the levy of small charges in the nature of statistical fees.

6. ad Article 37

Each Contracting Party shall consider whether certain restrictions or certain controls might not be dispensed with or relaxed at Customs offices en route in the case of transport operations covered by Chapter III of this Convention, in view of the safeguards afforded in the case of these operations by the system laid down in the Convention.

In witness whereof, the undersigned, being duly authorized thereto, have signed this Protocol.

Done at Geneva, this fifteenth day of January one thousand nine hundred and fifty nine, in a single copy in the English and French languages, each text being equally authentic.

LIST OF THE PRINCIPAL JURIDICAL TEXTS ON INTERNATIONAL ROAD TRANSPORT

A) ORGANIZATION OF THE UNITED NATIONS (UNO)

I. Road construction

— Declaration on the construction of main international traffic arteries, signed at Geneva on 16th September 1950 (*)

II. Road traffic and signalisation system

— Convention on road traffic, signed at Geneva on 19th September 1949. (*)

Protocol on road signs and signals, signed at Geneva on 19th September 1949. (*)

European Agreement supplementing the 1949 Convention on road traffic and the 1949 Protocol on road signs and signals, signed at Geneva on 16th September 1950.(*)

European Agreement on the application of article 23 of the 1949 Convention on road traffic, concerning the dimensions and weights of vehicles permitted to travel on certain roads of the Contracting Parties signed at Geneva on 16th September 1950.(*)

Agreement on signs for road works, amending the European Agreement of 16th September 1950 on road signs and signals, signed at Geneva on 16th December 1955. (*)

- Agreements on the lifting of restrictions on the freedom of the road (Geneva, in December 1947) (*) :

a) Freedom of transit for transport of goods by road ;

b) Freedom of transport of goods by road other than in transit ;

c) The most ilberal application of authorization systems for certain international transport by road ;

d) Freedom for international tourist traffic by road ;

e) The most liberal application of their authorization systems for all international passenger transport by road, and, in particular, for international tourist traffic services not falling under d).

- European Agreement on road markings, signed at Geneva on 13th December 1957.(*)

- Agreement concerning the adoption of uniform conditions of approval and reciprocal recognition of approval for motor vehicle equipment and parts, signed at Geneva on 20th March 1958. (*)

- Convention on transit trade of land-locked states, signed at New York, on 8th July 1965.
- Convention concerning customs facilities for touring, signed at New York on 4th June 1954. (*)
- Customs Convention on the temporary importation of commercial road vehicles, signed at Geneva on 18th May 1956. (*)

- Customs Convention on containers, signed at Geneva on 18th May 1956. (*)

- Customs Convention on the international transport of goods under cover of TIR carnets (TIR Convention) and Protocol of signature, signed at Geneva on 15th January 1959. (*)
- European Convention on customs treatment of pallets used in international transport, signed at Geneva on 9th December 1960. (*)

IV. Private law

- Convention on the contract for the international carriage of goods by road (CMR), signed at Geneva on 19th May 1956. (*)

V. Taxation

- Convention on the taxation of road vehicles engaged in international passenger transport, signed at Geneva on 14th December 1956. (*)
- Convention on the taxation of road vehicles engaged in international goods transport, signed at Geneva on 14th December 1956. (*)

VI. Social questions

- European Agreement concerning the work of crews of vehicles engaged in international road transport (AETR), signed at Geneva on 19th January 1962.

VII. Special transports

- European Agreement concerning the international carriage of dangerous goods by road (ADR), signed at Geneva on 30th September 1957.
- Agreement on special equipment for the transport of perishable foodstuffs and on the use of such equipment for the international transport of some of those foodstuffs, signed at Geneva on 15th January 1962.

B) COUNCIL OF EUROPE

- European convention on compulsory insurance against civil liability in respect of motor vehicles, signed at Strasburg on 20th April 1959.

C) INTERNATIONAL LABOUR ORGANIZATION (ILO)

- European Convention concerning the social security of workers engaged in international transport, adopted at Geneva on 9th July 1956. (*)

D) INTERNATIONAL INSTITUTE FOR THE UNIFICATION OF PRIVATE LAW (UNIDROIT), ROME

- Draft Convention on the contract for the international carriage of passengers and luggage by road.
- Draft Convention on the rights, obligations and liability of forwarding agents in international goods transport.
- Draft Convention on the contract for the international combined carriage of goods.

E) INTERNATIONAL ATOMIC ENERGY AGENCY (IAEA)

- Vienna Convention on civil liability for nuclear damage, signed at Vienna on 21st May 1963.

F) EUROPEAN NUCLEAR ENERGY AGENCY (ENEA)

- Convention on third party liability in the field of nuclear energy, signed at Paris on 29th July 1960.

(*) Conventions, protocols and agreements entered into force.

PAYS COUNTRIES LÄNDER	AUTRICHE (A)	BELGIQUE (B)	BULGARIE (BG)	SUISSE (CH)	TCHECOSLOVAQUIE (CS)	ALLEMAGNE occidentale (D)	ALLEMAGNE orientale (D)	DANEMARK (DK)	ESPAGNE (E)
Hauteur/Height/Höhe . . . . . m	3,8	4	4	4	3,8	4	4	3,8	4
Largeur/Width/Breite . . . . . m	2,5	2,5	2,5	2,5	2,5	2,5	2,5	2,5	2,5
Longueur/Length/Länge . . . . . m									
Autocar/Motorcoach/Autobus :									
à 2 essieux/axles/Achsen	12	12	11	12	12	12	11	12	12
à 3 essieux/axles/Achsen	12	12	12	12	12	12	12	12	12
Camion/Lorry/LKW :									
à 2 essieux/axles/Achsen	11	12	10	10	10	12	10	10	11
à 3 essieux/axles/Achsen	12	12	12	10	12	12	12	12	12
Remorques/Trailers/Anhänger :									
à 2 essieux/axles/Achsen	11	12	11	—	—	12	—	—	11
à 3 essieux/axles/Achsen	12	12	12	--	—	12	—	—	12
Véhicule articulé / Articulated vehicle / Sattelschlepper :									
à 3 essieux/axles/Achsen	15	15	14	14	22	15	14	14	16,5*
à 4 essieux/axles/Achsen	15	15	14	14	22	15	14	14	16,5
à 5 essieux/axles/Achsen	15	15	14	14	22	15	14	14	16,5
Train routier / Lorry with trailer / Lastzug	16,5*	18	18	18	—	18	18	18	16,5*
Poids/Weight/Gewicht . . . . . t									
par essieu/axle/Achse	10	13	10	10	10*	10*	—	10*	10
par essieu-tandem/double axle . . . Doppelachse	16*	20*	20	14	16*	16*	—	16*	16
Autocar/Motorcoach/Autobus :									
à 2 essieux/axles/Achsen	16	19	19	16	16	16*	14	15*	16
à 3 essieux/axles/Achsen	22	26	26	16	21,5	22*	18,5	15*	24
Camion/Lorry/LKW :									
à 2 essieux/Axles/Achsen.	16	19	19	16	—	16*	14	15*	16
à 3 essieux/axles/Achsen	22	26	26	16	—	22*	18,5	15*	24*
Remorques/Trailers/Anhänger :									
à 2 essieux/axles/Achsen	16	20	19	12	16*	16*	11	15*	16
à 3 essieux/axles/Achsen	22	26	26	12	24*	22*	16,5	15*	24*
Véhicule articulé / Articulated vehicle / Sattelschlepper :									
à 3 essieux/axles/Achsen	24	36	35	21	26	38	19,5	—*	24*
à 4 essieux/axles/Achsen	32	36	35	21	32	38	24	—*	32*
à 5 essieux/axles/Achsen	32	36	35	21	37,5	38	30	—*	32*
Train routier / Lorry with trailer / Lastzug	32	40	40	26*	—	38	—	—*	32*

328

ADMIS POUR LES VÉHICULES ROUTIERS EN EUROPE

PERMISSIBLE FOR ROAD VEHICLES IN EUROPE

GEWICHTSBEGRENZUNGEN FÜR STASSENFAHRZEUGE IN EUROPA

FRANCE (F)	ROYAUME-UNI (GB)	GRÈCE (GR)	HONGRIE (H)	ITALIE (I)	LUXEMBOURG (L)	NORVÈGE (N)	PAYS-BAS (NL)	PORTUGAL (P)	POLOGNE (PL)	ROUMANIE (R)	SUÈDE (S)	FINLANDE (SF)	TURQUIE (TR)	YOUGOSLAVIE (YU)	Desiderata de l'IRU
—	4,6*	3,8	4	4	4	—	4*	4	4	4	—	3,8	3,8	4	.
2,5	2,5	2,5	2,5	2,5	2,5	2,2*	2,5	2,45	2,5	2,75	2,5	2,5	2,5	2,5	.
11*	11	12	11	11	11	—	12	10,3	11	12	—	12	11	11	.
11*	11	12	11	11	12	—	12	10,3	11	12	—	12	11	—	.
11	11	10	10	10	10	—	11*	10	10	10	—	11	—	11	.
11	11	12*	11	11	12	—	11	10	11	12	—	11	—	12	.
11	7	—	—	7,5	—	—	11*	—	10	11	—	11	10	8	.
11	7	—	—	8	—	—	11	—	11	—	—	11	11	10	.
15	13	15*	14	14	14	—	15	12	14	—	—	14	14	15	15
15	13	15*	14	14	14	—	15	12	14	—	—	14	14	15	15
15	13	15*	14	14	14	—	15	12	14	—	—	14	14	15	15
18	18	18	18	18	20	—	18	14	18	20	—	18	18	18	18
13	9/10*	8	8	10	13	2*	10	10	8	10	10*	8	8	10	13
21	16/18*	14,5	14,5	14,5	20	—	16	16,5	14,5	16	16*	13	14,5	16	20
19	14-16*	14	16	15	19	—	16	15	16	16	—*	12,5	—	16	.
26	14-16*	20	20	19	26	—	—	20	21	24	—*	17,5	—	20	.
19	14-16*	14	16	14	19	—	—	15	14	16	—*	12,5	—	18	.
26	20-28*	20	20	18	26	—	—	20	21	24	—*	17,5	—	24	.
19	14-16*	—	16	14	—	—	—	15	14	20	—*	12,5	—	20	.
26	22*	—	20	18	—	—	—	15	21	—	—*	17,5	—	20	.
35	20-24*	32*	20	18	35	—	36	20	30	—	—	—	—	26	38
35	24-32*	32*	20	28	35	—	36	20	30	—	—	—	—	32	38
35	24-32*	32*	20	32	35	—	36	20	30	—	—	—	—	38	38
35	32	32*	36	—	40	—	40	30	—	40	—	—	—	40	40

AUSTRIA
* 18 m, if one of the vehicles was in service before 1st October 1962.

* 16 tons : distance of 1-2 m between axles.

BELGIUM
* Provided that the distance between the two axles is less than 1,60 m and the weight of the most heavily laden axle does not exceed 10 tons.

SWITZERLAND
* 21 tons : with trailer with 1 axle.

CZECHOSLOVAKIA
* 8 tons : trailer with 2 axles.
* Distance of 1,3 - 2 m between axles.

WESTERN GERMANY
* Saar : international transport : 13 tons : 1 axle, 21 tons : double axle, 19 tons : 2 axles, 26 tons : 3 axles and more

DENMARK
* On certain conditions in international road traffic.
* 15 t : maximum authorized weight of a vehicle or road train + 0,25 ton per 0,2 m distance between front and rear axle (omitting first 0,5 m from the calculation).

SPAIN
* By special permission.

FRANCE
* 12 m if the rear overhang does not exceed 6/10 of the length on the ground, nor the absolute length by more than 3,5 m.

UNITED KINGDOM
* Only to coaches.
* 9/10 tons on certain conditions.
* 16 tons : if axle spacing 1 m - 1,22 m.
* 18 tons : if axle spacing 1,22 m - 2,13 m.
* 14/15/16 : on certain conditions.
* 20 - 28 tons, 22 tons, 20 - 24, 24 - 32 tons : on certain conditions.

GREECE
* Main roads.

NORWAY
* Exceptions admitted : 2,35 m
* Exceptions admitted : 7 tons

NETHERLANDS
* From 1st June 1966.

SWEDEN
* Main roads (11 %).
* Maximum authorized weight of vehicles or road trains must not exceed 8,5 tons where the distance between front and rear axle is less than 2,2 m with an increase of 0,25 ton per 0,2 m additional distance between the axles, not counting first 2,2 m.

TURKEY
* Annex 7 of the 1949 Convention on road traffic.

VEHICLE TYPES

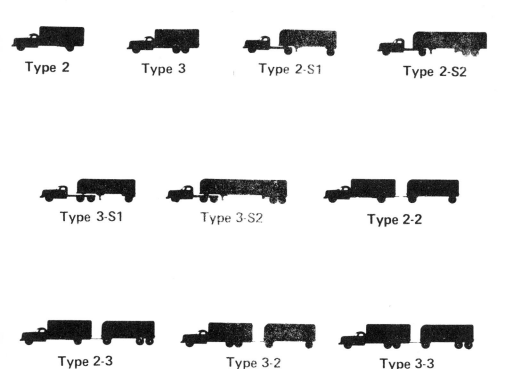

Type 2   Type 3   Type 2-S1   Type 2-S2

Type 3-S1   Type 3-S2   Type 2-2

Type 2-3   Type 3-2   Type 3-3

Notation

The figure shows silhouettes of most basic commercial vehicle types in regular operation as designated by code based
on axle arrangement. The first digit indicates the number
of axles of the truck or truck-tractor. The letter "S" indicates the number of axles on the semitrailer. Any digit
other than the first in a combination, when not preceded by
an "S", indicates a trailer and the number of its axles. For
instance, a 2-S2 combination is a two-axle truck-tractor
with a tandem-axle semitrailer. A 3-S1-2 combination is a
three-axle truck-tractor with tandem rear axles, a semi-
trailer with a single axle, and a trailer with two axles.

NORTH, CENTRAL & SOUTH AMERICA

COUNTRY PAIS	WIDTH ANCHO feet pies	meters metros	HEIGHT ALTURA feet pies	meters metros	Truck - Camiones feet pies	meters metros	Bus - Bus feet pies	meters metros	Truck & Semi-Trailer Camion con Remolque feet pies	meters metros	Other Combinations Otras Combinaciones feet pies	meters metros	Simple Simples pounds libras (1000)	metric tons ton. metricas	Tandem Tandem pounds libras (1000)	metric tons ton. metricas	2 pounds libras (1000)
Argentina		2.50		(2)		11.00		12.00		15.50		18.50 20.50 (3)		10.6		18.0	
Bolivia		2.50		3.80		10.00 15.30 (7)				22.00 (14)		24.00		6.0		6.0	
Brazil		2.60		4.00		11.00		11.00		17.50		17.50		10.0		16.0 (4)	
Canada (6)	8.0		12.5 14.5		33.0 to 40.0		33.0 to 40.0		50.0 85.0				18.0 to 20.0		24.0 to 32.0		20.0 to 30.0
Chile		2.50		3.80		10.00		11.00		14.00		18.00		8.00		15.0	
Colombia		2.45		4.00		10.65		10.65 12.20 (7)		13.70		15.25		8.2		14.5	
Costa Rica		2.50		3.80		10.00 11.00 (7)		11.00		14.00		–		8.0		14.5	
Ecuador		2.50		3.80		10.70				13.70		18.30		11.0		19.0	
Guatemala		2.50		3.80		10.00 11.00 (7)		11.00		14.00		14.00		8.0		14.5	
Guyana	7.5 (8)				27.5		27.5		30.0		30.0						
Honduras					No Regulations – No tiene Reglamento												
Jamaica (9)	8.0		10.5		26.0		30.0		33.0		52.0		17.92				27.0
Mexico (13)		2.50		4.15		11.60		11.60		(12)		18.30		9.0		14.5	
Netherland Antilles		2.50		3.50		10.00		11.00		14.00		18.00		No Regulations				
Nicaragua	8.2	2.50	12.5	3.80	36.0	10.00 11.00 (7)	36.0	11.00	46.0	14.00	46.0	14.00	17.6	8.00	32.0	14.5	
Panama (13)		2.50		4.10		11.00 12.00 (7)				15.25		15.25		10.0		16.4	
Paraguay		2.40		4.00		10.00		10.00		(15)		20.00		8.0		10.0	
Peru		2.50		4.00		10.50 12.00 (7)		10.50		15.00		15.00		11.0			
El Salvador		2.50		3.80		10.00 11.00 (7)		11.00		14.00		14.00		17.6 8.0		32.0 14.5		
Uruguay		2.50		4.00		10.70		10.70		15.30		18.30		10.0 (11)		14.4	
United States (10)	8.0 to 9.0		12.5 to 14.0		35.0 to 55.0		35.0 to 55.0		50.0 65.0		50.0 70.0		18.0 to 24.0		32.0 to 40.6		28.0 to 48.0
Venezuela		2.60		3.90		10.70 12.20 (7)		10.70 12.20 (7)		15.30		18.30		8.5		14.5	
AASHO Policy	8.5		13.5		40.0				55.0		65.0		20.0		32.0		
PanAm Hwy Congress Policy	8.2	2.5	12.5	3.8	33.0 to 36.0	10.0 to 11.0	36.0	11.00	46.0	14.00	72.0	22.00	17.6	8.0	32.0	14.5	

1. In accordance with vehicle type.
   De acuerdo con el tipo de vehículo.

2. Regulations vary with type of vehicle.
   El reglamento varía de acuerdo con el tipo de vehículo.

3. 18.50 meters for truck and trailer. 20.50 meters for truck-tractor with semi-trailer and trailer.
   18.50 metros para camión con remolque. 20.50 metros para camión con semi-remolque y remolque.

4. If the tandem axles are 1.34 meters or more apart, limit is 17.00 tons.
   Si los dos ejes están a una distancia de 1,34 metros o más, el límite es de 17,00 toneladas.

5. If the tandem axles are 1.34 meters or more apart, weight limit is the highest one.
   Si los dos ejes tandem están a una distancia de 1,34 metros o más, el límite de peso es el mayor.

6. Regulations vary by Province. Figures indicate range of variation in maximum limits.
   Los reglamentos varían de Provincia a Provincia. Las cifras indican las variaciones de los límites máximos.

7. Three axle vehicles have the highest limit.
   Vehículos de tres ejes tienen el límite mayor.

8. Except buses that can be 8.0 ft. wide.
   Excepto los buses que pueden tener 8,0 pies de ancho.

9. Special Permit may be granted for vehicles exceeding these limits.
   Permiso Especial puede concederse a vehículos que exceden estos límites.

AMERICA DEL NORTE, CENTRAL Y SUR

MAXIMUM GROSS WEIGHT (1) / PESO MAXIMO TOTAL

Values per vehicle type column: pounds libras (1000) / metric tons ton. métricas. Leftmost column: metric tons / ton. métricas.

metric tons ton. métricas	3	2-S1	2-S2	3-S1	3-S2	2-2	2-3	3-2	3-3
(Dependent on axle spacing — Según distancia entre ejes)									
11.0	14.0	14.0	14.0	14.0	14.0	14.0	20.0	20.0	20.0
15.0	21.0/22.0 (5)	25.0	31.0/32.0 (5)	31.0/32.0 (5)	37.0/39.0 (5)	35.0	40.0	40.0	40.0
	30.0 to 42.0	40.0 to 48.0	48.0 to 74.0	48.0 to 74.0	54.0 to 74.0	48.0 to 67.8	44.0 to 74.0	44.0 to 74.0	44.0 to 84.0
16.0	23.0	24.0	31.0	31.0	38.0	32.0	39.0	39.0	46.0
(Dependent on tire size and axle spacing — Depende del tamaño de llantas y distancia entre ejes)									
12.0	20.0	20.0	25.0	25.0	25.0	25.0	25.0	25.0	25.0
15.5	23.5	26.5	34.5	34.5	38.7	37.5	43.8	43.8	46.4
12.0	20.0	25.0	25.0	25.0	25.0	25.0	25.0	25.0	25.0
(Dependent on tire size and axle spacing — Depende del tamaño de llantas y distancia entre ejes)									
No regulations — No tiene Reglamento									
31.36	45.0	45.0	45.0	45.0	45.0				
14.0	19.5	23.0	28.5	28.5	34.0	32.0		37.5	
No tiene Reglamento No Regulations — No Tiene Reglamento									
12.0	18.5	20.0	26.5	26.5	33.0	28.0	34.5	34.5	
14.0	24.0	24.0	32.0	32.0	32.0	32.0	32.0	32.0	32.0
10.0	12.0	15.0	20.0	22.0	26.0	26.0	30.0	30.0	34.0
15.0	20.0	20.0	30.0	30.0	30.0	30.0	30.0	30.0	30.0
12.0	18.5	20.0	26.5	26.5	33.0	28.0	34.5	34.5	
15.0	19.4	25.0	29.4	29.4	32.9	32.9	32.9	32.9	32.9
	36.0 to 65.0	46.0 to 69.6	60.0 to 77.5	50.0 to 77.52	64.0 to 86.4	47.8 to 86.4	39.5 to 85.0	53.5 to 86.4	53.5 to 90.0
12.0	18.6	20.0			30.0	28.0			33.0
(Dependent on axle spacing and vehicle type — Depende de la distancia entre ejes y el tipo de vehículo)									
(Dependent on axle spacing — Según distancia entre ejes)									

10. Regulations vary by States. Figures indicate range of variation in maximum limits.
Los reglamentos varían de Estado a Estado. Las cifras indican las variaciones de los límites máximos.

11. Two-tire axle maximum load: 5.0 tons.
Four-tire axle maximum load: 10.0 tons.

Carga máxima en ejes de dos ruedas: 5,0 toneladas.
Carga máxima en ejes de cuatro ruedas: 10,0 toneladas.

12. Regulations vary with type of vehicle. Type 2-S1 is 14.00 m.; Type 2-S2 and 3-S1 is 14.50 m.; 3-S2 is 15.25 m.

Las regulaciones varían según el tipo de vehículo. El tipo 2-S1 tiene 14,00 m. El tipo 2-S2 y el 3-S1 tienen 14,50 m.; y el 3-S2 tiene 15,25 m.

13. Limits shown are for Class A highways.
Class B highways have lower limits.

Estos límites son para carreteras Clase A. Las Carreteras Clase B tienen límites más bajos.

14. Vehicle type 3-S2 has 24.00 m. of length as a limit.
El vehículo tipo 3-S2 tiene 24.00 m. de largo como límite.

15. Vehicle type 2-S1 has 10.00 m.; 2-S2 and 3-S1 have 12.00 m.; 3-S2 has 20.00 m.
El vehículo tipo 2-S1 tiene 10.00 m.; 2-S2 y 3-S1 tienen 12.00 m.; 3-S2 tiene 20.00 m.

The Equipment Interchange Association through its Special Intermodal Interchange Rules Committee devised the following procedures for the Interchange of Containers where the Motor Carrier industry was involved:

EQUIPMENT INTERCHANGE SCHEDULE No. 2

ITEM	INTERMODAL RULES AND REGULATIONS
	DEFINITIONS
	Equipment. As used herein the term "equipment" or the term "vehicle" shall be construed as including but not as limited to, those trailers, containers and other equipment, and components of any one thereof offered and used by subscribing carriers in interchange service.
	ACCEPTANCE AND RETURN OF EQUIPMENT
	1.1 Equipment will be accepted in intermodal interchange service only:
	a. If the name or *alpha carrier code (as published in the Directory of Registered Alpha Carrier Codes and Railroad and Private Car Company Reporting Marks), and equipment number of the owner are imprinted thereon.
	b. Under the terms of the contractual provisions set forth in ITEM S100.
	c. After inspection of the equipment by an authorized representative of the receiving carrier, and certification on the inspection report form set forth in ITEM 120, or other suitable form, that the person who made the inspection is competent and qualified to make such inspection and has been duly authorized to do so.
S20	d. By use of the Equipment Interchange Receipt and Safety Inspection Report set forth in ITEM 120, or other suitable form, duly completed and signed; and only if the equipment and the inspection meet the requirements of governmental agencies having jurisdiction thereof.
	e. If the tires are roadworthy, and if all applicable terms of ITEM S50 are observed.
	1.2 A carrier will accept the return of equipment from interchange service, ordinary wear and tear excepted, and in all cases as provided in paragraph "c" hereof, only:
	a. After inspection of the equipment by its authorized representative and certification to the effect that the person who made the inspection was competent and qualified to make such inspection and had been duly authorized to do so.

b. By use of the Equipment Interchange Receipt and Safety Inspection Report set forth in ITEM 120, or other suitable form, duly completed and signed; and only if the inspection meets the requirements of governmental agencies having jurisdiction thereof.

c. Subject to responsibility of the carrier initially acquiring the equipment for damages noted.

1.3 In the absence of agreements to the contrary, when equipment in interchange service is unloaded it must be returned promptly to the carrier from whom received at the point of original interchange in accordance with the Equipment Interchange Receipt and Safety Inspection Report.

Note: Provided it is adopted as official code designation by appropriate standardization organizations.

EQUIPMENT LOCATION REPORTS

S30

2.1 Carriers will maintain a register of equipment owned or held under long-term lease which register shall include the equipment number, a manufacturer's serial number and make.

2.2 The using carrier shall keep a record of each piece of interchanged equipment so as to indicate its present location.

2.3 When equipment leaves the line of the offering carrier the Equipment Interchange Receipt and Safety Inspection Report must be completed and distribution thereof made.

2.4 Each carrier will maintain a daily record showing, as to each carrier with which it conducts equipment interchange, and with adequate identification, equipment received from that carrier and equipment given to it.

EQUIPMENT ACCESSORIES

S40

3.1 All equipment to be interchanged must have (where applicable) the following accessories:

a. The SAE-ATA Recommended (7-Conductor) Electrical Connector Plug.

b. Flashing turn signals with not less than 12 square inches of lens surface.

c. Stop lights, tail light, clearance and marker lights as required by the Interstate Commerce Commission or other appropriate governmental agency.

d. Mud or rain flaps.

e. Attached to the exterior of the equipment, or as part of it, a waterproof container for necessary papers and documents to accompany the equipment.

f. Equipment offered in interchange may be refused by the receiving carrier if it does not comply with the foregoing requirements.

ITEM	INTERMODAL RULES AND REGULATIONS
S50	## TIRES 4.1 At the time of interchange of equipment the tires shall be thoroughly inspected: a. To determine if properly mated. b. To determine that all valve stems are provided with valve caps. c. To determine major defects, such as: (1) Separation of caps (2) Visible blisters (3) Visible Knots (4) Obvious signs of rim wear (5) Visible objects protruding from tire, such as nails, bolts, spikes. d. To determine that tires contain at least sufficient tread to assure safe travel for the complete round trip contemplated. 4.2 Any deficiency noted by inspection pursuant to paragraph 4.1 shall be corrected by the owner or by the carrier in possession in compliance with the provisions of ITEM S100, Section 5.4, and, in any event, prior to acceptance for interchange use. 4.3 At the time of inspection there shall be recorded on the Equipment Interchange Receipt and Safety Inspection Report the tire owner's name and number, if any, or manufacturer's name, and location of tire on the equipment. 4.4 The carrier in possession of foreign equipment in interchange service shall provide good maintenance to its tires and tubes, including proper inflation, the repair of flat tires, pulled valve stems, etc. 4.5 It is recommended that carriers mark all their company-owned tires on both sides. The marking should be placed immediately after the manufacturer's name or in the branding area provided on the tire.
S60	## ORDINARY MAINTENANCE 5.1 Interchanged equipment in a carrier's fleet is to be inspected, serviced, and lubricated in accordance with the using carrier's maintenance program.

REPAIRS

S70

6.1 By Owner or User. Defects in interchanged equipment noted at the time of inspection and prior to acceptance will be repaired at the expense of the owner. The user may cause such repairs to be made in compliance with the provisions of ITEM S100, Section 5.4.

6.2 By User. Repairs and other service adjustments occasioned after interchange will be absorbed by the user in accordance with the provisions of ITEM S100, Section 5.

6.3 Records To Be Kept. Where repairs are performed by user and charges assessed to owner, the user will prepare work order or job tickets covering repairs or other work required, and will furnish a copy to owner, suitably itemized.

ARBITRATION PROCEDURE

S80

When carriers are unable to agree unanimously on the settlement of a dispute respecting repairs to equipment or its maintenance or use, arising between two or more carriers who are subscribers to the Uniform Intermodal Interchange Rules, Regulations and Contractual Provisions, the following procedure shall be invoked in order to submit the dispute to arbitration. Disputes shall be submitted to the Intermodal Arbitration Panel of the Association for adjudication only in those instances where all the carriers in disagreement are subscribers to the Intermodal Interchange Rules, Regulations and Contractual Provisions.

(a) There shall be constituted within the framework of the Association an Intermodal Arbitration Panel, which shall consist of five members from each of the subscribing modes. Appointments to the Panel shall be made by the President of the Association and any member of the Association shall be eligible for appointment. Membership on the Panel shall continue until a member is replaced by appointment of a new member. Any member of the Intermodal Arbitration Panel shall be disqualified in a particular case if the carrier he represents is involved in the dispute. In such case, any three members of the Panel (each mode being represented) who are not involved in the controversy may be enlisted to settle the dispute. To the end that decisions be rendered as expeditiously as possible, whenever the number of arbitration dispute files referred to particular members of the Intermodal Arbitration Panel and still pending decision indicates that disposition of a new arbitration proceeding may be unduly delayed if submitted to those members, the carriers involved in the new dispute shall be notified thereof by the Managing Director of the Association, and requested to select, as provided in paragraph (c) of this ITEM, the required arbitrators from such members of the Panel as are not occupied with numerous other proceedings. The Managing Director of the Association, hereinafter sometimes referred to as the Director, shall serve as staff representative of the Association in the operation of the Intermodal Arbitration Panel, shall furnish forms, receive and distribute files, and shall otherwise undertake in all proper ways to assist the carrier and the Panel in the use of this arbitration procedure.

INTERMODAL RULES AND REGULATIONS

ITEM	
	ARBITRATION PROCEDURE—Concluded
(b)	To qualify a dispute for arbitration all carriers involved in the dispute shall execute the REQUEST FOR ARBITRATION form set forth in ITEM S90. The carrier initially desiring arbitration shall have the duty to notify all carriers against whom a charge may be made as a result of the Intermodal Arbitration Panel's decision by submitting to each four copies of the signed REQUEST FOR ARBITRATION form for signature. Upon signing the REQUEST FOR ARBITRATION form each such carrier shall return three signed copies to the carrier initiating the arbitration.
(c)	To submit a dispute to arbitration each side will appoint one member from the Intermodal Arbitration Panel and the two thus appointed shall select a third and neutral member. Such neutral member to be selected from a mode not involved in the dispute. If the carriers involved in the dispute are unable to agree on the slection of the neutral member the President of the Association shall make the appointment.
(d)	The carrier requesting arbitration shall prepare a brief written statement, setting forth the facts of the case, the carrier's contention as to proper settlement and the grounds for such position, and make request for such decision as is deemed appropriate. A copy of such brief written statement shall be sent by the initiating carrier to each other carrier involved in the arbitration and the statement shall show that such copies have been sent.
(e)	Within not more than fifteen days after service upon them of the initiating carrier's written statement of position provided in paragraph (d), each other carrier involved in the dispute may file reply, by letter to the Managing Director of the Association answering matter alleged in such initiating carrier's written statement. Copy of such reply must be sent to each other carrier involved in the arbitration and the reply shall show that such copies have been sent.
(f)	The entire dispute file, consisting of the complainant's contention, all papers provided in paragraph (d), and all documents in support thereof, shall be submitted by the complaining carrier by registered mail or prepaid express to the Managing Director of the Association with two copies of the executed REQUEST FOR ARBITRATION.
(g)	To promote orderly and efficient use of files all papers in the dispute file shall be numbered consecutively by the Managing Director, beginning with one (1) on the first paper, which shall be the executed REQUEST FOR ARBITRATION form. The consecutive numbers shall be placed in the upper right-hand corner of each paper and the file shall be presented in a neat condition.

(h) The Director of the Association shall make a record of the papers filed under paragraphs (d, e, and f) above, and will assign an official arbitration file number to the dispute submitted for arbitration. The Director will then transmit to one of the three members of the Panel the entire file with the exception of routine correspondence directed to him, accompanied by a letter of transmittal. The Director shall also keep a log of files referred to the members of the Intermodal Arbitration Panel. The letter of transmittal will indicate the order in which the Panel will receive and act upon the dispute, the order chosen by the Director being designed by him to achieve the greatest expedition.

(i) Each member of the Intermodal Arbitration Panel shall, in sequence, upon receipt of a dispute submitted, review all documents contained in the file and, as promptly as possible write an opinion clearly setting forth his definite conclusion as to the proper settlement of the dispute, giving his reasons for such opinion. The first member receiving the file shall, upon completing his opinion, forward the file to the second member indicated on the transmittal letter provided in paragraph (h), and shall send his signed opinion, in triplicate, to the Director. The second member shall, in turn, prepare his opinion and forward it, signed in triplicate, to the Director and forward the file to the third member who shall, in turn, prepare and submit his opinion, signed in triplicate, to the Director, along with the file.

If, in the opinion of a Panel member, an analysis of the facts submitted does not permit disposition, he may communicate directly with any party involved for the purpose of securing such additional information as may be required. All such correspondence must then become a part of the dispute file. If any such additional information is requested and obtained by a member of the Panel after an opinion has been written by another member or members, he shall prepare his opinion as provided above and the entire file shall be returned to the Director who shall forward it to the other member or members for possible reconsideration and amendment of the opinion already rendered. In other respects the procedure provided in the foregoing paragraph of this section shall be followed.

(j) When a decision in a dispute has been reached and the file returned to the Director, the Director shall send a copy of each Panel member's opinion to each party of record in the proceeding and return the original dispute file to the carrier submitting the dispute for arbitration. The Director shall keep a permanent record of each arbitration dispute, showing dates of transmittal to and between interested parties and containing a brief history of the proceeding, including copies of all opinions.

(k) A decision rendered by the Intermodal Arbitration Panel which has been concurred in by at least a majority (two members) of the Panel shall be final and binding upon the parties involved in the dispute.

The results of arbitration proceedings shall be published by the Association for the information of all subscribing carriers.

339

EQUIPMENT INTERCHANGE SCHEDULE No. 2

INTERMODAL RULES AND REGULATIONS

ITEM	

REQUEST FOR ARBITRATION

In consideration of the mutual agreements herein made, the undersigned carriers hereby request that the dispute identified below respecting repairs to equipment or its maintenance or use, or respecting compensation for the use of equipment or rules and regulations pertaining thereto, arising out of the use of the Intermodal Interchange Rules and Regulations in Equipment Interchange Schedule No. 2, be submitted to the Intermodal Arbitration Panel of the Association. It is agreed that the procedure established for said Intermodal Arbitration Panel by said Association shall govern the consideration, handling, and disposition of this dispute.

It is further agreed that the decision of the Intermodal Arbitration Panel shall be binding upon disputant carriers who are signatories hereto; and that remittance of any monies due any carrier party to this agreement resulting from decision of said Panel shall be made promptly upon receipt of the Panel's decision.

In accordance with the procedures established by the Association for appointment of arbitrators, plaintiff hereby appoints _____ and defendant hereby appoints _____ with the under-standing that the two arbitrators thus appointed will select a third and neutral arbitrator.

IN TESTIMONY WHEREOF, witness the hand and seal of the parties undersigned:

_____ (Seal)
Carrier

By: _____
Signature

_____
Official Title

Witness: _____   Date: _____

_____ (Seal)
Carrier

By: _____
Signature

_____
Official Title

Witness: _____

Date: _____

_____ (Seal)
Carrier

By: _____
Signature

_____
Official Title

Witness: _____

Date: _____

DISPUTE COVERED BY THIS AGREEMENT IS IDENTIFIED AS FOLLOWS:

Carrier	Dispute No.	Carrier	Dispute No.

Carrier	Dispute No.

EQUIPMENT INTERCHANGE SCHEDULE No. 2

ITEM	INTERMODAL CONTRACTUAL PROVISIONS
S100	**CONTRACTUAL PROVISIONS** 1. The subscribing carriers listed in Section R of this Schedule enter into this agreement governing their relationship with respect to the intermodal interchange of equipment, and to make this agreement operative respecting interchange of individual units of equipment will cause to be executed an Equipment Interchange Receipt and Safety Inspection Report form. The term equipment as used herein shall refer to any load-carrying vehicle without motive power. 2. At the time of interchange an authorized representative of each carrier shall execute, in multiple copies as the carriers may require, an Equipment Interchange Receipt and Safety Inspection Report form. The parties shall be bound by the notations on the Equipment Interchange Receipt and Safety Inspection Report. 3. Using Carrier—Responsibility and Liability. 3a. The subscribing carriers listed in the intermodal section of this Schedule shall be indicated by an asterisk so long as the following certification remains on file with the office of the Managing Director of the Equipment Interchange Association: (1) Form EIAL-2 certifying that the carrier has in effect and attached to his policies of liability insurance Uniform Endorsement Form EIAL-1 (or a revision thereof) covering his legal liability including any liability assumed under the contractual provisions of paragraph 3d (1) of this Item with limits of no less than \$250,000/\$500,000 for Bodily Injury and \$250,000 for Property Damage, or single limit Bodily Injury and Property Damage coverage of at least \$1,000,000. **Note:** To qualify for identification by an asterisk subscribing carriers who are self insured and are so recognized by the Interstate Commerce Commission or other appropriate Federal regulatory agencies, shall comply with this Section with respect to policies of insurance excess of self insured limits. 3b The carrier initially acquiring use of interchange equipment: (1) Shall complete promptly and expeditiously the use for which the equipment has been interchanged to it and promptly return the equipment to the terminal of the carrier from which received at the place received, unless otherwise agreed upon; (2) Shall be responsible for returning the equipment to the carrier from which it was received in the condition it was received, ordinary wear and tear excepted, notwithstanding that it may have interchanged such equipment to another connecting carrier;

(3) Shall be responsible to the carrier from whom it received the equipment for the performance of this agreement by itself and by all of other persons into whose possession such equipment may go until its proper return to the carrier from which it received the equipment;

(4) Shall have complete control and supervision of such equipment, and such equipment shall be operated under its common carrier responsibility to the public and public authority while in its possession; and the carrier initially furnishing the equipment shall have no right to control the detail of the work of any employe or agent operating or using said equipment during such time. Any person operating, in possession of, or using said equipment after the signing of said Equipment Interchanging Receipt and Safety Inspection Report, and until another such form is signed returning the equipment to the carrier entitled to receive it, is not the agent or employe of the carrier furnishing the equipment for any purpose whatsoever.

3c The carrier in possession of interchange equipment:

(1) Shall, when equipment is unloaded, assume the responsibility of loading such equipment back to the owning carrier in the following order of priority:

   a. With cargo to be moved by the owning carrier

   b. With cargo to the point of initial interchange or to an intermediate point, but in the direction of such initial point of interchange

   c. Return the equipment empty to the owning carrier.

3d Carrier liability for persons or property:

(1) For third persons or property of third persons:

   The Using Carrier, while in possession of interchange equipment, releases and agrees to defend and hold harmless the owner and any intermediate carrier or provider furnishing said equipment, from and against any and all loss, damage, liability, cost or expenses suffered or incurred by the owner, and any intermediate carrier or provider, arising out of or connected with injuries to or death of other persons or loss or damage to property of other persons arising out of the Using Carrier's use, operation, maintenance or possession of interchange equipment; except loss or damage to such interchange equipment, or cargo being transported therein or cargo being loaded or unloaded or held at terminal or transit points incident to transportation.

(2) For loss or damage to cargo:

   The carrier under whose operating authority interchange equipment is used shall be liable according to applicable law for loss or damage to, or delay of that property being transported therein, caused by or arising out of such carrier's use, operation, maintenance or possession of said equipment.

ITEM	
	INTERMODAL CONTRACTUAL PROVISIONS

CONTRACTUAL PROVISIONS—Continued

4. (Reserved for future use.)

5. Further remedies and responsibilities:

5.1 The carrier entitled to the return to it of equipment shall be entitled to all its lawful remedies.

5.2 Fuel used in providing temperature control shall be replaced by the using carrier at the time such equipment is returned. If the using carrier fails to replenish the fuel supply of the temperature control unit he may, unless otherwise agreed upon between the parties involved, be billed for the cost of fuel consumed.

5.3 When interchanged mechanical temperature control equipment that has moved unpackaged perishable commodities under refrigeration is unloaded it shall, unless otherwise agreed upon between the parties involved, be steam cleaned by the delivering carrier so as to be suitable for reloading other commodities.

5.4 In the event equipment offered for interchange shall require repairs before being received into interchange service, the offering carrier shall be responsible for the cost thereof and the receiving carrier may cause the repairs to be made. If the apparent cost of the foregoing repairs exceeds $25.00, the consent of the offering carrier shall be obtained by the receiving carrier before it causes the repairs to be made.

5.5 In the event interchanged equipment is damaged after being received into interchange service the carrier in possession at the time the damage occured shall, by repair, restore it to the condition in which it was received, and, in the event of failure of such carrier to make such repairs, it shall, nevertheless, be responsible for the cost thereof. If the apparent cost of the foregoing repairs exceeds $150.00, the consent of the owner shall be obtained by the using carrier before it causes the repairs to be made. Bills for repairs to damaged equipment shall, unless otherwise agreed upon, be rendered by the lessor within 120 days after the repairs have been completed.

5.6 Ordinary maintenance and other service adjustments occasioned by ordinary use in interchange will be:

(a) Absorbed by the user when cost thereof does not exceed $25.00;

(b) Billed to and borne by title owner in entirety when cost thereof would exceed $25.00;

(c) Authorized by the owner prior to commencement of repairs when estimated cost thereof would exceed $150.00;

(d) Billed to the owner (pursuant to subparagraphs (b) and (c)) within 120 days from the date the repairs were completed unless a different time period is agreed upon between the owner and the using carrier.

5.7 In the event equipment after being placed into interchange service is damaged, the providing carrier, while settlement therefor or repairs thereto are pending, shall promptly receive from the using carrier equipment of like condition, quality, and size, or, in lieu thereof, at its option, compensation equal to the applicable minimum daily rental in effect between the parties.

5.8 In the event equipment, including component parts, after being placed into interchange service is lost, stolen or totally destroyed, the providing carrier, while settlement therefor is pending, shall promptly receive from the using carrier equipment of like condition, quality, and size, or, in lieu thereof, at the option of the providing carrier, compensation equal to the applicable minimum daily rental in effect between the parties. Settlement for equipment lost, stolen or totally destroyed shall be made in the following manner:

(a) The using carrier shall promptly notify the providing carrier in writing that the equipment has been lost, stolen or totally destroyed, and that the using carrier will make settlement therefor on the basis herein provided. The accumulation of the per diem charges shall cease upon the providing carrier's receipt of such notice.

(b) The providing carrier, shall within thirty days after receipt of the notification provided in the foregoing paragraph, secure and furnish to the using carrier a written statement of the book value of the lost, stolen or totally destroyed equipment, including component parts.

(c) Book value shall be the basis for fixing the value of the equipment, and of component parts, such as re-frigeration units, chassis, LPG generators, etc. Book value shall be determined by utilizing the straight line depreciation method, with estimated ten-year life for the equipment and any component part, less ten percent salvage value.

(d) Settlement shall be made within 120 days after the using carrier has notified the providing carrier in writing that the equipment, including component parts, has been lost, stolen or totally destroyed.

6. The carrier providing equipment:

6.1 (1) Shall equip it with tires and tubes of proper size at the time of interchange. Thereafter, until the equip-ment is returned to the carrier furnishing it, repairs to tires and tubes shall be made by and at the expense of the carrier using the equipment at the time of such tire failure.

(2) In the event of a blow-out or total failure of a tire and/or tube the carrier in possession of the equipment shall furnish replacement tire and/or tube to return the equipment to the carrier from whom it was received (but may remove such replacement tires and/or tubes upon redelivery of the equipment to such carrier and in such event shall return the blown-out or unserviceable tires and/or tubes properly identified to the owner). In the event of failure to so return, payment therefor shall be made at the value thereof at the time of interchange.

INTERMODAL CONTRACTUAL PROVISIONS

ITEM	CONTRACTUAL PROVISIONS—Concluded
S100 (Concluded)	(3) If tires are ruined as a result of being run flat it will be the responsibility of the using carrier to replace the tires in kind or pay for the tires so ruined.
	(4) On tires and/or tubes that are lost or damaged (other than those covered above) the using carrier, within thirty days after the equipment is returned, is required to make settlement therefor by replacing the tires and/or tubes in kind or paying for the lost or damaged tires and/or tubes.
	(5) When a tire and/or tube is replaced by other than the owner of the equipment it must be properly mated and of similar quality and a report of such replacement must be made to owner of the equipment showing size, ply, brand and serial number of tire removed and applied.
	(6) When a tire and/or tube on equipment received in interchange service is considered unserviceable by the carrier in possession and is replaced by it, the unserviceable tire and/or tube must be held by the handling carrier for thirty days after notification to owner in the event the owner of the equipment desires return of the unserviceable tire and/or tube. In the event the owner desires return of unserviceable tire and/or tube, it must be returned to the point and in the manner designated by owner. Cost of transportation in such cases must be assumed by owner.
	6.2 When the using carrier offers to return interchanged equipment to the owner upon completion of the use for which the equipment was interchanged the owner shall, unless otherwise agreed upon, be required to accept the equipment so offered.
	6.3 Does not make any warranty or representation, express or implied, as to the fitness or condition of the equipment so interchanged, including tires and tubes, and the carrier acquiring the use thereof does so at its own risk.
	7. The provisions contained in ITEM S100, and said Equipment Interchange Receipt and Safety Inspection Report, or other suitable form, shall constitute the entire agreement between the parties hereto and no verbal amendment or modification thereof shall be permitted, except, that it may be supplemented by the respective parties in writing, as relates to compensation for the use of equipment.

Standard contract for Interchange of Equipment, including containers between various modes of transportation devised by the Equipment Interchange Association:

EQUIPMENT INTERCHANGE CONTRACT

1. The undersigned carriers enter into this agreement governing their relationship with respect to interchange of equipment, and to make this agreement operative respecting interchange of individual units of equipment will cause to be executed the Equipment Interchange Receipt and Safety Inspection Reports hereinafter mentioned. The term equipment as used herein shall refer to any load-carrying vehicle without power.

2. At the time of interchange an authorized representative of each carrier shall execute, in multiple copies as the carriers may require, a counterpart of the Equipment Interchange Receipt and Safety Inspection Report shown on p.353. The parties shall be bound by the notations on the Equipment Interchange Receipt and Safety Inspection Report.

3. Using Carrier — Responsibility and Liability

3a. Insurance Coverage Certified.

(1) The parties to this contract certify that each has in effect insurance covering its legal liability assumed under the provisions of paragraph 3c(1) of this contract, with limits of no less than $, /$____ for Bodily Injury and $____ for Property Damage, or single limit coverage of at least $____.

3b. The Carrier initially acquiring use of interchange equipment:

(1) Shall complete promptly and expeditiously the use for which the equipment has been interchanged to it and return the equipment to the terminal of the carrier from which received at the place received;

(2) Shall not permit the equipment to go out of its possession without permission of the owner in writing, as shown on the Equipment Interchange Receipt and Safety Inspection Report or otherwise in writing, and then only to the extent of written permission, and shall be responsible for the safe and timely return of the equipment to the carrier from which it was received, ordinary wear and tear excepted, notwithstanding that it may have had permission from the owner of such equipment to interchange such equipment to another connecting carrier;

347

(3) Shall be responsible to the carrier from whom it received the equipment for the performance of this agreement by itself and by all other persons into whose possession such equipment may go until its proper return to the carrier from which it received the equipment;

(4) Shall have complete control and supervision of such equipment, and such equipment shall be operated under its common carrier responsibility to the public and to public authority while in its possession; and the carrier initially furnishing the equipment shall have no right to control the detail of the work of any employee or agent operating or using said equipment during such time. Any person operating, in possession of, or using said equipment after the signing of said Equipment Interchange Receipt and Safety Inspection Report, and until another such form is signed returning the equipment to the carrier entitled to receive it, is not the agent or employee of the carrier furnishing the equipment for any purpose whatsoever.

3c. Carrier liability for persons or property:

(1) For third persons or property of third persons:

The Using Carrier, while in possession of interchange equipment, releases and agrees to defend and hold harmless the owner and any intermediate carrier or provider furnishing said equipment, from and against any and all loss, damage, liability, cost or expenses suffered or incurred by the owner, and any intermediate carrier or provider, arising out of or connected with injuries to or death of other persons or loss or damage to property of other persons arising out of the Using Carrier's use, operation, maintenance or possession of interchange equipment; except loss or damage to such interchange equipment, or cargo being transported therein or cargo being loaded or unloaded or held at terminal or transit points incident to transportation.

(2) For loss or damage to cargo:

The carrier under whose operating authority interchange equipment is used shall be liable according to applicable law for loss or damage to, or delay of that property being transported therein, caused by or arising out of such carrier's use, operation, maintenance or possession of said equipment.

4. How interchange is made and paid for:

4.1 Interchange may be made on an even exchange of equipment, or upon a compensation basis, as may be shown on the Equipment Interchange Receipt and Safety Inspection Report. If upon a compensation

348

basis, charges shall be as shown in the TABLE OF CHARGES on p. 352 hereof, subject to paragraph 4.2. Settlement shall be made at the end of each month or as otherwise agreed upon between the carriers. A day shall be considered a 24-hour period ending at 12:00 o'clock midnight, or a fraction of any such period.

4.2 (a) On all interchanged equipment the day of interchange and the first day after the day of interchange will be considered as days of grace during which time no charge will be made for the use of the equipment. Thereafter, full per diem will be assessed. Saturdays, Sundays and holidays will be excluded as chargeable days, and will not be counted when computing the free time allowance. As between carriers domiciled in the United States, holidays refer to those enumerated in labor contracts at point of interchange.

(b) On traffic moving between a point in the United States and a point in Canada, one additional day of grace will be allowed for the purpose of clearing customs.

(c) No charge will be assessed for such time as equipment is delayed at customs for reasons beyond control of the carrier in possession of the equipment at such customs point, provided adequate written notice of such delay is promptly given by such carrier in possession to the carrier from which the equipment was received at point of interchange.

(d) No charge for Canadian national holidays will be assessed for equipment of carriers domiciled in the United States while in Canada, and no charge for United States national holidays will be assessed for equipment of carriers domiciled in Canada while in the United States.

5. Further remedies and responsibilities:

5.1 The carrier entitled to the return to it of equipment shall be entitled to all its lawful remedies.

5.2 Fuel used in providing refrigeration shall be replaced by the using carrier at the time the mechanical refrigeration unit is returned. If the using carrier fails to replenish the fuel supply of the refrigeration unit he may, unless otherwise agreed upon between the parties involved, be billed for the cost of fuel consumed.

5.3 When an interchanged mechanically refrigerated unit of equipment that has moved unpackaged perishable commodities under refrigeration is unloaded, it shall, unless otherwise agreed upon between the parties involved, be steam cleaned by the delivering carrier.

5.4 In the event equipment offered by the owner for interchange shall require repairs before being received into interchange service, the owner shall be responsible for the cost thereof and the receiving carrier may

349

cause the repairs to be made. If the apparent cost of the foregoing repairs exceed _____ dollars, the consent of the owner shall be obtained by the receiving carrier before it causes the repairs to be made.

5.5 In the event interchanged equipment is damaged after being received into interchange service the carrier in possession at the time the damage occurred shall, by repair, restore it to the condition in which it was received, and, in the event of failure of such carrier to make such repairs, it shall nevertheless, be responsible for the cost thereof. If the apparent cost of the foregoing repairs exceed _____ dollars, the consent of the owner shall be obtained by the using carrier before it causes the repairs to be made. Bills for repairs to damaged equipment shall, unless otherwise agreed upon, be rendered within 90 days after the repairs have been completed.

5.6 In the event equipment is damaged after being placed into interchange service the providing carrier, while settlement therefor or repairs thereto are pending, shall promptly receive from the using carrier equipment of like condition, quality, and size, or, in lieu thereof, at its option, compensation equal to the applicable minimum daily rental in the TABLE OF CHARGES.

5.7 In the event equipment, including component parts, is lost, stolen or totally destroyed after being placed into interchange service the providing carrier, while settlement therefor is pending, shall promptly receive from the using carrier equipment of like condition, quality, and size, or, in lieu thereof, at the option of the providing carrier, subject to subparagraph (a) hereof, compensation equal to the applicable minimum daily rental in the TABLE OF CHARGES, on p. 352. Settlement for equipment lost, stolen or totally destroyed shall be made in the following manner:

(a) The using carrier shall promptly notify the providing carrier by registered mail, return receipt requested, that the equipment has been lost, stolen or totally destroyed, and that the using carrier will make settlement therefor on the basis herein provided. The accumulation of the per diem charges shall cease upon the providing carrier's receipt of such notice.

(b) The providing carrier shall, within thirty days after receipt of the notification provided in the foregoing paragraph, secure and furnish to the using carrier a written statement of the book value of the lost, stolen or totally destroyed equipment, including component parts.

(c) Book value shall be the basis for establishing the value of equipment, and component parts, such as refrigeration units, chassis, LPG generators, etc. To establish book value the purchase price, exclusive of trade-in and tire allowance, shall be used. The straight line depreciation method, with estimated eight year life for the equipment and any component part, less ten percent salvage value, shall be used.

350

(d) Settlement shall be made within 120 days after the using carrier has notified the providing carrier in writing that the equipment including component parts, has been lost, stolen or totally destroyed.

5.8 Ordinary maintenance and other service adjustments occasioned by ordinary use in interchange will be:

(a) Absorbed by the user when cost thereof does not exceed _____ dollars;

(b) Billed to and borne by title owner in entirety when cost thereof would exceed _____ dollars;

(c) Authorized by title owner prior to commencement of repairs when estimated cost thereof would exceed _____ dollars;

(d) Billed to title owner (pursuant to subparagraphs (b) and (c)) within ninety days from the date the repairs were completed unless a different time period is agreed upon between title owner and user.

6. The carrier providing equipment:

6.1 Shall equip it with tires and tubes of proper size at the time of interchange. Thereafter, until the equipment is returned to the carrier furnishing it, repairs to tires and tubes shall be made by and at the expense of the carrier using the equipment at the time of such tire failure. In the event of blow-out or total failure of a tire or tube, the carrier in possession of the equipment shall furnish replacement tires and tubes to return the equipment to the carrier from whom the equipment was received (but shall retain such replacement tires and tubes upon redelivery of the equipment to such carrier) and shall return the blown-out or unserviceable tire and tube, and the same make and type of rim that was on the equipment when the blow-out or tire failure occurred, with the equipment. If tires are ruined as a result of being run flat it will be the responsibility of the using carrier to replace or pay for the tire so ruined. On tires that are lost or damaged the using carrier, within thirty days after the equipment is returned, is required to make settlement therefor by replacing or paying for the lost or damaged tires.

6.2 When the using carrier offers to return interchanged equipment to the owner upon completion of the use for which the equipment was interchanged the owner shall, unless otherwise agreed upon, be required to accept the equipment so offered.

6.3 The carrier initiating the interchange shall, unless otherwise agreed upon, have the right to require equipment in exchange for that which he delivers into interchange service.

6.4 Does not make any warranty or representation, express or implied, as to the fitness or condition of the equipment so interchanged, including tires and tubes, and the carrier acquiring the use thereof does so at its own risk.

351

7. **THIS AGREEMENT** and said Equipment Interchange Receipt and Safety Inspection Report shall constitute the entire agreement between the parties and no verbal amendment or modification thereof shall be permitted. This agreement may be supplemented or amended only by a written agreement.

8. **THIS AGREEMENT** is for a period of one year from date and shall continue in effect from year to year. Either party to this agreement may terminate same as of any time by giving the other ten days' notice of such termination by registered United States mail addressed to the other party.

TABLE OF CHARGES

TYPE OF EQUIPMENT	Dollies	OUTSIDE LENGTH OF TRAILER							
		Under 32'		32' & Under 35'		35' & Over	
		M.R.	Other	M.R.	Other	M.R.	Other
Semi Trailer 1 Axle							
Semi Trailer 2 Axles							

EXPLANATION OF ABBREVIATIONS:

M.R.   Trailer equipped with mechanical refrigeration unit.

Other   Trailer not equipped with mechanical refrigeration unit.

EXECUTED AT .... this .... day of ...., 19..........

....
Name of Carrier   Name of Carrier

By .... By ....
Name and Title   Name and Title

....
Address   Address

352

COPY DISTRIBUTION

1 LESSOR'S CONTROL COPY (CARRIER FURNISHING TRAILER)   4 LESSEE'S CONTROL COPY (CARRIER RECEIVING TRAILER)
2 LESSOR'S TERMINAL COPY (CARRIER FURNISHING TRAILER)   5 LESSEE'S TERMINAL COPY (CARRIER RECEIVING TRAILER)
3 TO ACCOMPANY TRAILER AND TO BE RETURNED TO
LESSOR AT COMPLETION OF INTERCHANGE

EIA
INTERCHANGE ASSOCIATION EQUIPMENT

EQUIPMENT INTERCHANGE RECEIPT AND SAFETY INSPECTION REPORT

(This form may not be reproduced without the written permission of the Equipment Interchange Association 1616 P St. N.W., Washington, D.C.)

No.

POINT OF INTERCHANGE	DATE		O'CLOCK A.M. ☐ P.M. ☐	CROSS REF RECEIPT NO.

RECEIVED BY	BILLING ADDRESS			☐ EVEN EXCHANGE ☐ RENTAL

RECEIVED FROM		OWNER		OWNER'S UNIT NO

MAKE AND YEAR	TYPE	TRAILER LENGTH	LICENSE NO. AND STATE	IS REGIST TION CARD ON SEMI-TRAILER
		IN APPARENT GOOD ORDER EXCEPT AS NOTED		TRAILER IN EXCESS OF 12' 6"

TRAILER TO BE RETURNED TO CARRIER FROM WHOM RECEIVED PRIOR TO 12:00 MIDNIGHT (DATE)

Unless otherwise checked as subject to the paragraph immediately below, this interchange is made subject to the terms and conditions of the currently effective trailer interchange contractual provisions as contained in Equipment Interchange Schedule, and supplements thereto, the official copies of which are on file at the office of the executive secretary of the Equipment Interchange Association, 1616 P Street, Northwest, Washington, D.C. 20036 and copies of which have been supplied to the signatories hereto.
☐ The carriers making this interchange have signed a different and presently effective written interchange contract, that contract in all its terms shall constitute the contract between them

MARK CLEARLY ALL DAMAGE OR DEFICIENCY FOUND BY INSPECTION SYMBOL B· BRUISE · C· CUT · H· HOLE

INITIAL RECEIPT AND INSPECTION

LEFT SIDE

TOP

RIGHT SIDE

FRONT

REAR

FLOOR

BEGINNING HUB READING

RETURN RECEIPT AND INSPECTION

LEFT SIDE

TOP

RIGHT SIDE

FRONT

REAR

FLOOR

ENDING HUB READING

INTERSTATE COMMERCE COMMISSION REGULATIONS REQUIRE EACH PART LISTED TO BE INSPECTED
If not defective, use check mark. If defective, describe defect.

CLEARANCE LIGHTS	FRONT		REAR		STOP LIGHTS
SIDE MARKERS	LEFT		RIGHT		TAIL LIGHTS
REFLECTORS	LEFT SIDE	RIGHT SIDE			TURN SIGNALS
FLAPS			REAR		IDENT LIGHTS
		LANDING GEAR			SAE ATA 7-WAY PLUG
WHEEL LUGS	UNDER CARRIAGE				
WIRING	REAR END PROTECTION				SPRINGS

	HOSE	CONNECTIONS			TUBING
B R A K E S	AIR OR VACUUM LOSS				
	RELAY EMERGENCY VALVE	LININGS	OTHER DEFECTS		

	BRAND NO	CONDITION	POSITION	BRAND NO	CONDITION
T	R.O. FRONT		L.O. FRONT		
I	R.I. FRONT		L.I. FRONT		
R	R.O. REAR		L.O. REAR		
E	R.I. REAR		L.I. REAR		
S	SPARE		SPARE		

PROVIDING CARRIER

BY

ACQUIRING CARRIER

I hereby certify that on the date stated first above, I carefully inspected the equipment described above, that this is a true and correct report of the results of such inspection, and that possession of such equipment was taken on behalf of the acquiring carrier of the time, date and place indicated next above.

BY

I hereby certify that on the date stated above the person who made the inspection covered by this report was competent and qualified to make such inspection and was duly authorized to make such inspection and take possession of such equipment as a representative of the acquiring carrier.

BY

DATE	
TIME	
PLACE	

SIGNATURE OF OWNER PARTNER OR OFFICER OF ACQUIRING CARRIER

ACCOUNTING RECORD: TO BE USED UPON RETURN OF TRAILER TO COMPUTE CHARGES FOR RENTAL (REPAIRS, IF ANY, WILL BE BILLED SEPARATELY)

PER DIEM CHARGE (   )   DAYS @   (   ) $

1

LESSOR'S CONTROL COPY (Carrier Furnishing Trailer)

FORM NO T I R I (REV 10/1/66)   RECORDED FROM EQUIPMENT INTERCHANGE ASSOCIATION

ATLANTIC CONTAINER LINE, LTD.
26 BROADWAY
NEW YORK, NEW YORK 10004

STANDARD TRAILER INTERCHANGE CONTRACT WHERE TONNAGE
TRANSPORTED EITHER ORIGINATES AT OR IS ULTIMATELY
DESTINED TO POINTS OUTSIDE THE CONTINENTAL UNITED STATES

1. The undersigned enter into this agreement governing their relationship
with respect to lease of trailers, and to make this agreement operative
respecting lease of individual trailers will cause to be executed the interchange receipt and inspection reports hereinafter mentioned; provided, however,
that no provision in this contract shall be construed to increase the legal
liability of any party hereto. The term trailer as used herein shall refer
to any load carrying vehicle/or demountable container 20' or more in length
without power, except that power to operate heating or refrigerating units.

2. At the time of interchange an authorized representative of the undersigned shall execute, in multiple copies as the lessor may require, a trailer
interchange receipt and inspection report. The parties shall be bound by
the notations on the receipt and inspection report.

3. The lessee:

3.1 Shall complete promptly and expeditiously the use for which the
trailer has been leased to it and return the trailer to the terminal
of the carrier from which received, at the place received; or if
drawn from a pool other than at a port of the lessor, lessee shall
replace said trailer in kind with one drawn from the port of return
of said trailer; or such other place as the parties may designate.

3.2 Shall not permit the trailer to go out of its possession without
permission of the lessor in writing, and then only to the extent of
written permission, and shall be responsible for the safe and timely
return of the trailer to the lessor ordinary wear and tear excepted,
notwithstanding that it may have had the permission from the lessor
to lease or interchange such trailer to another party.

3.3 Shall be responsible to the lessor for the performance of this agreement by itself and by all other persons into whose possession such
trailer may go until its proper return to the lessor.

3.4 Shall have complete control and supervision of such trailer while
in its possession; and the lessor shall have no right to control
the detail of the work of any employee or agent operating or using
said trailer during such time. Any person operating, in possession
of, or using said trailer after the signing of said receipt and
inspection report and until such form is signed returning the trailer
to the lessor is not the agent or employee of the lessor for any
purpose whatsoever.

3.5 (a) Shall be responsible for the safety of any trailers provided by
the lessor under any provision of this contract, and for any damage
thereto or loss thereof while not in possession of the lessor.
(b) Shall be responsible for and shall defend, indemnify and hold
harmless the Carrier furnishing trailer and/or bogie wheels from
and against any and all loss, liability, damage, claims, fines,
demands and actions, and all expenses connected therewith, including
attorney's fees and costs for injury to persons (including death),
or damage to property or violations of statutes, arising in connection with their custody or use of trailer under this contract.

4. How interchange is made and paid for:

    4.1 Interchange may be made on a trailer-for-trailer exchange, or upon a compensation basis. If upon a compensation basis, charges shall be shown in the TABLE OF CHARGES included in this agreement subject to paragraph 4.2. Settlement shall be made at the end of each month or as otherwise agreed upon between the carriers. A day shall be considered a 24-hour period ending at 12:00 o'clock midnight, or a fraction of any such period.

    4.2 On all interchanged trailers the day of interchange and the following two days after the day of interchange will be considered as days of grace during which time no charge will be made for the use of the trailer. Thereafter, full per diem will be assessed. Saturdays, Sundays and Holidays will be excluded as chargeable days and will not be counted when computing the free time allowance. Holidays refer to those that are enumerated in labor contracts at the point of interchange.

5. The lessor shall be entitled to all its lawful remedies, and shall be entitled to receive from the lessee the special compensation shown in the table of charges herein until return of the trailer to lessor. In the event the trailer leased shall require repairs, the lessee shall cause the repairs to be made, provided, however, that consent of the lessor shall be first obtained if apparent cost of the repairs exceed $100. In the event a leased trailer is damaged, other than as provided for in the preceding sentence, the lessee shall, by repair, restore it to the condition in which it was received and, in the event of failure of lessee to make such repair, it shall, nevertheless, be responsible for all expenses in connection therewith.

    5.1 Fuel used in providing refrigeration shall be replaced by the using carrier at the time the mechanical refrigeration unit is returned. If the using carrier fails to replenish the fuel supply of the refrigeration unit he may, unless otherwise agreed upon between the parties involved, be billed for the cost of fuel consumed.

    5.2 When an interchanged mechanical refrigeration trailer that has moved unpackaged perishable commodities under refrigeration is unloaded, it shall, unless otherwise agreed upon between the parties involved, be steam cleaned by the delivering carrier.

6. The lessor:

    6.1 Shall equip trailer with tires and tubes of proper size. Thereafter, until the trailer is returned repairs to tires and tubes shall be made by and at the expense of the lessee. In the event of blowout or total failure of a tire or tube, lessee shall furnish replacement tires and tubes in order to return the trailer to the lessor (but shall retain such replacement tires and tubes upon re-delivery of the trailer to the lessor) and shall return the blown out or unserviceable tire and tube with the trailer. In the event of failure to so return, payment therefore shall be made at the value thereof at the time of original interchange, which in the absence of specific information to the contrary shall be $100.

    6.2 Does not make any warranty or representation, expressed or implied, as to the fitness or condition of the trailer so leased, including tire and tubes, and the lessee acquiring the use thereof does so at its own risk and inspection; provided, however, that only trailers considered to be in such mechanical condition as to be able to make the agreed tour without failure shall be leased but any repairs to vital parts such as brakes, wheel bearings, running gear, etc., necessary to complete tour shall be made by lessee and lessor billed when such repairs are in excess of $25.00 and lessor agrees to pay same. Such minor repairs as lights, latches, air connections, floor patching and any other individual repair costing no more than $25.00 shall be absorbed by lessee in possession. The lessee will be responsible for the labor expense in effecting repairs to refrigerating units. The lessor will be responsible for the cost of parts to refrigerating units when the old parts are returned to the lessor.

6.3 Shall equip trailer with State vehicle license plates, satisfactory mud flaps, working directional signal lights, clearance marker and stop lights, reflectors, and in compliance with part 193 of Motor Carrier Safety Regulations of the Interstate Commerce Commission.

7. THIS AGREEMENT and said trailer interchange receipt and inspection report shall constitute the entire agreement between the parties and no verbal amendment or modification thereof shall be permitted. This agreement may be supplemented or amended only by a written agreement.

8. TABLE OF CHARGES

		Normal Charges Per Day
A.	20 ft. trailer with S/A Bogie	$ 4.25
B.	20 ft. trailer with T/A Bogie	6.00
C.	20 ft. trailer in COFC Railroad Service	2.50
D.	40 ft. trailer with T/A Bogie	8.50
E.	40 ft. trailer in COFC Railroad Service	5.00
F.	40 ft. mech. reefer trailer/container with T/A Bogie	12.00

Charges per day for Excess Periods

1st day	$15.00
2nd day	18.00
3rd day	20.00
4th day	23.00
5th day & each day thereafter	25.00

Normal charge per day shall apply subject to provisions of paragraph 4.2 (Saturdays, Sundays and Holidays excluded) to use of the vehicle completed within the agreed normal period.

The agreed normal period shall be two (2) working days subject to provisions of paragraph 4.2 (excluding Saturday, Sunday and Holidays) or fraction thereof on round trip movements of up to 100 miles; four (4) working days subject to provisions of paragraph 4.2 (excluding Saturdays, Sundays and Holidays) or fraction thereof on round trip movements of up to 800 miles; six (6) working days subject to provisions of paragraph 4.2 on round trip movements up to 1600 miles; and one (1) additional working day for each additional 800 miles or fraction thereof. Where a two-way movement is involved, the agreed normal period will be increased by two (2) working days.

Mileage shall be computed on basis of latest available Rand McNally Maps, legal requirements as to routes to be followed, and complete routing of vehicles from departure to return to interchange point.

Where repairs to trailers are made under the provisions of Paragraph 5, the lessor shall be entitled to receive compensation from the lessee for each day a trailer remains out of operation because of the damage done to it. This cost or charge will be $6.30 and $9.75 per calendar day, including also Saturdays, Sundays and Holidays, for conventional and refrigerated trailers respectively.

9. Where excess trailer charges are caused by the acts of a shipper or consignee, the appropriate delay, demurrage or storage charges as outlined in the tariffs will be assessed by the carrier making the pick up or delivery. These charges will be billed

by the carrier direct to the shipper or consignee separate from the normal freight charges. The lessee will still be responsible to the lessor for any excess trailer charges caused by this type of delay. The lessee's relief will be from the shipper or the consignee. The excess trailer charges due to Atlantic Container Line, Ltd., shall in no way be dependent on the collection of any delay, demurrage or storage charges from the shipper or consignee.

10. THIS AGREEMENT is for a period of one year from date and shall continue in effect from year to year. Following the return and acceptance of all leased trailers and other leased equipment to lessor and full payment to lessor of charges and other expenses provided under this agreement, either party to this agreement may terminate same as of any time by giving the other ten days notice of such termination by registered or certified United States mail addressed to the other party at the address shown in this agreement or as changed by written notice.

UNITED STATES LINES, INC.
ORIGINAL EQUIPMENT INTERCHANGE RECEIPT AND INSPECTION REPORT

N°

VESSEL	CONTAINER NO.	CHASSIS NO.	BOGIE	TMT NO. (INTL. LIC. NO.)

TO CARRIER
(THIS PORTION OF REPORT IS TO BE COMPLETED AT TIME OF INTERCHANGE)
CARRIER RECEIVING CONTAINER AND/OR CHASSIS

TRACTOR NO.	LICENSE NO.	SEAL NO.

BOX (Circle One)	GROSS WEIGHT
LOADED EMPTY	

CONDITION OF CONTAINER AND CHASSIS

(REGULATIONS REQUIRE THAT EACH PART LISTED BE INSPECTED. IF NO EXCEPTIONS, USE CHECK (√) MARK).

BRAKES
LANDING GEAR
WHEEL LUGS
LIGHTS
UNDERCARRIAGE
DOORS
FLAPS

TIRES	POSITION	BRAND NO. & CONDITION	POSITION
	R.O.		L.O.
	R.I.		L.I.
	R.O.		L.O.
	R.I.		L.I.

IS AN EQUIPMENT DAMAGE REPORT BEING SUBMITTED HEREWITH? CHECK ONE: ☐ YES
IF YES, STATE REPORT NO._____   ☐ NO

INSPECTED BY_____
DATE AND TIME

LESSEE'S REPRESENTATIVES SIGNATURE (Driver, Etc.).
RECEIVED IN GOOD CONDITION, EXCEPT AS NOTED.

FROM CARRIER
(THIS PORTION OF REPORT IS TO BE COMPLETED UPON REDELIVERY)
CARRIER REDELIVERING CONTAINER AND/OR CHASSIS

TRACTOR NO.	LICENSE NO.	SEAL NO.

BOX (Circle One)	GROSS WEIGHT
LOADED EMPTY	

CONDITION OF CONTAINER AND CHASSIS

(REGULATIONS REQUIRE THAT EACH PART LISTED BE INSPECTED. IF NO EXCEPTIONS, USE CHECK (√) MARK).

BRAKES
LANDING GEAR
WHEEL LUGS
LIGHTS
UNDERCARRIAGE
DOORS
FLAPS

TIRES	POSITION	BRAND NO. & CONDITION	POSITION
	R.O.		L.O.
	R.I.		L.I.
	R.O.		L.O.
	R.I.		L.I.

INSPECTED BY_____
DATE AND TIME

LESSEE'S REPRESENTATIVES SIGNATURE (Driver, Etc.).
RECEIVED IN GOOD CONDITION, EXCEPT AS NOTED.

ACCOUNTING CODE
SUBSIDIARY ACCOUNT 660.21
CARRIER CODE NUMBER_____

DATE OF BILLING

Total Period of Interchange: Amount Due

........days @ $ per day .... $.........
........days @ $ per day .... $.........
........days @ $ per day .... $.........

$......... **Pay This Sum**

References

CHAPTER 1

Paper presented by Mr. James Henry and Mr. Henry J. Karsch of J. J. Henry & Co. at the Society of Naval Architects and Marine Engineers meeting at New York, Nov. 10-11, 1966, entitled, "Container Ships."

Talk given by Mr. C. D. Ramsden, President of PACECO, at the Europort '66 Congress at Amsterdam, Nov. 8, 1966.

The Journal of Commerce, May 13, 1968. Article "Van Marking Next Step in Standards."

Specifications for Cargo Containers-USASI MH 5.1-1965.

Aluminum Van Containers. Publication of Aluminum Company of America, 1963.

Containerisation International, Oct. 1967. Article "Steel Containers Are Gaining in Popularity."

Container News, Aug., 1967. Article "Plywood" by Robert P. Wenner, American Plywood Association.

Canadian Transportation, June 1965. Article "The New Look in Integrated Intermodal Containerization," by Mr. D. W. Francis.

Via Port of New York, Special Transatlantic Transport Prevue issue, 1967. Article "Booster for Freshness."

Containerisation International, Sept., 1967. Article "Refrigeration, the Trend Is Toward Mechanical Units."

CHAPTER 2

Transportation Facilitation Program. Prepared by Office of Facilitation, Department of Transportation.

Talk given by Alan S. Boyd, Secretary of Transportation, before the Committee on Commerce of the Senate on June 17, 1968.

Paper "The Role of Customs in the Handling of Unitized Cargoes" presented by Hugo Opazo, Latin American Free Trade Association at the First Inter-American Port Seminar in Bogota, Colombia, Mar. 25th-30th, 1968.

Container Transport, Sept., 1967, published by Mees & Hope and R. Mee & Zoonen, Holland.

Comprehensive Export Schedule. Section 379.1 (a) Exports by Water Carrier. Uniform Customs and Practice for Documentary Credits (1962 Revision). Publication of the International Chamber of Commerce. "Copyright by International Chamber of Commerce." Uniform customs and practice for documentary credits Vr. 222, has been published by the International Chamber of Commerce, 38, Cours Albert 1er, 75-Paris VIII

(United States Council of the ICC, Inc., 1212 Avenue of the Americas, New York, N.Y. 10036) in English-French, English-German, Spanish and Portuguese; translations in other languages will appear at a later date. This publication may be obtained from International Headquarters of the ICC and from the various National Committees.

Exporting Made Easy. Article "Financing Exports," by Mr. I. Paul Tesorero, Ass't. V. P., Int'l. Div., Greater Detroit Board of Commerce.

Talk given by Mr. Kurt S. Schalling at the 1st International Conference on Containerization held in Genoa, Italy, Oct. 19, 20, 21, 1967 entitled, "The Carrier's Liability on Combined Transportation and Insurance Problems."

Containerisation International, April, 1968. Article "Cargo Survival at Sea," by Hewlett R. Bishop.

Treasury Department, Bureau of Customs, "United States Treatment of Containers and Containerized Cargo."

Customs Convention on Containers. "Message from the President of the United States transmitting the Customs Convention on Containers together with the Protocol of Signature which forms an integral part thereof, done at Geneva on May 18, 1956."

Four Customs Conventions "Message from the President of the United States Transmitting—and on the International Transport of Goods under cover of T.I.R. carnets."

Presentation to Equipment Interchange Association concerning T.I.R. carnets, by Mr. Armour S. Armstrong, Chief, Transport Systems Division, Office of Facilitation, Department of Transportation, Sept., 1968.

CHAPTER 3

"Modern Ship Stowage," by Joseph Leeming, Bureau of Foreign and Domestic Commerce. Published by United States Government Printing Office, Washington, 1942.

"Suggested Methods for Loading, Blocking and Bracing of Freight in Closed Trailers for T.O.F.C. Service," Association of American Railroads, Oper. and Maint. Dept.

"Bonded Block and Palletized Method of Loading Commodities in Fibreboard Containers," Association of American Railroads, Oper. and Maint. Dept.

"Stowage of Sea-Going Containers—Advice to Shippers." National Cargo Bureau.

Containerisation International, July, 1968, Extract from paper read by J. S. Carter and I. J. Steel, Hays Int. Serv., London, at the International Container Symposium, London, May, 1968.

CHAPTER 4

Talk given by Mr. Gabriel Alter and Mr. H. J. Kirschning at the 1st Int. Conf. on Containerization in Genoa, Oct. 19, 20, 21, 1967, "Engineering Development of a Container System."

Report on Containerization, *National Joint Council on Materials Handling,* July, 1967.

CHAPTER 5

Containerisation International, Nov., 1967. Article "Gearing Up to Meet the Challenge."

Containerisation International, Dec., 1967. Article "Barge Ships—the Shipowner Compromise?"

Containerisation International, Feb., 1968. Article "Where Rail Helps Road."

Containerisation International, Mar., 1968. Article by Karsten Kieserling, "Truckers' Loss Won't Save Rail Debt."

Containerisation International, Jan., 1968. Article "Lash System Gets Underway."

Containerisation International, Apr., 1968. Articles "German Railways Prepare for Container and Trailer Traffic" and "Road Equipment for the Container Carrier."

Paper "Highway Requirements for Handling Unitized Cargoes," presented by José M. Zuñiga, Director of Engineering of the International Road Federation at the First Inter-American Port Seminar in Bogota, Colombia, Mar. 25th-30th, 1968.

May 1968 Publication of French Railways Ltd., "Direct Exports through British and French Rail."

Paper "Latin American Railways and Unitized Cargoes," presented by Operations Group of the Latin American Association of Railroads at the First Inter-American Port Seminar in Bogota, Colombia, Mar. 25th-30th, 1968.

Talk given by Mr. Paul Roshkind, Freight Transportation Planning, the Port of New York Authority, before the New York Society of Security Analysts, "Economic Implications of Containerization."

Containerisation International, July, 1968. Article "Containers in the System" by Mr. Peter Dawson.

Talk given by A. Lyle King, Director, Marine Terminals Department, the Port of New York Authority, before the Joint Industry/Government Conference on Unitization and Distribution, "Ports and Docks."

CHAPTER 6

The American Law of Ocean Bills of Lading, by Arnold W. Knauth; published by American Maritime Cases.

Handbook of International Road Transport, 5th Edition 1966, published by the International Road Transport Union.

International Convention concerning the Carriage of Goods by Rail, 1961 (CIM) as presented to Parliament by the Minister of Transport, Nov., 1963; published in London by Her Majesty's Stationery Office.

Limits of Motor Vehicle Sizes and Weights in North, Central and South America, as presented by José M. Zuñiga, Director of Engineering of the International Road Federation at the First Inter-American Port Seminar in Bogota, Colombia, Mar. 25–30, 1968.

Summary of Size and Weight Limits and Reciprocity authority (by regions) in effect as of Jan. 20, 1968, as prepared by Section on State Laws, Reciprocity and Taxation of American Trucking Associations, Inc.

International Standards Organization and United States of America Standards Institute.

GLOSSARY

Revised American Foreign Trade Definitions, "The American Law of Ocean Bills of Lading," by Arnold W. Knauth; published by American Maritime Cases.

Glossary of Terms

Bill of Lading.—An official detailed receipt given by a transport company to the person consigning goods, by which the company makes itself responsible for the safe delivery of the goods to the consignee.

Bogies.—Wheel units without chassis. Come with single and tandem axles.

Bolsters.—Frames which are utilized to handle container without wheels on railroad flatcars. At the present time, bolsters must be lifted and installed manually on conventional piggyback flatcars before such units can be used to handle containers. Railroads are developing railcars with retractable bolsters which will permit rapid conversion of a flatcar to handle either trailers or containers.

Bottom Lift.—By means of lifting tongs, overhead grippers or grabs to fasten onto the bottom rails of the container and effect lifting.

Breaking Bulk.—To "break bulk" is to commence to unload the cargo.

Broken Stowage.—The waste and loss of space caused by irregularity in the size and shape of packages.

Bulk.—Cargo is said to be stowed in bulk when it is stowed loose instead of being loaded into containers.

Cargo Container.—An enclosed, permanent, reuseable, nondisposable, weathertight shipping conveyance fitted with a minimum of one door.

Chassis.—A frame with wheels with devices for locking containers on. Comes in skeletal types, parallel frame types and perimeter frame types, among others.

Chock.—A piece of wood or other material placed at the side of a cask or package to prevent its rolling about or moving sideways.

COFC—Container on Flatcar.—Container only is put on railcar for transport.

Consolidate.—To receive cargo, combine it with other cargoes, and load.

Container Service Terms (in use by steamship carriers):

Pier to Pier.—The steamship company receives cargo on the pier and loads it into containers. The cargo is then taken out of containers at the pier of discharge.

Pier to House.—The steamship company receives cargo on the pier and loads it into containers. The cargo and container are delivered to consignee, after discharge, direct to consignee's facility.

House to Pier.—Cargo is loaded into container at shipper's facility, moves in container to pier and then overseas. Cargo is removed from container at overseas pier.

House to House.—Cargo is loaded into container at shipper's facility, moves in container to pier and then overseas. The cargo and container are delivered to consignee, after discharge, direct to consignee facility.

Customhouse Broker.—Acts as agent for the importer clearing inbound shipments through customs, arranging for entry, the payment of duties, the payment of collect freight charges, and the movement of the cargo to the door of the consignee.

Customs Entry.—To make a Customs entry it is necessary to produce a bill of lading and an invoice covering the merchandise. Customs entry may be made for consumption, for warehousing, for transportation to an interior point for the purpose of completing Customs clearance; for export to a foreign country or for transportation and exportation to a foreign country. Estimated duties must be deposited or secured by posting bond for payment.

Dock.—As used in trucking, that enclosed area of a truck terminal which is used for the handling of cargo on and off the trucks backed up to the doors.

Domestic Freight Forwarders and Carloaders.—Collect small and large shipments, consolidate them and ship them in carload and truckload lots. In addition to engaging trucking companies and railroads, they sometimes use the services of barge lines and utilize the inland waterways. Regulated by the Interstate Commerce Commission the domestic freight forwarder performs a through service, assuming full responsibility from point of receipt to the consignee.

Dunnage.—Loose wood or other material placed under and around the cargo to prevent damage and wedge it in place.

Export License.—Export merchandise declared by United States Department of Commerce as a requiring special authorization or "licensing" before being permitted to leave the United States.

Fifth Wheel.—That part of the tractor that engages the underside of the trailer or van and locks onto the "kingpin" with a manually operated spring lock device. Almost circular, "wheel like" looking which is reason for name "fifth wheel."

General License.—Merchandise leaving the United States that requires no special permission.

Intermodal (literally "between modes").—Used to denote ability of containers to change from rail to truck to ship in any order.

International Freight Forwarder.—From seller's plant site to the door of the consignee, the foreign freight forwarder handles every detail of the shipment and supervises its movement. He arranges for insurance, for transportation to the port. In the port he carries out the instructions of the letter of credit, he handles consular documentation, he prepares the export declaration, certificate of origin, import permit, export license. He books it aboard ship, prepares the bill of lading and the dock receipt and he advances the ocean freight.

Kingpin (as used in fastening trailers and vans to tractors).—A cylindrical projection from under the nose (front) end of the trailer or van used as a fastening device for the fifth wheel of the tractor.

Landing gear.—The "legs" with small wheels at the ends toward the forward end of the trailer or van that are manually lowered and locked before tractor is disengaged.

Lift-on/Lift-off.—Term applied to vessel with facilities for lifting containers without wheels onto and off the vessel. Facilities may be vessel powered or shore powered.

Loading with Bonded Block Method.—The principle of this method of loading is to build up the load in blocks, in which the cartons are bonded together into units by reversing each layer.

Nesting.—Fitting one article of cargo inside the other to economize space.

Non-Vessel Operating Common Carrier.—A "carrier" defined by Maritime Law, offering an international cargo transport service through the use of underlying carriers and under their own rate structure in accordance with tariffs filed with the Federal Maritime Commission in Washington. The rates filed are required to cover only the port-to-port portion. Specific authority for the NVOCC is given in the Code of Federal Regulations, Title 46, Chapter IV, Federal Maritime Commission Sub-Part B, entitled, "Regulations Affecting Maritime Carriers and Related Activities." General Order 4, Amendment 1, Section 510.21 (d) states:

> The term "non-vessel operating common carrier by water" means a person who holds himself out by the establishment and maintenance of tariffs, by advertisement, solicitation, or otherwise, to provide transportation for hire by water in interstate commerce as defined in the Act, and in commerce from the United States as defined in paragraph (b) of the section; assumes responsibility or has liability imposed by law for safe transportation of shipments; and arranges in his own name with underlying water carriers for the performance of such transportation whether or not owning or controlling the means by which such transportation is affected.

O.C.P.—Overland Common Point.—A special rate concession made by shipping lines serving the U. S. West Coast and by Rail Carriers service the U. S. West Coast for export and import traffic intended to benefit midwest shippers and importers by equalizing rates to and from other coastal areas and offering these midwest companies a comparable alternative. The steamship companies lower their rates and the railroads pick up the terminal charges which consist of handling charges, wharfage charges and carloading or unloading charges. The areas considered overland territories are east of and including North Dakota, South Dakota, Nebraska, Colorado and New Mexico.

Quay; Pier; Dock.—A berth is that part of a pier or quay used for vessel to tie-up. The dock is the area used to discharge or assemble cargo.

Rates.—In the transport of goods by vessel or truck or rail, rates are set up on two bases. They are known as "class" rates and "commodity" rates. A Class Rate is a rate stated, not on an article, but on a symbol which represents many articles.

A Commodity Rate is a rate stated on a specific commodity or description of traffic.

In a trade in which there are both class and commodity rates, an article on which a commodity rate is stated may come within a class description in the class rates. Because the existence of a commodity rate indicates that special consideration has been given to the article for rate purposes the commodity rate takes preference over the class rate.

Whether charges in a trade are based only on commodity rates or on both commodity and class rates, there is a catch-all description of traffic for shipments that come under no specific description in the classification or list of commodity rates. The description is usually "Cargo, N.O.S." (not otherwise specified). This is known as a general cargo rate and is usually higher than the rates on most commodities, for the carrier cannot know exactly what kinds of goods "not otherwise specified" will be offered for transportation.

The rate charged for transportation by a through route is called a "through" rate.

The rate charged for transportation over the line of one carrier is called a "local" rate.

When the rate is the sum of the local rate of the two or more carriers for their respective segments of the through route plus an amount to cover the cost of transfer at the transshipment points, such rates are called "combination through rates."

When the through rate is lower than the combination of local rates, it is called a "joint rate."

Proportional rates are those applying on traffic originating and/or beyond the points to which such rates apply.

Roll-on/Roll-off.—Term applied to vessel with facilities for the trailers being driven on and off the vessel by tractor power.

Schedule "A" No.—Commodity codes required for statistical classification of imports.

Schedule "B" No.—Commodity codes required for statistical classification on Export Declaration.

Shipper's Export Declaration ("SED.").—Document required by U. S. Customs for all cargo moving out of the United States.

TOFC—Trailer on Flatcar (also known as Piggyback).—Complete trailer or van is loaded on railcar for transport.

Tractor.—The motor unit (sometimes referred to as a "horse") that is used to pull trailers or vans. The complete coupled unit is called a "rig."

Trailer.—Used to describe a container together with a removable chassis or bogie.

Transcontainer.—A term used in Europe to describe the larger intermodal containers. Usually describes those upwards from 20 feet long.

Transmodalist.—Intermodal container operator that is not limited to one mode (such as a steamship company, a trucker or a railroad) but who has flexibility of all modes and ability to render a door-to-door service.

Truck (Motor).—A complete unit with motive power, cargo carrying area, and wheels all permanently attached. LKW (Landskraftwagon), Lorry.

Truck (Rail).—Wheel units without railcar.

Van (from "Caravan").—Used to describe the permanently attached box and wheels.

Vessel Terms:

Movements

Heave.—The vessel moves suddenly rapidly upward.

Pitch.—The downward plunge of a vessel's bow.

Roll.—The vessel moves from side to side more severely than when swaying, with the result that the top of the vessel moves greater distances from the center line.

Surge.—The vessel moves suddenly rapidly forward.

Surge.—The vessel moves suddenly rapidly forward.

Sway.—Vessel's movement forward or backward, or side to side, under normal conditions.

Yaw.—A movement of deviation from the direct course.

Locations

Aft.—Toward the after part of the vessel, or the stern.

Amidship.—In the center of a vessel, either with reference to its length or its breadth.

Athwart.—Across the width of a vessel. At right angles to the fore and aft line. This distance from one end to the other is known as the vessel's beam.

Fore.—Toward the forward part of the vessel, or the bow.

Hatch.—An opening in the deck of a vessel to allow cargo to be taken up and down to the holds. Cargo stowed directly in line with the hatch opening is said to be stowed in the "square" of the hatch.

Holds.—The cargo compartments of a vessel.

Wings.—The sides of the vessel's holds. Cargo stowed along the sides of the vessel is said to be stowed in the "wings."

Weights and Measures :

Definitions

Gross.—The weight of the cargo plus its packing.

Net.—The weight of the cargo alone.

Tare.—The weight of a packing box when empty, or container when empty.

Tonnages.—The word "ton" is a description connoting weight. However, the measurement of this weight can vary. In the United States, we have a long ton of 2240 pounds and a short ton of 2000 pounds. Where the metric system is used, as in Europe, a metric ton is 2204.6 pounds. In the Far East, Australia, Africa, a short ton is usually used.

Weight Ton, or Measurement Ton.—A measurement ton is 40 cubic feet. The assumption that 40 cubic feet equal one ton is believed to have originated in the Russian grain trade from the Black Sea, in which it was demonstrated by experience that one ton of Russian wheat required 40 cubic feet for stowing. Steamship rates outbound from the United States are usually quoted on a "weight" or a "measurement" basis, whichever brings the greater revenue. Cargo of such a nature that a long ton stows in less than 40 cubic feet is known as deadweight cargo and, in most instances pays freight on a weight basis. Cargo of which one long ton occupies 40 cubic feet or more is known as "measurement" cargo. The freight actually assessed by the line is known as a "freighted, or revenue ton." Cargo inbound from Europe is usually quoted on a weight, or "cubic meter" basis, a cubic meter being 35.4 cubic feet.

Hundredweight (CWT).—This unit is 1/20th of a ton. Since the tonnage type used can vary, as described above, the definition of hundredweight can vary.

Quintal (term similar to hundredweight).—It equals 100 pounds or 112 pounds or 100 kilograms, depending on the system used.

EQUIVALENTS

shipping ton (U. S.) .... 40 cubic feet
net, or short ton .... 2,000 pounds
metric, or kilo ton .... 2,204.6 pounds (1,000 kilos)
gross, or long ton .... 2,240 pounds (1,016 kilos)
1 millimeter .... 0.03937 inch
1 centimeter .... 0.3937 inch
1 meter .... 39.37 inches
1 kilometer .... 0.6214 mile
1 cubic foot .... 0.02832 cubic meter; 1728 inches
1 cubic meter .... 35.314 cubic feet

To find the volume of a cylinder, the formula $V = \pi r^2 h$ is used.
Volume = 3.1418 × the Radius Multiplied by itself × the Height

STANDARD TRADE TERMS (DEFINITIONS)

Clean Bill of Lading. A clean shipping document is one which bears no superimposed clauses expressly declaring a defective condition of the goods or packaging.

C. & F. (Cost and Freight)—(named point of destination). Under this term, the seller quotes a price including the cost of transportation to the named point of destination.

Under this quotation: Seller must (1) provide and pay for transportation to named point of destination; (2) pay export taxes, or fees or charges, if any, levied because of exportation; (3) obtain and dispatch promptly to buyer, or his agent, clean bill of lading to named point of destination; (4) where received-for-shipment ocean bill of lading may be tendered, be responsible for any loss or damage, or both, until the goods have been delivered into the custody of the ocean carrier; (5) where on-board ocean bill of lading is required, be responsible for any loss or damage, or both, until the goods have been delivered on board the vessel; (6) provide, at the buyer's request and expense, certificates of origin, consular invoices, or any other documents issued in the country of origin, or of shipment, or of both, which the buyer may require for importation of goods into country of destination and, where necessary, for their passage in transit through another country.

Buyer must (1) accept the documents when presented; (2) receive goods upon arrival, handle and pay for all subsequent movement of the goods, including taking delivery from vessel in accordance with bill of lading clauses and terms; pay all costs of landing, including any duties, taxes, and other expenses at named point of destination; (3) provide and pay for insurance; (4) be responsible for loss of or damage to goods, from time and place at which seller's obligations under (4) or (5) above have ceased; (5) pay the costs of certificates of origin, consular invoices, or any other documents issued in the country of origin, or of shipment, or of both, which may be required for the importation of goods into the country of destination and, where necessary, for their passage in transit through another country.

C. & F. Comments:

1. For the seller's protection, he should provide in his contract of sale that marine insurance obtained by the buyer include standard warehouse to warehouse coverage.
2. The comments listed under the following C. I. F. terms in many cases apply to C. & F. terms as well, and should be read and understood by the C. & F. seller and buyer.

C. I. F. (Cost, Insurance, Freight)—(named point of destination). Under this term, the seller quotes a price including the cost of the goods, the marine insurance, and all transportation charges to the named point of destination.

Under this quotation: Seller must (1) provide and pay for transportation to named point of destination; (2) pay export taxes, or other fees or charges, if any, levied because of exportation; (3) provide and pay for marine insurance; (4) provide war risk insurance as obtainable in seller's market at time of shipment at buyer's expense, unless seller has agreed that buyer provide for war risk coverage (5) obtain and dispatch promptly to buyer, or his agent, clean bill of lading to named point of destination, and also insurance policy or negotiable insurance certificate; (6) where received-for-shipment ocean bill of lading may be tendered, be responsible for any loss or damage, or both, until the goods have been delivered into the custody of the ocean carrier; (7) where on-board ocean bill of lading is required, be responsible for any loss or damage, or both, until goods have been delivered on board the vessel, (8) provide, at the buyer's request and expense, certificates of origin, consular invoices, or any other documents issued in the country of origin, or of shipment, or both, which the buyer may require for importation of goods into country of destination and, where necessary, for their passage in transit through another country.

Buyer must (1) accept the documents when presented; (2) receive the goods upon arrival, handle and pay for all subsequent movement of the goods, including taking delivery from vessel in accordance with bill of lading clauses and terms; pay all costs of landing, including any duties, taxes, and other expenses at named point of destination; (3) pay for war risk insurance provided by seller; (4) be responsible for loss of or damage to goods, or both, from time and place at which seller's obligations under (6) or (7) above have ceased; (5) pay the cost of certificates of origin, consular invoices, or any other documents issued in the country of origin, or of shipment, or both, which may be required for importation of the goods into the country of destination and, where necessary, for their passage in transit through another country.

C. & F. and C. I. F. Comments:

Under C. & F. and C. I. F. Contracts there are the following points on which the Seller and the Buyer should be in complete agreement at the time that the contract is concluded: (1) It should be agreed upon, in advance, who is to pay for miscellaneous expenses, such as weighing or inspection charges. (2) The quantity to be shipped on any one vessel should be agreed upon, in advance, with a view to the Buyer's capacity to take delivery upon arrival and discharge of the vessel, within the free time allowed at the port of importation. (3) Although the terms C. & F. and C. I. F. are generally interpreted to provide that charges for consular invoices and certificates of origin are for the account of the Buyer, and are charged separately, in many trades these charges are included by the Seller in his price. Hence, Seller and Buyer should agree, in advance, whether these charges are part of the selling price, or will be invoiced separately. (4) The point of final destination should be definitely known in the event the vessel discharges at a port other than the actual destination of the goods. (5) When ocean freight space is difficult to obtain, or forward freight contracts cannot be made at firm rates, it is advisable that sales contracts, as an exception to regular C. & F. or C. I. F. terms, should provide that shipment within the contract period be subject to ocean freight space being available to the Seller, and should also provide that changes in the cost of ocean transportation between the time of sale and the time of shipment be for account of the buyer. (6) Normally, the Seller is obligated to prepay the ocean freight. In some instances, shipments are made freight collect and the amount of the freight is deducted from the invoice rendered by the seller. It is necessary to be in agreement on this, in order to avoid misunderstanding which arises from foreign exchange fluctuations that might affect the actual cost of transportation, and from interest charges that might accrue under letter of credit financing. Hence, the Seller should always prepay the ocean freight unless he has a specific agreement with the Buyer, in advance, that goods can be shipped freight collect. (7) The Buyer should recognize that he does not have the right to insist on inspection of goods prior to accepting the documents. The Buyer should not refuse to take delivery of goods on account of delay in the receipt of documents, provided the Seller has used due diligence in their dispatch through the regular channels. (8) Sellers and Buyers are advised against including in a C. I. F. contract any indefinite clause at variance with the obligations of a C. I. F. contract as specified in these definitions. There have been numerous court decisions in the United States and other countries invalidating C. I. F. contracts because of the inclusion of indefinite clauses. (9) Interest charges should be included in cost computations and should not be charged as a separate item in C. I. F. contracts, unless otherwise agreed upon, in advance, between the Seller and Buyer; in which case, however, the term C. I. F. and I (Cost, Insurance, Freight, and Interest) should be used. (10) In connection with insurance under C. I. F. sales, it is necessary that Seller and Buyer be definitely in accord upon the following points: (a) The character of the marine insurance should be agreed upon in so far as being W.A. (With Average) or F.P.A. (Free of Particular Average), as well as any other special risks that are covered in specific trades, or against which the Buyer may wish individual protection. Among the special risks that should be considered and agreed upon between Seller and Buyer are theft, pilferage, leakage, breakage, sweat, contact with other cargoes, and others peculiar to any particular trade. It is important that contingent or collect freight and customs duty should be insured to cover Particular Average losses, as well as Total Loss after arrival and entry but before delivery. (b) The Seller is obligated to exercise ordinary care and diligence in selecting an underwriter that is in good financial standing. However, the risk of obtaining settlement of insurance claims rests with the Buyer. (c) War risk insurance under this term is to be obtained by the Seller at the expense and risk of the Buyer. It is important that the Seller be in definite accord with the Buyer on this point, particularly as to the cost. It is desirable that the goods be insured against both marine and war risk with the same underwriter, so that there can be no difficulty arising from the determination of the cause of the loss. (d) Seller should make certain that in his marine or war risk insurance there be included the standard protection against strikes, riots and civil commotions. (e) Seller and Buyer should be in accord as to the insured valuation, bearing in mind that merchandise contributes in General Average on certain bases of valuation which differ in various trades. It is desirable that a competent insurance broker be consulted, in order that full value be covered and trouble avoided.

Ex (Point of Origin)—"*Ex Factory*," "*Ex Mill*," "*Ex Mine*," "*Ex Plantation*," "*Ex Warehouse*," etc. *(named point of origin)*. Under this term, the price quoted applies only to the point of origin, and the seller agreed to place the goods at the disposal of the buyer at the agreed place on the date or within the period fixed.

Under this quotation:

Seller must

(1) bear all costs and risks of the goods until such time as the buyer is obliged to take delivery thereof;

(2) render the buyer, at the buyer's request and expense, assistance in obtaining the documents issued in the country of origin, or of shipment, or of both, which the buyer may require either for purposes of exportation, or of importation.

Buyer must

(1) take delivery of the goods as soon as they have been placed at his disposal at the agreed place on the date or within the period fixed;

(2) pay export taxes, or other fees or charges, if any levied because of exportation;

(3) bear all costs and risks of the goods from the time when he is obligated to take delivery thereof;

(4) pay all costs and charges incurred in obtaining the documents issued in the country of origin, or of shipment, or of both, which may be required either for purposes of exportation, or of importation at destination.

Ex Dock—"*Ex Dock (named port of importation)*." Under this term, seller quotes a price including the cost of the goods and all additional costs necessary to place the goods on the dock at the named port of importation, duty paid, if any.

Under this quotation:

Seller must

(1) provide and pay for transportation to named port of importation;

(2) pay export taxes, or other fees or charges, if any, levied because of exportation;

(3) provide and pay for marine insurance;

(4) provide and pay for war risk insurance, unless otherwise agreed upon between the buyer and seller;

(5) be responsible for any loss or damage, or both, until the expiration of the free time allowed on the dock at the named port of importation;

(6) pay the costs of certificates of origin, consular invoices, legalization of bill of lading, or any other documents issued in the country of origin, or of shipment, or of both, which the buyer may require for the importation of goods into the country of destination and, where necessary, for their passage in transit through another country;

(7) pay all costs of landing, including wharfage, landing charges, and taxes, if any;

(8) pay all costs of customs entry in the country of importation;

(9) pay customs duties and all taxes applicable to imports, if any, in the country of importation, unless otherwise agreed upon.

Buyer must

(1) take delivery of the goods on the dock at the named port of importation within the free time allowed;

(2) bear the cost and risk of the goods if delivery is not taken within the free time allowed.

Ex Dock Comments:

This term is used principally in United States import trade. It has various modifications, such as "Ex Quay," "Ex Pier," etc., but it is seldom, if ever, used in American export practice.

F.A.S. (Free Along Side). "*F.A.S. Vessel (named port of shipment).*" Under this term, the seller quotes a price including delivery of the goods alongside overseas vessel and within reach of its loading tackle.

Under this quotation:

Seller must

(1) Place goods alongside vessel or on dock designated and provided by, or for, buyer on the date or within the period fixed; pay any heavy lift charges, where necessary, up to this point;

(2) provide clean dock or ship's receipt;

(3) be responsible for any loss or damage, or both, until goods have been delivered alongside the vessel or on the dock;

(4) at the buyer's request and expense, render assistance in obtaining the documents issued in the country of origin, or of shipment, or of both, which the buyer may require either for purposes of exportation, or of importation at destination.

Buyer must

(1) give seller adequate notice of name, sailing date, loading berth of, and delivery time to, the vessel;

(2) handle all subsequent movement of the goods from alongside the vessel:

(a) arrange and pay for demurrage or storage charges, or both, in warehouse or on wharf, where necessary;

(b) provide and pay for insurance;

(c) provide and pay for ocean and other transportation;

(3) pay export taxes, or other fees or charges, if any, levied because of exportation;

(4) be responsible for any loss or damage, or both, while the goods are on a lighter or other conveyance alongside vessel within reach of its loading tackle, or on the dock awaiting loading, or until actually loaded on board the vessel, and subsequent thereto;

(5) pay all costs and charges incurred in obtaining the documents, other than clean dock or ship's receipt, issued in the country of origin, or of shipment, or of both, which may be required either for purposes of exportation, or of importation at destination.

F.A.S. *Comments:*

1. Under F.A.S. terms, the obligation to obtain ocean freight space, and marine and war risk insurance, rests with the buyer. Despite the obligation on the part of the buyer, in many trades the seller obtains ocean freight space, and marine and war risk insurance, and provides for shipment on behalf of the buyer. In others, the buyer notifies the seller to make delivery alongside a vessel designated by the buyer and the buyer provides his own marine and war risk insurance. Hence, seller and buyer must have an understanding as to whether the buyer will obtain the ocean freight space, and marine and war risk insurance, as is his obligation, or whether the seller agrees to do this for the buyer.

2. For the seller's protection, he should provide in his contract of sale that marine insurance obtained by the buyer include standard warehouse to warehouse coverage.

F.O.B. (Free on Board). "F.O.B. (named inland carrier at named inland point of departure)." Under this term the price quoted applies only at inland shipping point, and the seller arranges for loading of the goods on, or in, railway cars, trucks, lighters, barges, aircraft, or other conveyance furnished for transportation.

Under this quotation:

Seller must

(1) place goods on, or in conveyance, or deliver to inland carrier for loading;

(2) provide clean bill of lading or other transportation receipt, freight collect;

(3) be responsible for any loss or damage, or both, until goods have been placed, in or on, conveyance at loading point, and clean bill of lading or other transportation receipt has been furnished by the carrier;

(4) at the buyer's request and expense, render assistance in obtaining the documents issued in the country of origin, or of shipment, or of both, which the buyer may require either for purposes of exportation, or of importation at destination.

Buyer must

(1) be responsible for all movement of the goods from inland point of loading, and pay all transportation costs;

(2) pay export taxes, or other fees or charges, if any, levied because of exportation;

(3) be responsible for any loss or damage, or both, incurred after loading at named inland point of departure;

(4) pay all costs and charges incurred in obtaining the documents issued in the country of origin, or of shipment, or of both, which may be required either for purposes of exportation, or of importation at destination.

"F.O.B. (named inland carrier at named inland point of departure) Freight Prepaid To (named point of exportation)." Under this term, the seller quotes a price including transportation charges to the named point of exportation and prepays freight to named point of exportation, without assuming responsibility for the goods after obtaining a clean bill of lading or other transportation receipt at named inland point of departure.

Under this quotation:

Seller must

(1) assume the seller's obligations as under F.O.B. above, except that under (2) he must provide clean bill of lading or other transportation receipt, freight prepaid to named point of exportation,

Buyer must

(1) assume the same buyer's obligations as under F.O.B. above, except that he does not pay freight from loading point to named point of exportation.

"F.O.B. (named inland carrier at named inland point of departure) Freight Allowed To (named point)." Under this term, the seller quotes a price including the transportation charges to the named point, shipping freight collect and deducting the cost of transportation, without assuming responsibility for the goods after obtaining a clean bill of lading or other transportation receipt at named inland of departure.

Under this quotation:

Seller must

(1) assume the same seller's obligations as under F.O.B. above, but deducts from his invoice the transportation cost to named point.

Buyer must

(1) assume the same buyer's obligations as under F.O.B. above, including payment of freight from inland loading point to named point, for which seller has made deduction.

"F.O.B. (named inland carrier at named point of exportation)." Under this term, the seller quotes a price including the costs of transportation of the goods to named point of exportation, bearing any loss or damage, or both, incurred up to that point.

Under this quotation:

Seller must

(1) place goods, on, or in, conveyance, or deliver to inland carrier for loading;

(2) provide clean bill of lading or other transportation receipt, paying

all transportation costs from loading point to named point of exportation; (3) be responsible for any loss or damage, or both, until goods have arrived in, or on, inland conveyance at the named point of exportation; (4) render the buyer, at the buyer's request and expense, assistance in obtaining the documents issued in the country of origin, or of shipment, or of both, which the buyer may require either for purposes of exportation, or of importation at destination.

Buyer must

(1) be responsible for all movement of the goods from inland conveyance at named point of exportation;

(2) pay export taxes, or other fees or charges, if any, levied because of exportation;

(3) be responsible for any loss or damage, or both, incurred after goods have arrived in, or on, inland conveyance at the named point of exportation;

(4) pay all costs and charges incurred in obtaining the documents issued in the country of origin, or of shipment, or of both, which may be required either for purposes of exportation, or of importation at destination.

"F.O.B. Vessel (named port of shipment)." Under this term, the seller quotes a price covering all expenses up to, and including, delivery of the goods upon the overseas vessel provided by, or for, the buyer at the named port of shipment.

Under this quotation:

Seller must

(1) pay all charges incurred in placing goods actually on board the vessel designated and provided by, or for, the buyer on the date or within the period fixed;

(2) provide clean ship's receipt or on-board bill of lading;

(3) be responsible for any loss or damage, or both, until goods have been placed on board the vessel on the date or within the period fixed;

(4) render the buyer, at the buyer's request and expense, assistance in obtaining the documents issued in the country of origin, or of shipment, or of both, which the buyer may require either for purposes of exportation, or of importation at destination.

Buyer must

(1) give seller adequate notice of name, sailing date, loading berth of, and delivery time to, the vessel;

(2) bear the additional costs incurred and all risks of the goods from the time when the seller has placed them at his disposal if the vessel named by him fails to arrive or to load within the designated time;

(3) handle all subsequent movement of the goods to destination:

  (a) provide and pay for insurance;

  (b) provide and pay for ocean and other transportation;

(4) pay export taxes, or other fees, or charges, if any, levied because of exportation;

(5) be responsible for any loss or damage, or both, after goods have been loaded on board the vessel;

(6) pay all costs and charges incurred in obtaining the documents, other

than clean ship's receipt or bill of lading, issued in the country of origin, or of shipment, or of both, which may be required either for purposes of exportation, or of importation at destination.

"*F.O.B. (named inland point in country of importation).*" Under this term, the seller quotes a price including the cost of the merchandise and all costs of transportation to the named inland point in the country of importation.

Under this quotation:

Seller must

(1) provide and pay for all transportation to the named inland point in the country of importation;

(2) pay export taxes, or fees or charges, if any, levied because of exportation.

(3) provide and pay for marine insurance;

(4) provide and pay for war risk insurance, unless otherwise agreed upon between the seller and buyer;

(5) be responsible for any loss or damage, or both, until arrival of goods on conveyance at the named inland point in the country of importation;

(6) pay the costs of certificates of origin, consular invoices, or any other documents issued in the country of origin, or of shipment, or of both, which the buyer may require for the importation of goods into the country of destination and, where necessary, for their passage in transit through another country;

(7) pay all costs of landing, including wharfage, landing charges, and taxes if any;

(8) pay all costs of customs entry in the country of importation;

(9) pay customs duties and all taxes applicable to imports, if any, in the country of importation.

Buyer must

(1) take prompt delivery of goods from conveyance upon arrival at destination;

(2) bear any costs and be responsible for all loss or damage, or both, after arrival at destination.

Comments on All F.O.B. Terms. In connection with F.O.B. terms, the following points of caution are recommended:

1. The method of inland transportation, such as trucks, railroad cars, lighters, barges, or aircraft should be specified.

2. If any switching charges are involved during the inland transportation, it should be agreed, in advance, whether these charges are for account of the seller or the buyer.

3. The term "F.O.B. (named port)," without designating the exact point at which the liability of the seller terminates and the liability of the buyer begins, should be avoided. The use of this term gives rise to disputes as to the liability of the seller or the buyer in the event of loss or damage arising while the goods are in port, and before delivery to or on board the ocean carrier. Misunderstandings may be avoided by naming the specific point of delivery.

4. If lighterage or trucking is required in the transfer of goods from

the inland conveyance to ship's side, and there is a cost therefor, it should be understood, in advance, whether this cost is for account of the seller or the buyer.

5. The seller should be certain to notify the buyer of the minimum quantity required to obtain a carload, a truckload, or a barge-load freight rate.

6. Under F.O.B. terms, excepting "F.O.B. (named inland point in country of importation)," the obligation to obtain ocean freight space, and marine and war risk insurance, rests with the buyer. Despite this obligation on the part of the buyer, in many trades the seller obtains the ocean freight space, and marine and war risk insurance, and provides for shipment on behalf of the buyer. Hence, seller and buyer must have an understanding as to whether the buyer will obtain the ocean freight space, and marine and war risk insurance, as is his obligation, or whether the seller agrees to do this for the buyer.

7. For the seller's protection, he should provide in his contract of sale that marine insurance obtained by the buyer include standard warehouse to warehouse coverage.

Negotiability and Assignability. It may be said at the outset that *negotiability* is used in the sense that the bill of lading, drawn to order and endorsed in blank with the name of the consignor, and sold for value to a bona fide purchaser who has no notice of any infirmity in the goods or in document, confers on the purchaser not merely all the rights which the consignor possessed, but a right possibly superior, to the extent that the purchaser is not bound by any concealed or undisclosed knowledge in the possession of the consignor. Thus a bona fide purchaser of an endorsed "order" bill of lading which is "clean" is not bound by any arrangements between the consignor and the carrier as to the goods being actually in bad order and covered by some sort of a "letter of indemnity."

Assignability, on the other hand, means only that the purchaser acquires such rights—disclosed or concealed—as the seller has. "Order" bills are seldom assigned, although a restricted transfer is possible. "Straight" bills of lading on the contrary are only transmissible by assignment; they cannot be negotiated, because by definition they are not drawn to "order." A business man wishing to avoid the broader perils of a negotiable bill of lading should use the "straight" form, keeping control more securely in his own hands.

"Order" Bill of Lading. In these days, almost all ocean bills of lading are "order" bills of lading; they state that the carrier, shipowner, charterer or master and/or ship will deliver the goods at the port of destination not merely to the named consignee, but to his order. The word "order" means that the bill of lading is more than the ship's receipt for the goods, more than the contract to carry the goods; it possesses, by reason of the words "to the order of" a named party, a third and highly legal and commercially important characteristic—namely, it becomes a document of title. In the United States the Pomerene Act expressly states in section 3 that "a bill in which it is stated that the goods are consigned or destined to the order of any person named in such bill is an order bill."

And for good measure the statute adds:

> "Any provision in such a bill, or any notice, contract, rule, regulation, or tariff that it is nonnegotiable shall be null and void and shall not affect its negotiability within the meaning of this *Act* unless upon its face and in writing agreed to by the shipper."

In some countries the same proposition is established by case law.

Thus the legal ownership—the property interest in the goods described can be transferred from the named consignee to any other persons whatsoever, and by them to still other persons, without any of these persons ever seeing the goods or having the goods in their physical possession.

This transfer of possession is accomplished initially merely by the written signature of the named consignee—his name written on the bill of lading. Once so endorsed, the bill may pass from hand to hand, and needs no further endorsement until the holder presents it to the carrier at the port of destination and demands his goods.

This means that all the endorsees and holders for value are legally entitled to rely upon the tally and upon the statements of "apparent (good) order and condition" in the bill as true; they may in the United States hold the carrier and also (except under various chartering circumstances) hold the ship liable for their loss if the statements are not true.

Index

On the following pages you will find full descriptions of recently published CORNELL MARITIME PRESS books and descriptions of back list titles on related subjects.

TANKER

TANKER PERFORMANCE AND COST, **Measurement, Analysis and Management** is written to help the tanker operator and manager, large or small, to arrive, from his own fleet operating statistics, at meaningful expressions of demand for his vessels, the supply of ships to meet this demand, the performance of vessels so employed, and the costs arising from these operations.

Answers are now available to questions of tanker performance which have long lain dormant for lack of any adequate approach to their measurement. It now becomes possible and practical to measure performance of tankers in a standard term, to compare the performance of one tanker with that of another even under widely different circumstances and on different trade routes, to accumulate the performance of groups of tankers for expressions of fleet performance, and to measure tanker costs in a standard unit.

The Tanker Transportation Unit is introduced as a measurement of tanker demand, supply, performance and cost and provides the means of integration and ready relation of all of these factors.

Tanker Performance and Cost analyzes and puts a finger on the flaws inherent in tanker

128 Pages   6 × 9" Format

" A much needed reference volume on the economics and statistics involved in the tanker divisions of the shipping and oil industries. Information covered can be used as a yardstick for management in measuring and analyz-

Performance and Cost

By ERNEST GANNETT

tonnage measurement methods in common usage and provides the basis for a step-by-step adoption of improved measurement and reporting techniques that any operator can put to immediate use.

Particular factors necessary to implement measurement techniques are identified, and means for the derivation of these from the raw material of any fleet are suggested.

The approach to total performance measurement with the concept of Effective Deadweight is unique and yields performance (and cost) data truly significant over a wide range of vessel sizes, voyages, and operating conditions.

Ernest Gannett is well suited to write a book of the nature of **Tanker Performance and Cost,** for he has had a career of over twenty-five years in the tanker management field where his association with problems of the nature outlined and solved in this book was on a firsthand and day-to-day basis.

His fifteen years of actual sea experience preceding his tanker management career have given Mr. Gannett an appreciation of maritime activities and tanker operations from both the seafarer's and the landsman's viewpoint.

Indexed $6. 50

ing tanker industry essentials. Many examples and tables of performance are found throughout the text. "(K. F.)
NEW TECHNICAL BOOKS
New York Public Library
November 1969

STEAMSHIP ACCOUNTING
By Philip C. Cheng, Ph. D., C. A.

This book is designed to deal with the peculiar accounting techniques, practices, and problems of the industry and does not represent an attempt to catalog every conceivable financial and accounting question in the industry, since it is believed that a comprehensive treatment of the more critical areas would be more useful and constructive. Several important special considerations and critical analyses are explored, with a discussion of the pros and cons of various suggested solutions and the author's views as to the best solutions.

The investigation may frequently prove useful in the analysis of the income tax problems ordinarily encountered by steamship companies. Since acceptable tax practices are not always consistent with good accounting principles, the results in this respect would sometimes be disappointing. Maritime industry taxation is somewhat unique and differs significantly from the taxation of any other industry. This is particularly true of the federal taxation of subsidized steamship companies. The income tax problems are discussed as they arise throughout the book.

Any author, particularly one writing about steamship accounting, instinctively tries to reach the largest possible audience. An effort has been made, therefore, to develop this work in such a way that it will appeal not only to accountants but to other people involved with the steamship business. This book can be used in the classroom, in the practitioner's library, and in the training programs of steamship companies.

Steamship Accounting encompasses the subject matter of both shipping operations and accounting and is designed to foster a better understanding of the relationship between the two. Emphasis is placed upon the application of accounting principles and accounting theory to the steamship business.

192 Pages $10.00

$10.00

SHIPMASTER'S HANDBOOK
ON SHIP'S BUSINESS

By Ben Martin, Master Mariner

There has, for some time, been a real need for an up-to-date, practical guide book to assist the shipmaster in the preparation and completion of the large amount of complicated paper work which is required in the operation of the present-day merchant vessel. The primary object of *Shipmaster's Handbook on Ship's Business* is to provide a useful source of information and ready reference regarding the many phases of ship's business and paper work with which the master must be concerned.

The paper work necessary for the efficient operation of the modern cargo vessel may appear onerous at first, but it will be noted, after a voyage or two, that the work will be less troublesome and difficult if it is started as early as possible on the voyage and as much is done toward completion as available information and details permit. This is especially true of making up the payrolls and other accounting items. There will be occasions, however, when some paper work, involving reports and official documents, must be done on short notice. The details of such reports and documents may not be known in advance so to attempt their preparation early could lead to serious errors. Much of the paper work throughout the voyage is for the home office or charterer and many special forms are issued by both. These differ in various companies but the information required is essentially the same. There are over 61 sample forms in the appendix.

Shipmaster's Handbook on Ship's Business is principally for the newly appointed master of a merchant vessel but it may also be of assistance to more experienced masters as a check in completing the paper work required; and much of the information herein will be found useful by masters and mates of foreign flag vessels calling at U.S. ports.

320 Pages   Indexed

NAUTICAL CALCULATIONS EXPLAINED

Klinkert & White

Nautical Calculations Explained is unique in its field in that it concentrates the quantitative aspects of the navigator's profession into a single volume. It contains three hundred fully-worked calculations illustrated by one-hundred-and eighty line drawings. There are over three-hundred-and-seventy additional examples for exercise with answers covering no less than twelve subjects pertinent to the work of navigating officers aboard ship.

The book constitutes an important work of reference and forms a portable course of study for cadets in training and professional navigators seeking advancement. Most theoretical topics of a quantitative nature are discussed and these led to step-by-step solutions of a large and varied number of practical problems which deck officers encounter either aboard ship or at examinations.

Nautical Calculations Explained will be required reading for anyone sitting for his Master's Certificate, or for any navigational examination. It will be indispensable for every professional navigator, every nautical college and shipping organisation, and on every ship afloat.

616 Pages $14. 00   (November 1969)

Related Books

Shipping: Operations and Management—Chartering and Freight Rates

CHARTERING AND CHARTER PARTIES

by H. B. COOLEY. A basic book on chartering—not case histories. For owners, charterers, owners of cargo on chartered vessels, their representatives, and the lawyer with little experience in maritime law.

160 Pages, 6" x 9", (1947). $6.00

THE MARITIME INDUSTRY: Federal Regulation in Establishing Labor and Safety Standards

by RUDOLF W. WISSMANN. 400 Pages, 6" x 9", Illustrated, (1942). $6.00

MERCHANT MARINE POLICY

by HOWARD C. REESE, PH.D., Editor. Foreword by WARREN G. MAGNUSON, Chairman, Senate Committee on Commerce.

Proceedings of the Symposium of the Fifteenth Ocean Shipping Management Institute—Marvin L. Fair, Director. Held at The American University, Washington, D.C.

Contents: How Serious is the Situation of the U.S. Merchant Marine? — What is the Military Need for a U.S. Merchant Marine for the Last Third of the 20th Century? — What Could and Should Be Done Through Collective Bargaining in the Maritime Industry? — What the Shipping Industry Can and Should Do; — Appendix.

224 Pages, 6" x 9", (1963). $7.00

MERCHANT SHIPS: A Pictorial Study

by J. H. LA DAGE with SAUERBIER, STEINER, HIRSCHKOWITZ and FIORE

This greatest collection of organized information about ships in book form has been described as a work on which superlatives can be bestowed without hesitation. By the use of 1160 photographs and 150,000 well chosen words of text, the factual story of merchant ships is unfolded in a most instructive and interesting manner. From the forecastle head to the tail shaft and propeller, all types of Merchant Ships are dissected as by a skilled surgeon, with their innermost parts laid bare with camera and pen.

If it is on a ship or part of a ship, the odds are that it is pictured and described here. All descriptions are concise and to the point, making clear for all, the functions of each part, job, space and marking, on ships seen the world over.

Merchant Ships: A Pictorial Study is a "must-own" volume for all in any phase of the shipping business and the interested layman.

512 Pages, 8½" x 11", Illustrated, **Second Edition** (1968). $15.00

NUCLEAR SHIP PROPULSION

by HOLMES F. CROUCH; see under *Marine Engineering*. $10.00

OCEAN FREIGHT RATES

by WILLIAM L. GROSSMAN. The three geographical phases of ocean transportation are covered: Foreign ocean transportation, coastwise and intracoastal, and non-contiguous.

Contents: Freighting Units — Ship's Option — Class, Commodity, and Volume Rates — Through Rates and Other Special Rates — Additional and Accessorial Charges — Charter Parties — The Goal of Rate Making — Out of Pocket Cost — Value of Service — The Conference System — The Dual Rate System — Conference Rate Procedure — Rate Making for Group Profit — Purposes of Rate Regulation — History and Scope of Rate Regulation — Common, Contract and Private Carriage — Reasonableness: General Rate Level and Specific Rates — Unjust Discrimination — Tariffs — Volatility of the Market in Foreign Ocean Transportation — Billing and Payment — Appendices — Index.

224 Pages, 6" x 9", Illustrated, (1956). $6.00

OCEAN FREIGHTS AND CHARTERING

by C. F. H. CUFLEY. The scope of this book covers all the subjects of which those who charter and operate ships should have a working knowledge.

Contents: The Size of the Problem and Basic Knowledge Required of a Freight Market

Operator — The Practice of Shipbroking — Charterparties and Bills of Lading: Definitions and Basic Provisions — The Construction of a Charterparty: Parts I & II — The Hague Rules and Clause Paramount — Merchant Vessels: Parts I & II — The Classification of Merchant Ships — Cargo Movements and the Development of World Merchant Shipping — The Pattern of Sea-Borne Trade — Tramp Ship Cargoes — Stowage Factors of Cargoes — Freight Market Behaviour — The Economics of Shipowning in Relation to Chartering and Management Policies — Appendix: Evolution of Merchant Ships, 1892-1962 — Index.

462 Pages, 5½" x 8½", Illustrated, (1962). $12.00

OCEAN SHIPPING

by STEWARD R. BROSS. A practical book on the general subject of ocean transportation written to serve a twofold purpose: a. As a practical handbook for steamship operators and their employees, foreign freight traffic managers, freight forwarders, and ships' officers. b. As a guide for students of transportation, exporters and importers, and the many other groups and individuals vitally concerned with international commerce and shipping.

Contents: Introduction — Ocean Trade Routes — Types of Ocean Vessels — Marine Insurance — Tramp Shipping — Ship Brokers — Agents — Liner Company Organization and Administration — The Marine Department — Terminal Operations — Stevedoring and Ship Stowage — Chartering — Freight Traffic — Bills of Lading — Damage Claims — Passenger Traffic — Accounting — Military Sea Transportation Service — International Relations — Appendices: Carriage of Goods by Sea Acts, U.S. British, Australian and Canadian — The Harter Act — York/Antwerp Rules, 1950 — Index.

352 Pages, 6" x 9", Illustrated, (1956). $7.00

SHIP MANAGEMENT: A Study In Definition and Measurement

by RODNEY M. ELDEN. The basic objective of *Ship Management* is to identify and depict the many problems of ship management, and then, through analysis and synthesis, to measure and identify the degree of controllability inherent in each problem. In this way, those key aspects which should have real import and significance to the owners and managers of ships are revealed.

Contents: The Conservator — Labor — Food — Maintenance — Materials — Insurance — Definition — Measurement — Tables, Charts & Bibliography — Index.

128 Pages, 6" x 9", (1962). $4.00

COORDINATED TRANSPORTATION

E. GROSVENOR PLOWMAN, Editor. There is an increasing realization of the possibilities of coordination of transportation in rendering a more economic and more expedited service. In many ways it is a new frontier in transportation development. The development of the container, the demountable truck-trailer body and other means of transferring unit loads has stimulated recent interest in coordination. This book presents the subject of transport coordination by means of about 40 excerpts from the papers presented at five workshop sessions conducted between 1964 and 1966 as part of the *Transportation and Logistics Program of the School of Business Administration of The American University* under the general direction of Professor Marvin L. Fair. As thus developed, this book presents the dimensions and potentials of coordination, its demand and supply characteristics and requirements, its economic and legal environment, institutional barriers, and the impacts of new cargo handling methods and of applied research. A pioneering work on the subject of coordinated transportation and containerization as it affects all forms of transportation . . . . Sea, Air, Truck and Rail. The papers are the work of outstanding management men in the shipping industries and officials of the several governmental regulatory bodies involved with the movement of the goods by land, sea and air.

320 Pages, 6" x 9", (Mid-1968). $7.50

TRANSPORTATION MANAGEMENT

by H. B. COOLEY. An invaluable guide to modern management of air, water and truck transportation companies. For those in traffic, operating, treasurer's and comptroller's departments.

Contents: Organization — Traffic Department — Operating Department — Treasurer's Department — Comptroller's Department — Personnel Relations — Wage Systems and Incentives — Purchasing Stores and Repair — Budgeting — Standard Costs — Reports — Index.

224 Pages, 6" x 9", (1946). $6.00

VESSEL VOYAGE DATA ANALYSIS: A Comparative Study

by KIM J. LOROCH. The comparative vessel performance model described in this work attempts to set up a system which can be used to guide, record and measure the accomplishments that may be forthcoming from a well organized flow, presentation, analysis, and interpretation of the mass of operating vessel data that is or should be made available not only to those who operated the vessels, but also to those who design and build them, who engage in research and development, as well as various maritime agencies on governmental and international levels.

Better organization leads to better control, resulting in greater efficiency and utilization of equipment. The benefits of a standardized system and forms are offered as an important by-product.

Contents: Introduction — Voyage Data Survey Response — The Method — The Method in Practice — Economic and Practical Implications — Proposal for an International Research Clearing House Devoted to Vessel Data — Appendices — Extensive Bibliography — Index.

176 Pages, 6" x 9", Illustrated, (1965). $10.00

Marine Insurance—Export/Import

EXPORT/IMPORT TRAFFIC MANAGEMENT AND FORWARDING

by ALFRED MURR, M.B.A., DR. JUR. In this new edition every section affected by changes in law, general orders and trade practice has been revised and updated. The foreign consular regulations reflect the current regulations of all countries and include those of the many emerging countries. Throughout the text there are numerous corrections and additions reflecting the many changes that have affected the export/import and forwarding business over the past ten years. This handbook is a comprehensive coverage of the subject prepared by a man of broad practical business and teaching experience in the field. It has been designed to serve those in the business as guide in their daily work, and to serve as text in business schools and colleges in courses dealing with foreign commerce and international trade. This comprehensive coverage is contained in 37 chapters under the following ten general headings:

The Forwarder's Place in International Transportation and Foreign Trade — Legal Principles Applicable to Foreign Freight Forwarding — Ocean Shipping — Marine Insurance — Export and Import Banking — Procedures in Handling of Export Shipments — Customs Brokerage — Accessorial Services and Trade Promotion — The Business Organization of the Foreign Freight Forwarder — Revenue — Plus 23 Appendices.

607 Pages, 6" x 9", (2nd Ed. 1967). $10.00

GENERAL AVERAGE AND THE YORK/ANTWERP RULES, 1950 American Law and Practice

by LESLIE J. BUGLASS. **Contents:** The Ancient Maritime Law of General Average — General Principles — General Average Sacrifices — General Average Expenditures — Salvage Charges — Basis of Contribution — Security of Contribution — Liability of Cargo to Contribution — Application to Policies of Insurance — Bibliography of the York/Antwerp Rules — Appendices: A: The York/Antwerp Rules, 1950 — B: Rules of Practice of the Association of Average Adjusters of the U.S. — C: Form of Average Agreement, Underwriter's Guarantee and Deposit Receipt — D: Summary of Principal Differences between the York/Antwerp Rules, 1950 and American Law and Practice — Index.

160 Pages, 5½" x 8½", (1959). $4.00

MARINE INSURANCE CLAIMS: AMERICAN LAW AND PRACTICE

by LESLIE J. BUGLASS. Deals in concise form with the fundamentals of marine insurance with particular regard to claims. Whenever possible, American authorities are cited.

Contents: List of Cases Cited — Introduction to Marine Insurance — Total Losses — Particular Averages — General Averages — Salvage Expenses and Sue and Labor Charges — Liabilities to Third Parties — Subrogation and Double Insurance — Marine Insurance Claims on the Great Lakes — Bibliography.

Appendices: American Institute (Hulls) Form of Policy — Rules of Practice of Association of Average Adjusters of the U.S. — Resolutions and Rules of Practice of the Great Lakes Protective Association — English Marine Insurance Act — York/Antwerp Rules, 1950 — Merritt-Chapman & Scott Corporation's Form of Salvage Agreement — Index.

160 Pages, 6" x 9", Illustrated, (1963). $4.00

MARINE INSURANCE DIGEST

by HUGH A. MULLINS and LESLIE J. BUGLASS. A "Digest of the Law and Practice of Marine Insurance for the Layman," with "An Analysis of the American Institute Time (Hulls) Form of Policy."

320 Pages, 5½" x 8½", (2nd Ed. 1959). $6.00

THE OCEAN FREIGHT FORWARDER, THE EXPORTER AND THE LAW

by GERALD H. ULLMAN. In this volume Gerald H. Ullman, counsel for the New York Foreign Freight Forwarders and Brokers Association, Inc., sets the applicable laws, decisions and agency interpretations so that the area of legal responsibility may be expeditiously determined. Where the law or agency action is in a developmental stage, the trend is indicated and the arguments, pro and con, are provided. If a decision in the author's opinion is erroneous or questionable or an agency regulation or interpretation of doubtful validity, he furnishes his reasons for so thinking so that future conduct of the parties concerned may be weighed accordingly. Whenever possible, procedures have been recommended to avoid legal pitfalls which experience has shown are continually encountered by forwarders and shippers.

Ocean freight forwarders, shippers, ocean carriers and their attorneys will find this text to be a handy reference to applicable judicial and regulatory authorities pertinent to the shipper-forwarder-carrier relationship.

Contents: The Role of the Forwarder, His Legal Status and Jurisdiction Over His Activities — Liability of the Forwarder to Carrier and Shipper — Obligations of the Forwarder and Shipper Under the Shipping Act — Qualifying for a Forwarder License — Forwarder's Obligations to Shippers and Carriers Under Commission Regulations — The Forwarder, Shipper and the Dual Rate Law — Responsibilities of Forwarder and Shipper under U.S. Export Control Laws — Extracts from the Shipping Act, 1916 — Organization Chart of the Federal Maritime Commission — General Order 4 of Federal Maritime Commission — Forms of Shipper's Credit Agreement — Form of Section 15 Agreement Between Forwarders in Different Ports.

144 Pages, 6" x 9", (1967). $5.00

Ports: Administration and Operation

FREE PORTS AND FOREIGN-TRADE ZONES

by DR. RICHARD S. THOMAN. **Contents:** Introduction — Part I. The Free Port: Definition, Distribution, and Historical Development — Part II. Free Ports of the German Federal Republic — Part III. The Free Ports of Scandinavia — Part IV. The North European Free Port and Alternative Systems — Part V. The Foreign Trade Zones of the United States: A European Heritage — Part VI. Free Ports and Free Trade Zones: A Summary Review — Appendices: German Federal Republic, Denmark, Sweden, United States — Bibliography — Index.

250 Pages, 6" x 9", Illustrated, (1956). $7.00

PORT ADMINISTRATION IN THE UNITED STATES

by MARVIN L. FAIR. **Contents:** Water Ports — Port Administration: Development and Functions — Federal Activities Related to Port Administration — The Port Authorities of the United States — Port Authorities: Organization, Procedures, Trends — Port Management in the United States — Port Development — Planning of General Cargo Terminal Facilities — Port Jurisdiction and Supporting Facilities — Port Finance — Some Aspects of Port Operation — Selling the Port — Port Association — Index.

232 Pages, 6" x 9", Illustrated, (1954). $7.00

PROGRESS IN CARGO HANDLING—Volume 3: International Cargo Handling Coordination Association

by R. P. HOLUBOWICZ, Editor. This volume presents in full the 30 papers and attendant discussion of the ICHCA, 1961 General Assembly and Technical Conference, held in New York City, September 1961. The papers cover the latest technological developments in these general fields: Shipborne Breakbulk Cargo, Containerization of International Commerce; Port and Cargo Handling in Recently Emergent Economies; Cargo Handling in Airborne Commerce.

240 Pages, 6" x 9", Illustrated, Indexed, (1962). $10.00

Tugs and Towing—Fishing Boats

A GUIDE TO FISHING BOATS AND THEIR GEAR

by CARVEL HALL BLAIR and WILLITS D. ANSEL. A recognition guide to commercial fishing vessels and their equipment. Describes and pictures the fishing gear and the craft used in the world fisheries. The introductory chapter describes the basic methods of a commercial fisherman and explains the equipment he uses. Each succeeding chapter discusses and illustrates one of the major types of fishermen: trawlers, hook and liners, gill netters, seiners, harpooners, support ships, research vessels, and small craft. A Glossary explains terms not covered elsewhere, and a complete bibliography refers to books, reports and periodical articles containing further information.

Prepared to serve the professional mariner as well as the amateur seaman, and the landlubber as a "look-her-up book." It is a modest attempt to do for fishing boats what Mr. Jane, Mr. Talbot Booth and their successors have done for naval vessels and merchant ships.

Contents: Introduction — Methods of fishing and types of gear — Trawlers — Hook and Line — Gill netters — Purse seiners — Harpooners — Fish factories and mother ships — Research ships and others — Small boats — Glossary — Bibliography — Index.

160 Pages, 9″ x 6″, Illustrated, (1968). $5.00

TUGS, TOWBOATS AND TOWING

by EDWARD M. BRADY. This book embraces a wide range of subjects directly concerned with tug and towboat operations, and methods of suitable "hookups," including various types of towing gear, and gives practical directions for connecting the gear for the various types of tugs and tows. *Tugs, Towboats and Towing* is a most practical reference manual covering American practice, River; Harbor; Lakes; Inland River; Coastal and Ocean-going; as well as British and European practices. Only those design features that affect tug and towboat maneuverability have been touched upon with the thought in mind to acquaint operating personnel with some of the more important aspects affecting maneuverability. The working parts of tugs and towboats are fully illustrated and there are diagrams showing the various hookups as well as numerous illustrations of the various types of vessels.

Contents: Types of Tugs and Towboats — Tug and Towboat Construction and Design — Towing Theory — Inland Towing — Offshore Towing — Miscellaneous Towing Operations, Techniques and Hazards — Rescue Towing — Glossary — Bibliography — Index.

256 Pages, 6″ x 9″, (1967). $10.00

Seamanship—Ship Handling—Deck Officers' Guides—Ship's Business—Stowage—Cargo Loss and Damage

AMERICAN MERCHANT SEAMAN'S MANUAL—Fifth Edition

by F. M. CORNELL and A. C. HOFFMAN. The complete handbook for merchant seamen. This printing contains corrections made in 1967; and the Rules of the Road — effective September 1, 1965. Covers every phase of good seamanship and all navigation necessary to prepare for 3rd Mate's license.

Contents: Organization Tree of a Merchant Freight Vessel and Passenger Vessel with a Departmental Breakdown of Duties and Responsibilities — Introduction — Marlinespike Seamanship — Wire and Wire Splicing — Blocks and Tackle — Deck Seamanship — Cargo Stowage and Handling — Canvas Work — Paints and Painting — Ground Tackle — Steering Gear — Examination Guide for Lifeboatman and Able Seaman — Handling Small Boats Under Oars — Handling Small Boats Under Sail — Signals — Tankerman's Guide — Construction and Stability — Navigation — Weather — Ship Sanitation — First Aid — The Rules of the Road — Safety — Consular Regulations Affecting Relief of Seaman — Laws Pertaining to Merchant Seaman — General Information — Index.

880 Pages, 5¼″ x 7½″, Illustrated, (1967). $10.00

BLUE BOOK OF QUESTIONS AND ANSWERS FOR THIRD MATES—Ocean, Coastwise, Yachts & Limited and Mineral and Oil Industry

by W. A. MacEwen.

160 Pages, 6″ x 9″, Illustrated, (1968). $4.00

BLUE BOOK OF QUESTIONS AND ANSWERS FOR SECOND MATE, CHIEF MATE AND MASTER—Ocean, Coastwise, Yachts & Limited Mineral and Oil Industry

by W. A. MacEwen. These two volumes answer the specimen examinations for all grades as published in Specimen Examiniations for Merchant Marine Deck Officers, *CG-101*.

The questions and answers are representative examples, answered in the required manner, covering the subjects on which the examinations are based.

320 Pages, 6″ x 9″, Illustrated, (1969). $6.00

DICTIONARY OF COMMODITIES CARRIED BY SHIP

by Pierre Garoche. Presents useful information regarding the many classes and items of goods, substances, and merchandise carried by ship, in handy dictionary form. Description is in English. Corresponding name is given in French and Spanish or in language of country of origin.

384 Pages, 5½″ x 8½″, (1952). $6.00

MERCHANT MARINE OFFICERS' HANDBOOK—Fourth Edition

by E. A. Turpin and W. A. MacEwen. Revised and corrected with the addition of much new material, this edition in a new enlarged format is intended to serve as a modern well-indexed practical reference for everyday use on shipboard, and to give the essential information required for the Masters' and Mates' examinations. It contains more needed information required by Deck Officers and Cadets than ever compiled in any single volume.

Contents: Everyday Labors of a Ship's Officer — Instruments and Accessories Used in Navigation — Piloting — Tides and Currents — Sailings — Celestial Navigation — Meteorology — Cargo — Shiphandling — Cargo Gear — Ground Tackle — Signals — Rules of the Road — Radar and Rules of the Road — Ship Construction, Maintenance and Repairs — Ship Calculations and Stability — Fire — Emergencies — U.S. Navigation Laws and Ship's Business — Engineering for Deck Officers — Steering Gear, Windlasses, Capstans and Gypsies — First Aid and Ship Sanitation — Rules and Regulations for Deck Officer's Licenses — Commissions in the U.S. Navy Reserve — Mathematics — Tables and Useful Information — Index.

896 Pages, 5″ x 8″, Illustrated, (1965). $10.00

MODERN SHIPS: Elements of Their Design, Construction and Operation— Second Edition

by John H. La Dage. A reference text for Ship's Officers, Cadets, Naval Architects, Marine Surveyors and anyone interested in the efficient building and operation of ships.

Contents: Principal Dimensions and Characteristics — Modern Types of Ships — Tonnage Measurement — Classification — Freeboard and Load Lines — Strength of Materials and Ships — Lines, Offsets and the Mold Loft — Riveting and Welding — Tanks, Bilges, and Piping Systems — Turning and Steering — Launching — Drydocking — Ship's Calculations — The Ship in Waves — Resistance and Powering — Propellers and Propulsion — Ship Trials — Index.

392 Pages, 6″ x 9″, Illustrated, (1965). $7.00

SHIP HANDLING IN NARROW CHANNELS

by C. J. Plummer. In this new edition special consideration is given to the new grants of the waterways . . . . the supertankers and large bulk carriers.

Contents: Making Suction an Asset — Anchoring — Mooring — Best Trim for Maneuvering — Using Tugs Advantageously — Use of Anchor to Maneuver — Conclusion.

112 Pages, 5¼″ x 7½″, Illus. (2nd Ed. 1966). $3.00

SHIP'S BUSINESS and CARGO LOSS AND DAMAGE—2nd Edition

by M. E. McFarland. This new combined volume contains revised editions of Captain McFarland's *Ship's Business* and *Cargo Loss and Damage* which were formerly published as separate titles.

Contents: CARGO LOSS AND DAMAGE — The Representative — The Surveyor & the Survey — The Report — The Assured, Third Parties and the Underwriter — The Claim, The General Average — Appendices — Index.

Contents: SHIP'S BUSINESS — Relating to Business with Customhouse, Government Officials, etc. — Relating to Business with Agents, Charterers, Charter Parties, Bills of Lading, etc. — Casualties, Damages, Personal Injuries and Illness, Surveys and Surveyors, Repairs and Repair Contracts — Bibliography — Index.

160 Pages, 5½″ x 8½″, Illustrated, (1963). $5.00